D0982032

Probability Modeling and Computer Simulation

An Integrated Introduction with Applications to Engineering and Computer Science

THE DUXBURY SERIES IN STATISTICS AND DECISION SCIENCES

Applications, Basics, and Computing of Exploratory Data Analysis, Velleman and Hoaglin
Applied Regression Analysis and Other Multivariable Methods, Second Edition, Kleinbaum and Kupper
Classical and Modern Regression with Applications, Myers
A Course in Business Statistics, Second Edition, Mendenhall
Elementary Statistics for Business, Second Edition, Johnson and Siskin
Elementary Statistics, Fifth Edition, Johnson
Elementary Survey Sampling, Third Edition, Scheaffer, Mendenhall, and Ott
Essential Business Statistics: A Minitab Framework, Bond and Scott
A First Course in Linear Regression, Second Edition, Younger
Fundamental Statistics for Human Services and Social Work, Krishef
Fundamental Statistics for the Behavioral Sciences, Howell
Fundamentals of Biostatistics, Second Edition, Rosner
Fundamentals of Statistics in the Biological, Medical, and Health Sciences, Runyon
Introduction to Contemporary Statistical Methods, Second Edition, Koopmans
Introduction to Probability and Mathematical Statistics, Bain and Engelhardt
Introduction to Probability and Statistics, Seventh Edition, Mendenhall
An Introduction to Statistical Methods and Data Analysis, Third Edition, Ott
Introductory Statistics for Management and Economics, Second Edition, Kenkel
Linear Statistical Models: An Applied Approach, Bowerman, O'Connell, and Dickey
Management Science: An Introduction, Davis, McKeown, and Rakes
Mathematical Statistics with Applications, Third Edition, Mendenhall, Scheaffer, and Wackerly
Minitab Handbook, Second Edition, Ryan, Joiner, and Ryan
Minitab Handbook for Business and Economics, Miller
Operations Research: Applications and Algorithms, Winston
Probability Modeling and Computer Simulation, Matloff
Probability and Statistics for Engineers, Second Edition, Scheaffer and McClave
Probability and Statistics for Modern Engineering, Lapin
Quantitative Management: An Introduction, Second Edition, Anderson and Lievano
Quantitative Models for Management, Second Edition, Davis and McKeown
Statistical Experiments Using BASIC, Dowdy
Statistical Methods for Psychology, Second Edition, Howell
Statistical Thinking for Behavioral Scientists, Hildebrand
Statistical Thinking for Managers, Second Edition, Hildebrand and Ott
Statistics for Management and Economics, Fifth Edition, Mendenhall, Reinmuth, Beaver, and Duhan
Statistics: A Tool for the Social Sciences, Fourth Edition, Ott, Larson, and Mendenhall
Time Series Analysis, Cryer
Time Series Forecasting: Unified Concepts and Computer Implementation, Second Edition, Bowerman and O'Connell
Understanding Statistics, Fourth Edition, Ott and Mendenhall

Probability Modeling and Computer Simulation

An Integrated Introduction with Applications to Engineering and Computer Science

Norman S. Matloff
University of California—Davis

PWS-KENT Publishing Company
Boston, Massachusetts

20 Park Plaza
Boston, Massachusetts 02116

PWS-KENT Publishing Company is a division of Wadsworth, Inc.

Library of Congress Cataloging-in-Publication Data

Matloff, Norman S.
Probability modeling and computer simulation.

Includes index.
1. Probabilities. 2. Mathematical statistics.
3. Probabilities—Computer simulation. 4. Mathematical statistics—Computer simulation. I. Title.
QA273.M385 1988 519.2 87-32792
ISBN 0-534-91854-9

Printed in the United States of America.
88 89 90 91 92—10 9 8 7 6 5 4 3 2 1

Sponsoring Editor *Michael Payne*
Production Coordinator *Elise Kaiser*
Production/Composition *Publication Services*
Interior Design *Ellie Connolly*
Cover Design *Julie Gecha*
Cover Printer *John P. Pow Company*
Text Printer/Binder *Haddon Craftsmen, Inc.*

Preface

This text is designed for a calculus-based, one-term introductory course in probability and statistics. The intended audience is students majoring in Engineering and Computer Science, with most of the examples and exercises drawn from applications to these areas. The applications are definitely realistic; they are *not* merely conversions to Engineering/Computer Science contexts of hackneyed problems from other fields. On the other hand, the text does not at all take a "cookbook" approach—*it emphasizes principles and concepts*, not formulas.

A vital feature of the text is that it also serves to introduce computer simulation methodology, an indispensable tool in modern applications. In real-world applications, the simpler systems can be modeled mathematically, but more complicated cases must be handled through computer simulation. Thus, the point of view taken is that since both mathematical and simulation methods are *used* together in practice, it is desirable to also *teach* them together.

Thus, simulation methodology is incorporated throughout the presentation. The primary emphasis is still on probability and statistics, rather than on comprehensive treatment of simulation. Simulation is integrated into the examples and exercises in a natural manner as a *supplement* to the mathematical aspects of probability and statistics, to give the student both an additional practical tool, and much deeper insight.

The pattern followed in examples and exercises is the same as that cited above in real-world applications: a mathematically tractable problem is dealt with mathematically, and then an intractable extension of the problem is analyzed using simulation. Exercises are designated either "M" or "S," indicating whether the problem is to be analyzed mathematically or through simulation. Most exercises are designated *both* M and S (i.e., both solutions are requested); in such exercises, the two solutions serve as checks on each other, feedback that students find very valuable.

The author and others who have class-tested the text have found that the use of simulation in the course not only provides the students with an additional tool of practical value, but also turns out to have a very substantial beneficial effect on the student's intuitive understanding of the concepts. For example, students in introductory probability courses often have trouble distinguishing between the concepts $P(A \text{ and } B)$ and $P(A \mid B)$. The author has observed that this confusion is resolved when the students write simulation programs to compute the two quantities. Similarly, confusion about sampling concepts, about the distribution of $\overline{X}$ for example, is also dispelled quite rapidly when a student writes simulation programs.

In essence, the use of simulation serves to add a "laboratory" component that complements the mathematical aspects of the course, a feature particularly appreciated by students in Engineering and Computer Science. This makes the concepts much more concrete by enabling the students to actually see the effects of the Central Limit Theorem, for example, "in action." And, as mentioned before, simulation allows exploration of concepts at a deeper level, supplementing what the students learn in working the mathematical exercises.

On the other hand, the simulation aspects of the book are meant as a supplement to the mathematical aspects, not as a replacement for them. Indeed, although the presentation style is informal—it does not follow a formal definition/theorem/proof format—the text is still precise and mathematically complete, and it is substantially more mathematically sophisticated than are most "Engineering Statistics" texts.

Considerable time—more than in most other texts—is spent on developing the motivations and goals of the concepts and discussing the modeling process, instead of on stark presentation of formulas. The goal is to prepare the students to *use* this material, correctly and creatively, after they finish their education, rather than to learn just enough to get some "partial credit" in course examinations.

Prerequisite background for the course is a year of calculus, and one course in computer programming; no previous background in probability, statistics, or simulation is assumed. Typical timing for such a course is the junior year, but sophomore or senior levels might be more appropriate at some schools.

Since the inclusion of computer programming is unusual for a text in probability and statistics, some comments to the instructor on this subject would be useful, especially to those instructors who have little or no background in simulation. These instructors will be surprised to discover that students adapt to the idea of simulation programming quite rapidly indeed. All the programs follow one of a few general formats, which in fact are formalized in "templates" such as Program 2.4.3; thus the programming is fairly simple. Many of the author's students in this course enter with minimal background in programming, typically a single introductory course taken one or two years previously—yet they still adapt very quickly after the first homework assignment or two. If the instructor is uncertain of the students' programming skills, he or she might wish to give outlines of some of the homework programs to the students for the first assignment or two, or might have students work in groups or in pairs. Again, extensive class testing of the book has shown that such aids rapidly become unnecessary as the course progresses.

The programming language used for examples in this book is Pascal. This choice was made because it is currently the most widely taught language, and because of its clarity of expression. However, if the instructor and/or students know only some other language, typically FORTRAN, this will not present any problem. The instructor can have the students read the text's examples in Pascal and still work the simulation exercises in FORTRAN if they wish. Reading the Pascal examples is easy for "non-speakers" of Pascal, since one of the primary design goals in developing the Pascal language was its English-like readability. A short introduction to Pascal for those who know FORTRAN is provided in an Appendix; a few minutes spent on this will be quite sufficient for reading the Pascal programs in this book (the sole exception being Program 8.2.2).

A number of plans may be followed in teaching from this book. A course intended to cover only basic probability and statistics might exclude Chapter 8, which introduces stochastic process. On the other hand, courses that emphasize probability more than statistics can include Chapter 8, but either skip or cover very lightly Chapters 5, 6, and 7, which deal with statistics. Similarly, there is flexibility of coverage within a given chapter; for example, some instructors may wish to skip some of the more sophisticated topics in Chapter 4, such as transform methods.

As with any book, constructive criticism from users and reviewers of this text has been of very substantial benefit. The author would like to thank Alan Fenech, Neil Willitts, and Ron Pruitt, who class-tested earlier versions of the manuscript, and the reviewers: David Ellis, Richard S. Kleber, Linda C. Malone, Bradley Skarpness, Paul Speckman, Stephen Vardeman, Lyn R. Whitaker, and Peter C. Wollan. The text has benefited greatly by efforts of the PWS-KENT staff, Michael Payne and Elise Kaiser. And I must include a special expression of gratitude and love to my wonderful wife, Gamis Yu; all of my energy and inspiration derive from her.

Norman Matloff
Davis, California

Contents

APPENDIXES

CHAPTER 1

Randomness, Mathematics, and Computers

The field of probability and statistics may be viewed as a branch of applied mathematics in which intuitive ideas of randomness are made precise in formal mathematical models. The resulting methods are of great practical value in almost any field involving quantitative, or even qualitative, data.

These methods vary in mathematical complexity. On the one hand, a simple model known as the **normal** (or **Gaussian**) distribution is the basis for a very substantial portion of commonly used statistical procedures. On the other hand, there are many stochastic optimization problems that are so difficult to solve mathematically that an alternative approach known as **computer simulation** must be used for their solutions.

Although the field of probability and statistics uses modern tools from mathematics and computer science, the history of this area is surprisingly long. Applications of probability theory extend back to the seventeenth century (if analysis of gambling games can be considered an "application"), and probabilistic methods were being used in the design of telephone systems as early as 1909. Also around that time, statistical methods were beginning to play a vital role in agricultural research. Even simulation, which today is considered to depend completely on the use of computers, was proposed in primitive form in 1907.

Today the list of applications of probability, statistics, and simulation is virtually endless. Here are just a few examples:

- Machine recognition and synthesis of speech has a range of applications, including industrial automation technology and the construction of "reading" machines for the blind. How can machine processing of speech be done?
- How can one analyze the fluctuations in the height of a river? How can one determine the necessary capacity in building a new dam?
- In a certain interactive computer system, what is the maximum number of terminals that can be attached before mean response time becomes unacceptably slow? How would this number be affected by a given increase in the amount of memory in the system?
- What would be the effects of installing various kinds of traffic signals at a busy intersection?
- What are the important predictors of air pollutant levels? How can they be used to identify the major contributants to pollution?
- An office automation system is to be designed for combined communication of text, voice, and image data. What system of transmission should be used to achieve the lowest possible mean delay?
- A business or government agency needs to determine the population prevalence of a certain attribute that is sensitive in nature. How can it determine the overall rate without violating the privacy of individuals?
- A manager of a large system of warehouses wishes to optimize the operation of the system, with respect to questions such as: What is the best schedule for replenishing inventories? How many trucks should the operation include? In what order should the various distribution points be visited?
- A manufacturer of medical equipment is designing a computer system for monitoring patients in hospital intensive care wards. It is vitally important that such equipment have as low a probability of failure as is technologically possible. How should this be done?
- In television coverage of election returns, how can the final result be predicted from a very small sample of early returns?
- A tire manufacturer is comparing treadwear for two kinds of tires by conducting wear tests on 50 tires of each kind. How accurate will the information based on this sample be? Will a sample of 50 be enough?
- In a large computer database, how should the data be arranged in order to minimize the average data access time?

Questions such as those above can be analyzed using the mathematical and computer simulation tools developed in this book, given a thorough knowledge of the field of application itself.

The subject of statistics can be learned at several different levels of mathematics. This can be seen in the variety to be found in statistics textbooks. On the one hand, there are a number of "statistical methods" texts, which do not use calculus or computer programming. Although these texts present a few formulas that are useful in some simple situations, they do not cover sufficient material for many engineering applications (e.g. most of the examples above).

On the other hand, there are a number of "mathematical statistics" texts, which are theoretical in nature and use a theorem/proof format. Such theory can be rather elegant, and such books are needed for those who wish to study it. However, this text is focused primarily on applications. It does not stress the mathematical aspects of the subject, in the sense that it does not emphasize theory. The treatment here is informal in nature, using only "concrete" mathematics—calculus.

An extremely important feature of this text is its integration of computer simulation methods into its presentation. While many real-world problems in engineering and management can be solved easily with very simple mathematical models, there are also many problems too complex to analyze mathematically. In these cases, computer simulation is an absolutely vital tool.

A simulation program mimics the operation of the system under study, in much the same way a wind tunnel helps aeronautical engineers by simulating the effects of air movement in actual flight. (Indeed, aeronautical engineers use a form of computer simulation too.) While mimicking the operation of the system, the simulation program collects data which enable us to calculate quantities of interest, (e.g., mean data access time).

The philosophy of this text is that since mathematical and simulation tools are *used* together, it makes sense to *teach* these two tools together in an integrated manner. This is also the basis on which the homework exercises are designed, with some problems requiring mathematical solutions, and some others requiring simulation solutions. Many problems require both solutions. For example, Part (a) of a problem may treat a simple case, which can be done mathematically. Then Part (b) may ask for the solution of a more complicated, mathematically intractable problem for which a simulation solution is necessary. However, even the mathematically tractable problems usually ask for simulation solutions as well; this interaction not only gives additional practice in using simulation, but also has an extremely valuable effect on the student's power to apply the concepts of statistical modeling.

FURTHER READING

Tung Au, Richard Shane, and Lester Hoel, *Fundamentals of Systems Engineering: Probabilistic Models*, Addison-Wesley, 1979.

Richard Barlow and Frank Proschan, *Statistical Theory of Reliability and*

Life Testing: Probability Models, Holt, Reinhart and Winston, 1975. Reprinted by To Begin With, Silver Spring, Md., 1981.

W. Biles and J. Swain, *Optimization and Industrial Experimentation*, Wiley, 1980.

Roy Billinton and Ronald Allan, *Reliability Evaluation of Engineering Systems: Concepts and Techniques*, Pitman, 1983.

William Cochran, *Sampling Techniques*, Wiley, 1977.

Ove Ditlevsen, *Uncertainty Modeling with Applications to Multidimensional Civil Engineering Systems*, McGraw-Hill, 1981.

Charles Haan, *Statistical Methods in Hydrology*, Iowa State University Press, 1977.

Owen Hanson, *Design of Computer Data Files*, Computer Science Press, 1982.

Harold Larson and Bruno Shubert, *Probability Models in Engineering Sciences*, Wiley, 1979.

Michael Pidd, *Computer Simulation in Management Science*, Wiley, 1984.

L. Rabiner and R. Schafer, *Digital Processing of Speech Signals*, Prentice-Hall, 1978.

Edward Reingold, Jurg Nievergelt, and Narsingh Deo, *Combinatorial Algorithms: Theory and Practice*, Prentice- Hall, 1977.

Charles Sauer and K. Mani Chandy, *Computer Systems Performance Modeling*, Prentice-Hall, 1981.

Mischa Schwartz, *Computer-Communication Network Design and Analysis*, Prentice-Hall, 1977.

Jack Sklansky and Gustav Wassel, *Pattern Classifiers and Trainable Machines*, Springer-Verlag, 1981.

R. Syski, *Introduction to Congestion Theory in Telephone Systems*, Oliver and Boyd, 1960.

Kishor Trivedi, *Probability and Statistics, with Reliability, Queueing and Computer Science Applications*, Prentice-Hall, 1982.

CHAPTER 2

Fundamental Concepts of Probability

2.1 INTUITIVE NOTIONS

We all have an intuitive idea of the word **probability.** If we say that the probability of heads for a coin is 1/2, we mean that if we toss the coin repeatedly, then in the long run, 50% of the tosses will result in heads. In this section we will elaborate on this idea. We start with some *informal* definitions, based on this intuitive concept (more formal definitions will be given in Section 2.5).

Informal Definition 2.1.1

Suppose we have an experiment that we perform repeatedly, recording the result of each repetition, and we have an event—a possible outcome of the experiment, denoted by A. Let $n(A)$ denote the number of times A occurs among the first n repetitions of the experiment. Then the probability of A, denoted by $P(A)$, is

$$\lim_{n\to\infty} \frac{n(A)}{n} \tag{2.1.1}$$

In the coin example above, our "experiment" consists of tossing a fair coin once, and A represents the event that we get a head. Then $n(A)$ is the number of heads we obtain in the first n of infinitely many tosses, and thus the limiting value of $n(A)/n$ will be 1/2. Note that in applications of probability we don't actually do the experiment more than once, but the interpretation of probability is based on the long-run results we would get if we were to repeat the experiment indefinitely. Computer simulation, a very valuable practical tool, is also based on this notion. Note also that the word "experiment" is not to be taken with the meaning of a laboratory experiment.

In probability computations, we are very often interested in various interactions between events:

Informal Definition 2.1.2

Let A and B denote two events associated with some experiment, and again consider the long-run results from repeating the experiment infinitely many times.

a. Let $n(A \text{ and } B)$ denote the number of repetitions of the experiment which have the property that both A and B occur on the same repetition among the first n times the experiment is performed. Then $P(A \text{ and } B)$ is

$$\lim_{n\to\infty} \frac{n(A \text{ and } B)}{n} \qquad \textbf{(2.1.2)}$$

b. Let $n(A \text{ or } B)$ denote the number of repetitions having the property that at least one of the events A and B occurs among the first n repetitions. Then $P(A \text{ or } B)$ is

$$\lim_{n\to\infty} \frac{n(A \text{ or } B)}{n} \qquad \textbf{(2.1.3)}$$

c. $P(B|A)$, read as "the conditional probability of B, given A," is

$$\lim_{n\to\infty} \frac{n(A \text{ and } B)}{n(A)} \qquad \textbf{(2.1.4)}$$

The intuitive meaning of $P(B|A)$ is the long-run proportion of the times that B occurs, among those times in which A occurs.

To make these notations more concrete, consider the following example.

Example 2.1.1

A box contains two trick coins. Coin 1 is heavily weighted on the "tails" side, so much so that $P(\text{heads}) = 0.9$. Coin 2 is the opposite, with $P(\text{tails}) = 0.9$. Our experiment is to choose a coin at random from the box, and toss that coin once. Let B denote the event that the result is heads. From the symmetry of the situation, it should be clear that the result of the experiment is equally likely to be heads or tails—if we don't know which coin was chosen. This is written as $P(B) = 0.5$. However, if the two coins are easily distinguishable and we know that we chose Coin 1, we would say there is a 90% chance of getting heads, which we write as $P(B|A) = 0.9$, where A denotes the event that Coin 1 is chosen. On the other hand, $P(A \text{ and } B)$ means the likelihood that our experiment results in both A and B occurring; to find this probability, note that if the experiment were to be done infinitely many times, half of the repetitions would result in A occurring, and 90% of these B would occur, so that 45% of all repetitions would result in both A and B occurring. Thus $P(A \text{ and } B)$ is 0.45.

Definition 2.1.3

The set of all possible outcomes of an experiment is called the **sample space** for that experiment.

Example 2.1.2

Suppose our experiment consists of rolling two dice. Then the sample space is the set of 36 ordered pairs in Figure 2.1. For example, the ordered pair (2,5) means that the first die came up "2" and the second die was a "5."

Note that some sample spaces have infinitely many points. Suppose we roll a die until a "2" appears. Then the sample space may be written

$$S = \{2,\ N2,\ NN2,\ NNN2,\ \ldots\} \tag{2.1.5}$$

where N denotes "non-2." For example, $NN2$ means that we rolled two non-2's (say, a 4 and a 1), and then rolled a 2 on the third try.

(1,1) (1,2) (1,3) (1,4) (1,5) (1,6)
(2,1) (2,2) (2,3) (2,4) (2,5) (2,6)
(3,1) (3,2) (3,3) (3,4) (3,5) (3,6)
(4,1) (4,2) (4,3) (4,4) (4,5) (4,6)
(5,1) (5,2) (5,3) (5,4) (5,5) (5,6)
(6,1) (6,2) (6,3) (6,4) (6,5) (6,6)

Figure 2.1 Sample space for the experiment of rolling two dice.

Readers should keep in mind that many probability problems are solved much more easily if they first take the simple step of writing down a precise description of the sample space.

2.2 COMPUTING PROBABILITIES: UNIFORM SAMPLE SPACES

A **uniform** sample space is one in which all outcomes are equally probable. For example, if the dice used for Figure 2.1 are fair (i.e., each of the six sides has probability 1/6), then all 36 outcomes of the experiment are equally probable. This section is devoted to computation of probabilities in such situations.

At first glance, this may seem to be a trivial matter of counting. For example, suppose we want to find the probability of obtaining a sum of 4 from tossing two fair dice. Since each of the 36 pairs in Figure 2.1 is equally likely, the probability of getting a sum of 4 will then be $x/36$, where x is the number of pairs in Figure 2.1 which have a sum of 4. It is easy to list these pairs:

$$A=\{(1,3),(2,2),(3,1)\} \tag{2.2.1}$$

where we have used the letter A to denote the event of getting a sum of 4. Here $x=3$, so $P(A)=3/36=1/12$.

In general, for a uniform space,

$$P(A)=\frac{\#(A)}{\#(S)} \tag{2.2.2}$$

where #(G) denotes the number of elements in a set G. In the above example, $\#(A)=3$ and $\#(S)=36$.

Thus, computation of probabilities for uniform spaces is "simply" a matter of counting. However, as you might guess from the quotation marks in the last sentence, in some problems the counting process may not be a simple matter at all. Consider the following example.

Example 2.2.1

Suppose we select five cards at random from a standard 52-card deck. What is the probability of getting exactly two hearts among our five cards?

To solve this problem, let us first state the sample space. Recalling that the sample space is the set of all possible outcomes of the experiment, we see that in this case the sample space will be the set of all possible five-card hands. Thus, the sample space here will consist of a set of 5-tuples, just as the sample space in Example 2.1.2 consisted of 2-tuples, or pairs.

$$S=\{(\text{AC}, \text{4D}, \text{3H}, \text{7S}, \text{KC}),\ (\text{9H}, \text{5C}, \text{9C}, \text{3D}, \text{2S}), \ldots\} \qquad \textbf{(2.2.3)}$$

Here 4D means the 4 of diamonds, KC the king of clubs, and so on. Thus the "point" (AC,4D,3H,7S,KC) represents the case in which our five-card hand happens to consist of the ace of clubs, 4 of diamonds, 3 of hearts, 7 of spades, and the king of clubs. Note that in this problem we don't care about the order in which we obtain the cards. For example, (AC,4D,3H,7S,KC) and (7S,4D,AC,KC,3H) are considered the same and do not have separate entries in Equation (2.2.3).

The counting difficulties to which we alluded above are already quite clear in the example; direct computation of $\#(S)$ by simply writing down a list of all elements in S is out of the question (in fact, we will find below that there are actually more than 2 million hands listed in S).

Since counting by direct enumeration is not possible in many problems, we need to develop indirect techniques of counting. These techniques are known collectively as **combinatorial methods.** The methods we will use all derive from the following simple observation.

The Multiplication Principle

> Suppose a process consists of two stages. Let n_1 denote the number of ways to do Stage 1, and suppose that regardless of which way is used for Stage 1, the number of ways to do Stage 2 once Stage 1 is done is n_2. Then the total number of ways to do the entire process is $n_1 n_2$.

This simple principle is the basis for the solution of most counting problems. To introduce its use, we consider the following example.

Example 2.2.2

Suppose we wish to know how many two-letter "words" may be formed without duplicate letters (for example, "cg" and "za" are included but "cc" is not). We will count words with the same letters in different orders as different words (e.g., "by" and "yb" count as two separate words). To use the Multiplication Principle, note that the process of forming a two-letter word consists of two stages: Stage 1 is the action of choosing the first letter, and Stage 2 is the choice of the second letter. n_1 is of course equal to 26, and n_2 is 25, since there will be only 25 choices for Stage 2 once Stage 1 is done. Thus, the Multiplication Principle implies that there are (26)(25) ways to do the two-stage process here; that is, there are 650 possible words.

The possible results of the two-stage process in the last example are shown in Figure 2.2. The figure also shows why the numbers 26 and 25 are

— ab ac ad ae af ag ah ai aj ak al am an ao ap aq ar as at au av aw ax ay az
ba — bc bd be bf bg bh bi bj bk bl bm bn bo bp bq br bs bt bu bv bw bx by bz
ca cb — cd ce cf cg ch ci cj ck cl cm cn co cp cq cr cs ct cu cv cw cx cy cz
da db dc — de df dg dh di dj dk dl dm dn do dp dq dr ds dt du dv dw dx dy dz
ea eb ec ed — ef eg eh ei ej ek el em en eo ep eq er es et eu ev ew ex ey ez
fa fb fc fd fe — fg fh fi fj fk fl fm fn fo fp fq fr fs ft fu fv fw fx fy fz
ga gb gc gd ge gf — gh gi gj gk gl gm gn go gp gq gr gs gt gu gv gw gx gy gz
ha hb hc hd he hf hg — hi hj hk hl hm hn ho hp hq hr hs ht hu hv hw hx hy hz
ia ib ic id ie if ig ih — ij ik il im in io ip iq ir is it iu iv iw ix iy iz
ja jb jc jd je jf jg jh ji — jk jl jm jn jo jp jq jr js jt ju jv jw jx jy jz
ka kb kc kd ke kf kg kh ki kj — kl km kn ko kp kq kr ks kt ku kv kw kx ky kz
la lb lc ld le lf lg lh li lj lk — lm ln lo lp lq lr ls lt lu lv lw lx ly lz
ma mb mc md me mf mg mh mi mj mk ml — mn mo mp mq mr ms mt mu mv mw mx my mz
na nb nc nd ne nf ng nh ni nj nk nl nm — no np nq nr ns nt nu nv nw nx ny nz
oa ob oc od oe of og oh oi oj ok ol om on — op oq or os ot ou ov ow ox oy oz
pa pb pc pd pe pf pg ph pi pj pk pl pm pn po — pq pr ps pt pu pv pw px py pz
qa qb qc qd qe qf qg qh qi qj qk ql qm qn qo qp — qr qs qt qu qv qw qx qy qz
ra rb rc rd re rf rg rh ri rj rk rl rm rn ro rp rq — rs rt ru rv rw rx ry rz
sa sb sc sd se sf sg sh si sj sk sl sm sn so sp sq sr — st su sv sw sx sy sz
ta tb tc td te tf tg th ti tj tk tl tm tn to tp tq tr ts — tu tv tw tx ty tz
ua ub uc ud ue uf ug uh ui uj uk ul um un uo up uq ur us ut — uv uw ux uy uz
va vb vc vd ve vf vg vh vi vj vk vl vm vn vo vp vq vr vs vt vu — vw vx vy vz
wa wb wc wd we wf wg wh wi wj wk wl wm wn wo wp wq wr ws wt wu wv — wx wy wz
xa xb xc xd xe xf xg xh xi xj xk xl xm xn xo xp xq xr xs xt xu xv xw — xy xz
ya yb yc yd ye yf yg yh yi yj yk yl ym yn yo yp yq yr ys yt yu yv yw yx — yz
za zb zc zd ze zf zg zh zi zj zk zl zm zn zo zp zq zr zs zt zu zv zw zx zy —

Figure 2.2 All two letter words without repeated letters.

multiplied, rather than added: Each of the 26 rows in Figure 2.2.1 contains 25 words, so there are (26)(25) words in all.

The Extended Multiplication Principle

> The Multiplication Principle extends to multi-stage processes. In a k-stage process in which there are n_i ways to do Stage i once Stages 1 through $i-1$ have been done ($i = 2, 3, \ldots, k$), there are $n_1 n_2 \cdots n_k$ ways to do the entire process.

Example 2.2.3

Consider the three-letter version of Example 2.2.2. The Extended Multiplication Principle tells us that there are a total of (26)(25)(24) = 15600 possible three-letter words with no repeat letters.

Example 2.2.4

Let us look at another variation of the previous examples. How many five-letter words are possible by rearranging the letters p, l, a, c, and e? For example, two possibilities are "place" and "ceapl." Think of the process of forming such a word as consisting of five stages. Stage 1 consists of choosing the first letter for the five-letter word from among the five original letters p, l, a, c, and e; Stage 2 represents the choice of the second letter of the word we are forming from among the four letters remaining at that time and so on through Stage 5. We see from the Extended Multiplication Principle that the total number of words possible is (5)(4)(3)(2)(1) = 120.

The reasoning in Example 2.2.4 is easily generalized:

The Permutation Formula

> The total number of possible arrangements of n objects is $n(n-1)(n-2)\ldots(2)(1) = n!$

We now need just one more tool to be able to solve the card problem in Example 2.2.1. Again, we will develop it by considering an example first.

Example 2.2.5

Suppose we choose five (distinct) letters at random from the alphabet, not caring about the order in which they are chosen. How many sets of letters are possible? To answer this question, note first that if we did care about the order, then the problem would be similar to Examples 2.2.2 and 2.2.3, and the answer would be $(26)(25)(24)(23)(22) = 7{,}893{,}600$. Of course, this number is much larger than the number in which we are actually interested here, since each particular set of five letters is represented many times among the 7,893,600 words—once for each arrangement, or ordering, of the given five letters. This last remark is the key to finding the number of *unordered* sets: Example 2.2.4 shows that *any* set of five letters has 120 possible arrangements. Thus, since the number 7,893,600 represents all possible arrangements of all possible sets, and each set contributes 120 arrangements to this total, there must be $7{,}893{,}600/120 = 65{,}780$ different sets.

The principle we discovered in Example 2.2.5 is a counterpart of the Permutation Formula.

The Combination Formula

The total number of unordered subsets consisting of k distinct objects drawn from a set of size n is

$$n(n-1)(n-2)\ldots[n-(k-1)]/k! = \frac{n!}{k!(n-k)!} \qquad \textbf{(2.2.4)}$$

We denote the quantity in Equation (2.2.4) by $\binom{n}{k}$.

We now are in a position to solve the problem posed in Example 2.2.1. The probability of getting exactly two hearts in our five-card hand was pointed out in Equation (2.2.2) to be $\#(A)/\#(S)$, where A is the set of all possible five-card hands having exactly two hearts, and the S is the set of all possible five-card hands of any composition. To find the sizes of these two sets, we note first that the Combination Formula is immediately applicable to finding $\#(S)$, since we are drawing 5 cards from a set of 52 cards; we thus have $\#(S) = \binom{52}{5}$.

Finding $\#(A)$ involves both the Combination Formula and the Multiplication Principle: Consider the process of choosing a five-card hand with exactly two hearts to be a two-stage process. Stage 1 consists of choosing the two hearts from the 13 available, and Stage 2 consists of choosing the three non-hearts from the 39 available. By the same line of reasoning that we used in finding $\#(S)$, we see that n_1 is $\binom{13}{2}$ and n_2 is equal to $\binom{39}{3}$. Thus we can solve Example 2.2.1.

$$P(\text{exactly two hearts}) = \frac{\binom{13}{2}\binom{39}{3}}{\binom{52}{5}} = 0.274 \qquad \textbf{(2.2.5)}$$

Moreover, during our analysis we also developed some tools that are applicable to a large class of problems, such as in the following example.

Example 2.2.6

You bought three tickets in a lottery, for which 60 tickets were sold in all. There will be five prizes given. What is the probability that you win at least one prize?

To solve this problem, first note that it is easier to find the probability that you win no prize, and then subtract this number from 1.0 to get the probability of at least one prize; otherwise we would have to find three separate probabilities—of getting exactly one prize, of getting exactly two, of getting exactly three—and then add these three numbers.

Now to set up our sample space S, recall that the sample space is supposed to be a list of all possible outcomes of the experiment. What is the "experiment" in this case? Imagine that it is the day of the drawing, and there is a box containing the stubs of the 60 tickets sold; included among the 60 are the stubs for the three tickets that you bought. The experiment here will then be the random drawing of five stubs from the 60 in the box. Thus, we may take our sample space S to be the set of all possible subsets of 5 tickets from the 60 sold.

Since any of these subsets is equally as likely as any other (assuming that the tickets are fairly drawn), we see that the probability of winning no prizes is equal to $\#(A)/\#(S)$, where A is the collection of all five-ticket sets in which none of the five is among the three that you bought.

$\#(S)$ is obviously equal to $\binom{60}{5}$ but what about $\#(A)$? Actually, this number can be found simply too: The set A consists of groups of five tickets chosen from the 57 that you didn't buy, so $\#(A)$ is $\binom{57}{5}$. Thus

$$P(\text{win at least one prize}) = 1 - \frac{\binom{57}{5}}{\binom{60}{5}} = 0.233 \tag{2.2.6}$$

Most combinatorial problems have solutions which are variations of those in the examples above. We present one more example below.

Example 2.2.7

Two five-person committees are to be formed from your group of 20 people. In order to foster communication between the two committees, it is required that they both have the same chairperson. However, no further overlap is allowed. The membership is to be assigned randomly. What is the probability that you are chosen for exactly one committee?

This can be thought of as a three-stage process: Choose 4 people as non-chairs for Committee 1; choose the common chair; and choose the 4 non-chairs for Committee 2. There are $\binom{20}{4}$ ways to do the first stage. Once that is completed, there will be only 16 people left from which to choose, so there will be $\binom{16}{1}$ ways to do this stage of the process. Then there will be 15 people left for the last stage, making $\binom{15}{4}$ ways to choose. Thus by the Multiplication Principle, there will be

$$\binom{20}{4}\binom{16}{1}\binom{15}{4}$$

points in the sample space S.

The set A consists of points in S at which you are chosen as a non-chair. This set in turn can be broken down into sets A_1 and A_2, where A_i denotes the set of points at which you are chosen as non-chair of Committee i, $i = 1, 2$.

First look at A_1. There are $\binom{19}{3}$ ways to choose the other three non-chairs from among the 19 people eligible (not 20, since you are already chosen). This will leave 16 people who have not been chosen for anything yet, so there will be $\binom{16}{1}$ possible choices for the chair. After the chair is chosen, there will be 15 people left, of which we must choose four for the non-chairs of Committee 2; there are $\binom{15}{4}$ ways to do this. Thus

$$\#(A_1) = \binom{19}{3}\binom{16}{1}\binom{15}{4} \tag{2.2.7}$$

We could then use the same reasoning to find an expression for $\#(A_2)$. However, a faster way is to notice that the symmetry of the situation implies that $\#(A_1) = \#(A_2)$. Thus, the probability that you are chosen for exactly one committee is

$$\frac{2\binom{19}{3}\binom{16}{1}\binom{15}{4}}{\binom{20}{4}\binom{16}{1}\binom{15}{4}} = 0.4 \tag{2.2.8}$$

2.3 COMPUTING PROBABILITIES: NONUNIFORM SPACES

There are many probability problems that cannot be solved simply by counting. The reason for this is that the formula

$$P(A) = \frac{\#(A)}{\#(S)}$$

which we used in the last section depended on the fact that, in the examples presented there, all elements of S were equally likely. There are many problems in which this is not true. In this section we will discuss methods for solving problems of this type.

The basic technique is usually to break an event down into two or more subevents whose probabilities are easily obtainable, usually from simple intuition. In this section we will present some rules which are often helpful in this process, but first we need two important definitions.

Definition 2.3.1

Two events A and B are said to be **mutually exclusive** if they cannot both occur on the same repetition of the experiment, that is, if $P(A \text{ and } B)=0$. Similarly, a group of events $E_1, E_2, \ldots$ is said to be mutually exclusive if $P(E_i \text{ and } E_j)=0$ for any distinct i and j.

For example, suppose we draw a five-card hand from a standard deck of cards. Then if we let $A=\{$we obtain exactly one heart$\}$ and $B=\{$we obtain exactly two hearts$\}$, A and B are clearly mutually exclusive; A might happen on one repetition of the experiment and B might occur on some other repetition, but they cannot happen on the same repetition. On the other hand, if we change B to the event $\{$we obtain exactly two diamonds$\}$, then A and B are not mutually exclusive.

Definition 2.3.2

We say that two events A and B are **independent** if knowledge that one has occurred on a particular repetition of the experiment does not change our assessment of the probability that the other occurred, that is, if $P(A|B)=P(A)$ and $P(B|A)=P(B)$. Similarly, we say that a group of events $E_1, E_2, \ldots$ is independent if the probability of any one of them is the same, with or without the knowledge of whether the others occurred.

For example, suppose I toss a coin twice in the next room, where you cannot see whether the result is heads or tails. Let $A=\{$first toss is a head$\}$ and $B=\{$second toss is a head$\}$. If I ask you for your assessment of the probability of A, you would of course say 1/2; but what would your assessment be if I gave you a hint by informing you that the second toss resulted in a head (i.e., B occurred)? Would your assessment of the probability of A change? Of course not; you would still say that there is a 50% chance that A occurred. Thus A and B are independent events.

On the other hand, consider Example 2.1.2. Let $A=\{$sum of the dice is 2$\}$, and $B=\{$the first die is a 1$\}$. Here $P(A)=1/36$, but $P(A|B)=1/6$, so A and B are not independent events. This shows the nonindependence according to the definition, but the reader should note that independence is usually not checked by definition. In most cases, independence or nonindependence should be intuitively clear, as it is in this case.

The reader should keep in mind that the terms **mutually exclusive** and **independent** are *relational* in nature; they describe relations among events.

It is nonsense to ask whether a single event, say A, is mutually exclusive.

We now present some rules used in computing probabilities. They are likely to appear rather abstract to you at first, but they will be illustrated by a series of examples. After going through the examples carefully, you will see that these rules are actually quite simple, and nothing more than common sense.

Rule 1: Finding Probabilities Involving "or"

If A and B are mutually exclusive events, then

$$P(A \text{ or } B) = P(A) + P(B) \qquad \textbf{(2.3.1)}$$

Otherwise, a more general formula must be used:

$$P(A \text{ or } B) = P(A) + P(B) - P(A \text{ and } B) \qquad \textbf{(2.3.2)}$$

If $E_1, E_2, \ldots$ are mutually exclusive events, then

$$P(E_1 \text{ or } E_2 \text{ or } \ldots) = P(E_1) + P(E_2) + \cdots \qquad \textbf{(2.3.3)}$$

[An analog of Equation (2.3.2) for Equation (2.3.3) exists, but it is too cumbersome to be generally useful and is not presented here.]

Note that the "correction factor," $P(A \text{ and } B)$ in Equation (2.3.2), is necessary to avoid double-counting in the cases in which A and B overlap. For example, let $A = \{\text{the first die is a } 2\}$ and $B = \{\text{the sum of the two dice is } 4\}$ in Example 2.1.2. $P(A) = 6/36$ and $P(B) = 3/36$. $P(A \text{ or } B) = P[(2,1),(2,2),(2,3),(2,4),(2,5),(2,6),(1,3),(3,1)] = 8/36$, which is not the same as $P(A) + P(B)$. The latter sum, equal to 9/36, is too large because the point (2,2) is implicitly counted twice; thus, adjustment by subtracting $P(A \text{ and } B) = P(2,2) = 1/36$ provides the correct answer.

Rule 2: Finding Probabilities Involving "and"

If A and B are independent events, then

$$P(A \text{ and } B) = P(A)P(B) \qquad \textbf{(2.3.4)}$$

Otherwise,

$$P(A \text{ and } B) = P(A)P(B|A) = P(B)P(A|B) \qquad \textbf{(2.3.5)}$$

Also, for a set of independent events $E_1, E_2, \ldots,$

$$P(E_1 \text{ and } E_2 \text{ and } \ldots) = P(E_1)P(E_2) \cdots \qquad \textbf{(2.3.6)}$$

Both equations in (2.3.5) can be put to good use in computing complicated probabilities. To see why they are correct, observe that in the notation of Section 2.1, we have

$$P(A)P(B|A) = \left[\lim_{n\to\infty} \frac{n(A)}{n}\right]\left[\lim_{n\to\infty} \frac{n(A \text{ and } B)}{n(A)}\right]$$

$$= \lim_{n\to\infty} \frac{n(A \text{ and } B)}{n}$$

$$= P(A \text{ and } B)$$

Thus Equation (2.3.5) can be used to find an "and" probability from a conditional probability (and one unconditional probability). However, in some cases we might have the reverse situation, in which the "and" probability is known, or can be found easily, and the conditional probability is the one we would like to find. Equation (2.3.5) is thus sometimes useful in another form:

Rule 3: Finding Conditional Probabilities

$$P(B|A) = \frac{P(A \text{ and } B)}{P(A)} \qquad \textbf{(2.3.7)}$$

The next rule is one that we have already encountered, in the solution of Example 2.2.6.

Rule 4: Simplifying through Complementation

$$P(A) = 1 - P(\text{not } A) \qquad \textbf{(2.3.8)}$$

Now let us apply these rules in the following example.

Example 2.3.1

Urn 1 contains four blue marbles and two yellow ones, while in Urn 2 the mix is three blue and five yellow. A marble is drawn at random from Urn 1, and then put into Urn 2. A marble is then drawn at random from Urn 2. What is the probability that the second marble drawn is blue?

To solve this problem, we first set up some notation: Let B_1 denote the event that the first marble we draw is blue, and let B_2 denote the event that the second marble is blue. Similarly, let Y_1 and Y_2 denote the corresponding events, in which we draw a yellow marble on the first or second draw.

Our first step will be to *break big events down into small events,* in preparation for using Rule 1. This approach is extremely common; thus you should always keep it in mind as a possible solution to many other problems. We write

$$P(B_2) = P(B_1 \text{ and } B_2 \textbf{ or } Y_1 \text{ and } B_2) \tag{2.3.9}$$

That is, we ask ourselves, "How can B_2 happen?" If we are to break down the event B_2 (again, we remind you that this is a common first step), the natural breakdown seems to depend on the outcome of Stage 1 of the experiment; that is, we break down an event concerning Stage 2 by asking what outcomes are possible for Stage 1.

Next, note that the events

$$\begin{aligned} E_1 &= B_1 \text{ and } B_2 \\ E_2 &= Y_1 \text{ and } B_2 \end{aligned} \tag{2.3.10}$$

are mutually exclusive. This may seem a little strange at first, since their definitions both involve B_2. However, if E_1 occurs, it is impossible for E_2 to occur and vice versa, so the two events are mutually exclusive. Thus Rule 1 gives us

$$P(B_2) = P(B_1 \text{ and } B_2) + P(Y_1 \text{ and } B_2) \tag{2.3.11}$$

Now from Rule 2, we have

$$\begin{aligned} P(B_1 \text{ and } B_2) &= P(B_1)P(B_2|B_1) \\ &= \frac{4}{6} \cdot \frac{4}{9} \end{aligned} \tag{2.3.12}$$

[Note that if B_1 occurs, there then will be a total of 9 marbles in Urn 2, 4 of which are blue; thus $P(B_2|B_1)$ is equal to 4/9.]

Similarly,

$$\begin{aligned} P(Y_1 \text{ and } B_2) &= P(Y_1)P(B_2|Y_1) \\ &= \frac{2}{6} \cdot \frac{3}{9} \end{aligned} \tag{2.3.13}$$

Thus

$$P(B_2) = \frac{4}{6} \cdot \frac{4}{9} + \frac{2}{6} \cdot \frac{3}{9} = \frac{11}{27} \tag{2.3.14}$$

There is one more rule to discuss before going on to more examples.

Rule 5: Bayes' Rule

$$P(B|A) = \frac{P(B)P(A|B)}{P(B)P(A|B) + [1 - P(B)]P(A|\text{not } B)} \quad \textbf{(2.3.15)}$$

[subject to the technical conditions that $P(A) > 0$, $P(B) > 0$, and $P(B) < 1$].

More generally, let A and $E_1, E_2, \ldots$ be events of positive probability. Suppose $E_1, E_2, \ldots$ are mutually exclusive of each other (but not necessarily of A), and that

$$E_1 \cup E_2 \cup \cdots = S$$

meaning that

$$P(E_1 \text{ or } E_2 \text{ or } \cdots) = 1$$

(recall that S denotes the entire sample space of the experiment). Then

$$P(E_j|A) = \frac{P(E_j)P(A|E_j)}{P(E_1)P(A|E_1) + P(E_2)P(A|E_2) + \cdots} \quad \textbf{(2.3.16)}$$

The reasoning behind Rule 5 is as follows. First, by applying Rule 3 twice,

$$P(B|A) = \frac{P(B \text{ and } A)}{P(A)} = \frac{P(B)P(A|B)}{P(A)}$$

Now we can decompose the denominator, $P(A)$, using the "break big events down into small events" approach mentioned above: The event A can occur either with B or without B, so

$$P(A) = P(A \text{ and } B) + P(A \text{ and not } B)$$

Then by using Rule 3 once more, this last equation becomes

$$P(A) = P(B)P(A|B) + [1 - P(B)]P(A|\text{not } B)$$

which yields Equation (2.3.15). Equation (2.3.16) is derived in the same way.

We can now also answer another question in Example 2.3.1.

Example 2.3.2

Recall the marbles-in-urns example above. Suppose we know that the second marble drawn was blue; what are the chances that the first marble chosen was blue? Here we use Rule 5: Since the event A in Rule 5 is the event which is known to have occurred, in this case A should be B_2; the event B in Rule 5 should be B_1. Then we can compute the desired probability, $P(B_1|B_2)$, from Equation (2.3.15), as

$$\frac{P(B_1)P(B_2|B_1)}{P(B_1)P(B_2|B_1) + P(Y_1)P(B_2|Y_1)} = \frac{\frac{4}{6}\cdot\frac{4}{9}}{\frac{4}{6}\left(\frac{4}{9}\right) + \frac{2}{6}\left(\frac{3}{9}\right)} = \frac{8}{11} \qquad \textbf{(2.3.17)}$$

It should be noted that Bayes' Rule plays a fundamental role in the areas of pattern recognition and artifical intelligence. The applications are too complex for use in detailed examples here, but to get an idea of the role of Bayes' Rule, consider the following example.

Example 2.3.3

Suppose we are designing a machine to do automatic recognition of blood types. Here the event B from Equation (2.3.15) might be the event that the blood is of Type O, while A is the event that the blood sample is observed by the machine to have a certain set of characteristics. In this context, we know A has occurred, and we want to know the probability that B has occurred. If we find this probability to be high, then we might want to do further manual checks to verify it; if not, we would consider other blood types. Bayes' Rule will then give us this conditional probability, in terms of other probabilities we would know in this context, for example P(B), the *overall* prevalence of Type O in the human population.

The next example will not only illustrate the rules presented above, but also give us another opportunity to use the combinatoric techniques we developed in the last section.

Example 2.3.4

Suppose we toss a fair coin five times. What is the probability that we get exactly two heads?

Again we need to break down the original problem into more easily handled pieces. In this case, we break the problem down according to the possible orders of appearance of the two heads and three tails:

$$\text{exactly 2 heads out of 5 tosses} = \text{HHTTT or HTHTT or THTHT or} \ldots$$

where, for example, the notation HHTTT means that the first two tosses resulted in heads, and the third, fourth, and fifth tosses were tails. Thus, from Rule 1 concerning "or" we see that P(exactly 2 heads out of 5 tosses) is equal to

$$P(\text{HHTTT}) + P(\text{HTHTT}) + P(\text{THTHT}) + \cdots \tag{2.3.18}$$

As noted previously, successive tosses of a coin are independent, so

$$\begin{aligned} P(\text{HHTTT}) &= P(H_1 \text{ and } H_2 \text{ and } T_3 \text{ and } T_4 \text{ and } T_5) \\ &= P(H_1)P(H_2)P(T_3)P(T_4)P(T_5) \\ &= (0.5)^5 \end{aligned} \tag{2.3.19}$$

from Rule 2 (note that H_1 denotes the event of getting a head on Toss 1, etc.).

The same reasoning shows that each of the terms in Equation (2.3.18) is equal to $(0.5)^5$, so our final answer in that equation will be $x(0.5)^5$, where x is the number of terms. To find the value of x, note that the number of terms in (2.3.18) is equal to the number of ways to choose two of five slots

— — — — —

to be filled with the letter "H." One way is H _ _ H _; the other three slots are automatically filled with "T": H T T H T. From the material in the last section, we know that

$$x = \binom{5}{2} = 10$$

Thus

$$P(\text{exactly 2 heads out of 5 tosses}) = 10(0.5)^5 = \frac{5}{16} \tag{2.3.20}$$

It should be noted that for any given problem, there may be many different ways to solve it, such as in the following case.

Example 2.3.5

In the last section, the answer to Example 2.2.7 (involving committee selection) had a fairly simple form, but the reasoning involved was somewhat delicate. Here is a different approach, which some people might consider easier: Let A be the event that you are picked for some committee, and let B be the event that you are chosen to be a non-chair. Then

$$\begin{aligned} P(B) &= P(A)P(B|A) \qquad \text{(Rule 2)} \\ &= [1 - P(\text{not } A)]\,P(B|A) \qquad \text{(Rule 4)} \end{aligned} \tag{2.3.21}$$

Now $P(\text{not } A)$ is the probability you are not chosen for any committee, so that the event "not A" means

you are not chosen on the first pick, and
you are not chosen on the second pick, and

. . .

you are not chosen on the ninth pick

Thus

$$P(\text{not } A) = \frac{19}{20} \cdot \frac{18}{19} \cdots \frac{11}{12} = \frac{11}{20}$$

Meanwhile, if you are chosen to be on some committee, you will be one of 9 such people, exactly one of which will be chair. Thus there is probability 8/9 that you are not chosen as chair (given that you are chosen for committee membership). Then the probability which we are seeking is found by substituting into Equation (2.3.21), giving

$$P(B) = \left(1 - \frac{11}{20}\right)\frac{8}{9} = 0.4$$

consistent with the answer we found by purely combinatorial methods in Section 2.2. [By the way, this is a good opportunity for you to review the difference in meaning between conditional and unconditional probability. You should make sure you understand the difference in interpretation between $P(B)$ and $P(B|A)$ in this example.]

We now present several more examples of the techniques illustrated above. This text has been written from an applications-oriented point of view. In keeping with this philosophy, we will use a number of examples from applied contexts, almost always from fields in engineering and computer science. This is of the utmost importance, since many new engineers find that they have insufficient *practical* insight into probability methods. Examples based on dice and cards are fun and simple to state, but they do not do much to develop insight or the ability to formulate probabilistic models. Actually, as you will see below, applied problems are not very difficult; the mathematical formulation of the problem—the "translation" from the English statement of the problem—takes some time, but with practice is easily done.

On the other hand, to give highly detailed applications-oriented examples would mean spending too much time explaining the application field itself. Thus, we have tried to strike a balance between academic and applied examples. And in the case of the applied examples, we have aimed for simplicity, avoiding the deeper and more complex applications, but nevertheless giving you a feel for the ways in which probability modeling is used in the real world.

You will find that the probabilistic principles used below are the same as those used above, and that these examples are just as easy as the ones discussed previously. In fact, you will probably be surprised to see how easy they are.

Example 2.3.6

Traffic engineering makes heavy use of probabilistic analysis. Here is a simple introductory example: Suppose you are driving along a small street and come to a busy highway, which you wish to cross. The question is how long you will have to wait to do so.

Let us suppose it takes 4 seconds to cross the highway safely; that is, there must be a 4 second gap between cars in order to make the crossing. For simplicity, let us divide time into 2 second slots, with the first slot beginning at the moment you arrive at the highway. Assume that for each time slot, the probability that a car passes by on the highway during that slot is 0.3; assume also that the slots are independent. Note that the time you need to cross the highway, measured in terms of time slots, is 2 slots.

Let S_1 be the first 2 second time slot after you arrive at the highway; then S_2 will denote the next time slot, S_3 the third, and so on. Let us find $P(A)$, where A denotes the event that you complete your crossing by a time at most 10 seconds after you arrive at the highway (i.e., that you finish crossing the highway by the end of the fifth slot).

As usual, we break big events down into small events. Write $P(A)$ as

$$P(A) = P(A_2) + P(A_3) + P(A_4) + P(A_5) \qquad \textbf{(2.3.22)}$$

where A_i denotes the event that you finish crossing the highway at the end of the slot S_i.

We will compute $P(A_4)$ here; the other terms are computed similarly. Again, we ask "How can the event happen?" Well, A_4 will occur precisely under the following conditions:

A car passes during S_2, but no cars pass during S_3 and S_4.

Note that it doesn't matter what happens during S_1 in this particular case. Note also that the condition that the slot S_2 must contain a car is crucial; otherwise, you could finish earlier than S_4, which then would not be relevant to the probability we are currently trying to determine, $P(A_4)$.

Thus, $P(A_4)$ is equal to the probability that S_2 contains a car and S_3 contains no car and S_4 contains no car. By Rule 2, and the assumed independence of the time slots, we have

$$P(A_4) = (0.3)(0.7)(0.7) = 0.147$$

After similar computations for the other terms in Equation (2.3.22), we then could find P(A), and we would be done (the details are left to the reader).

In this example, you might ask about our assumptions and data: Where,for example, do we get the value of 0.3 as the probability that a 2 second time slot contains a passing car? The answer is of course that one must go to the field and collect data. For example, if we watched the highway for 1000 time slots, and found that 30% of those contained cars, we would use 0.3 as the probability of a slot being filled. Issues of data collection and analysis are covered in Chapters 6 and 7 of this text.

The reader might also wonder if there is a general formula to $P(A_k)$, for general k, as opposed to our approach above, which involved "manually" listing all possibilities. This was easy for the case done above, $P(A_4)$, which involved only one possibility. $P(A_5)$ is a bit more complex, involving *three* possibilities: CCCNN, NCCNN, CNCNN, where "C" means that a car passes during a particular slot, and "N" means that no car passes during the slot. (NCCNN means no car in the first slot, cars in the second and third slots, and no cars in the fourth and fifth slots.)

$P(A_6)$ has even more possibilities, and so on. It would be nice to have a general formula, rather than having to try to list all possibilities and then add up the corresponding probability terms. Actually, such a formula can be obtained through the use of more sophisticated methods. We do this in the Appendix at the end of this section.

Example 2.3.7

Reliability engineering plays an absolutely crucial role in many applications. Projects such as airplanes and nuclear power plants form obvious examples, but other examples are numerous. For instance, equipment which monitors a patient in a hospital's intensive care unit must have as low a probability of failure as we can possibly achieve.

Consider a device (e.g., mechanical or electrical) in virtually constant use. Suppose it tends to have a short lifespan; to make the example simple, let us suppose that the device fails after either 1 month, 2 months, or 3 months, with probabilities 0.1, 0.7, and 0.2, respectively. Let us consider the value of having spare parts; specifically, suppose we have one spare copy of the device. What is the probability that the two copies last us a total of at least 4 months?

Again, we break big events down into small events; we note that the two-component system will last at least 4 months if it lasts exactly 4 months, or it lasts exactly 5 months, or it lasts exactly 6 months. Due to the "or's" here, Rule 1 and Equation (2.3.3) apply here, since the 3 small events are mutually exclusive (e.g., if the system lasts exactly 5 months, it does *not* last exactly 4 months, or exactly 6 months).

Now that we have used the "break big events down into small events" trick, we will use the other common trick for probability problems: We ask how an event happened. For example, how can the system last exactly 5 months? Well, either the original device lasts 2 months, and the spare lasts 3 months, or the original lasts 3 months and the spare lasts 2. (You should make sure there are no other possibilities.) Due to the or and and nature of the problem, we can use Rules 1 and 2:

$$P(\text{system lasts exactly 5 months}) = (0.7)(0.2) + (0.2)(0.7) = 0.28$$

(Note that we are assuming that the two components have independent lifetimes, which is a reasonable assumption; if by some chance the original component happens to be one with a short lifetime, this will neither increase nor decrease the chance that the spare has a short lifetime.) Proceeding similarly for the 4 month and 6 month cases, you should check that the probability that the system lasts at least 4 months is

$$(0.7)^2 + 2(0.7)(0.2) + (0.2)^2 = 0.81$$

Example 2.3.8

A concept that arises often in electrical engineering and computer science is that of a **bit**. This concept is quite simple: A bit is a digit in the base-2 form of a number (thus it is called a binary digit, or bit). For example, consider the number 5. Its base-2 representation is 101, since

$$5 = 1(2^2) + 0(2^1) + 1(2^0)$$

Then the bits for the number 5 are 1, 0, and 1. More precisely, we can refer to the bits according to their position in the sum-of-powers-of-2 expansion above: We say that the 2^2- bit is 1, the 2^1-bit is 0, and the 2^0-bit is 1.

The base-2 form of a number is very convenient because in electronic devices 1 can be represented by a high voltage and 0 by a low voltage, or in a telephone we can use a high-pitched tone for 1 and a low-pitched tone for 0. These can be used for storing numbers (e.g., storing account balances in a bank's disk file—see Example 2.3.9), or in telecommunications (e.g., transmitting account balances from one branch of a bank to another). For instance, to send the number 5 from one branch to another via a phone link, we would send a high pitch (1), then a low pitch (0), and then another high pitch (1), signifying the number 101, or 5.

A problem that arises is that there can be errors in bits; for example, transmission errors may occur. Suppose the error rate is 1%. In some applications (e.g. bank balances), even this seemingly small error rate is considered too high. To achieve a lower rate, a number of methods have been developed. One technique is **Triple Modular Redundancy** (TMR). In the bank balance example, we could store three copies of account balances, in 3 separate disk units. Then if, for instance, 2 of the copies agree with each other, but the third is different, we know one of them is wrong—probably the third copy, although we cannot be certain. This is known as a **voting** scheme, since X, Y, and Z are "voting" as to what the true bit value is.

For example, let X, Y, and Z be three copies of the same bit position (e.g., 3 copies of the 2^2-bit). What if we find that $X=1$ and $Y=1$, but $Z=0$? It looks as if Z is in error, and we conclude that the true bit was a 1. Of course, there is still a chance that we are wrong, but at least this TMR scheme will make our overall error rate much lower than 1%.

Let us see how much lower it will be. The error rate r will be the probability that at least two of X, Y, and Z are in error (if none or only one is in error, the voting scheme will provide us with the correct value of the bit). How can this happen? Here are the ways:

a. X and Y in error, Z not in error

b. X and Z in error, Y not in error

c. Y and Z in error, X not in error

d. X, Y, and Z all in error

Again, we will assume that the three copies act independently; an error in one copy neither increases nor decreases the chance of another copy being in error. Then situation (a) has probability $(0.01)^2(0.99)$, as do (b) and (c). Situation (d) has probability $(0.01)^3$. Thus

$$r = 3(0.01)^2(0.99) + (0.01)^3 = 0.000298.$$

In other words, the TMR scheme has reduced the error rate from 1% to only about 0.03%, an excellent reduction.

A number of error detection/correction schemes such as the TMR method shown above are in common use, for example, the **parity** method illustrated in Exercise 2.20.

Example 2.3.9

Many computer applications deal with large databases. Examples are credit card records, airline reservation systems, and parts databases for computer-aided engineering systems. Due to the potentially large volumes of data involved, fast access of the data depends on thorough analysis of the data storage devices, and often also on analysis of the communications networks connected to the database. Both aspects are probabilistic in nature. In this example, we will introduce the data storage concerns.

Computers store data on round, flat **disks**, very similar in appearance to a phonograph record (Figure 2.3). A disk is divided into a set of concentric rings called **tracks**. Each track is then

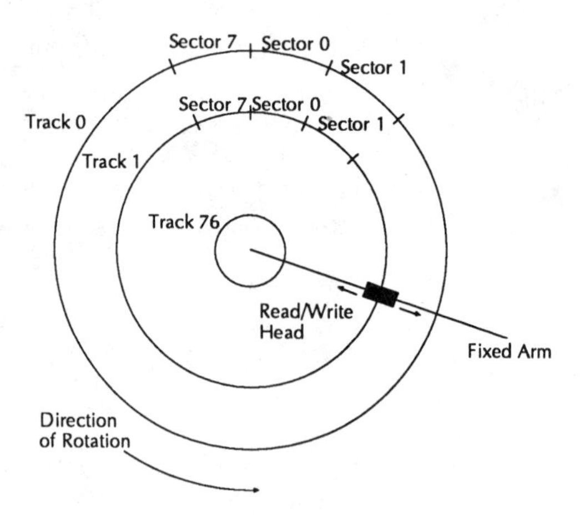

Figure 2.3 A disk with 77 tracks and 8 sectors.

subdivided into storage areas called **sectors**. The actual access to the disk is done by a **read/write head**, which can move back and forth along a fixed arm. The disk itself rotates underneath this arm, and the read/write head either senses (reads) the stored data or changes (writes) it as a sector passes by the head. In Figure 2.3, we have shown a disk with 77 tracks, numbered 0 through 76, with 8 sectors per track, numbered 0 through 7.

A program in the computer (the **operating system**) maintains a list which states the location on the disk at which each data item is stored. This list must state in which track, and which sector within that track, any given data item is stored. Thus, the list might indicate, for example, that the data for Mary Jones's credit card are stored in Sector 2 of Track 51.

Suppose a billing program running on the computer wishes to access Ms. Jones's record, and suppose the read/write head is currently positioned over Track 22. First the head must be moved to Track 51, a movement of $51-22=29$ tracks; this movement is called a **seek**. Then the head must wait for Sector 2 to rotate around so that it will pass under the read/write head, at which time the actual access of this record will occur; this period of time is called the **rotational delay**.

For concreteness, suppose the speed at which the head moves along the arm is 3.2 milliseconds per track, so that the seek in the above example will take $(29)(3.2)=92.8$ ms. Suppose also that the disk makes one rotation every 30 ms. This means that the rotational delay can be anywhere from 0 to 30 ms, with an average value of 15 ms. The access itself takes 1.2 ms.

The system is certainly probabilistic in nature. Access requests arrive at random times, and the requests are for random data items; that is, we cannot predict which item will be requested next. Following is an example of a very simple computation.

Suppose the disk has 76 tracks, numbered Track 0 (the innermost) through Track 75 (the outermost). Assume that all tracks are accessed with equal frequency, (i.e., the probability that Track i is requested is 1/76 for $i=0, 1, \ldots, 75$) and that successive requests are independent. Suppose the read/write head is currently positioned above Track 20. What is the probability that the total seek time for the next two requests is less than 50 ms?

Again, we break big events down into small events. Let A denote the event that the combined seek will take less than 50 ms, and for $i=0, 1, \ldots, 76$, let T_i denote the event that the *first* of the next two access requests is for a data item in Track i. Then

$$P(A)=P(A \text{ and } T_0)+P(A \text{ and } T_1)+\cdots+P(A \text{ and } T_{76}) \qquad \textbf{(2.3.23)}$$

Let us discuss a typical term in Equation (2.3.23), say $P(A$ and $T_{26})$.

$$P(A \text{ and } T_{26}) = P(T_{26})P(A|T_{26}) = \frac{1}{76}P(A|T_{26})$$

Thus we must find $P(A|T_{26})$.

Again, we ask how this can happen. How can the total seek time be less than 50 ms, if the first seek is to Track 26? First, note that the first seek will take $(26-20)(3.2)=19.2$ ms. Thus the total seek time will be less than 50 ms if the next seek takes less than $50-19.2=30.8$ ms. During 30.8 ms, $30.8/3.2=9$ (rounded down to the nearest integer) tracks can be traversed, so we can go as far inward as Track $26-9=17$, or as far outward as Track $26+9=35$.

In other words, $P(A|T_{26})$ is equal to the probability that the second access request is for a data item in one of the $35-17+1=19$ tracks between Tracks 17 and 35, inclusive. Since each track has probability 1/76, we now see that $P(A|T_{26})=19/76=1/4$. The other terms in Equation (2.3.23) can be found similarly.

A General Formula for $P(A_k)$

Recall that at the end of Example 2.3.6, we discussed the possibility of finding a general formula for $P(A_k)$. We will find such a formula here, not so much for the sake of that example as to illustrate some more sophisticated techniques, which are applicable to many other kinds of problems.

Let p_k denote $P(B_k)$, $k=2, 3, \ldots$, where B_k represents the event that you are not able to cross the road within the first k time slots. Note that

$$\begin{aligned} P(A_k) &= P(B_{k-1}) - P(B_k) \\ &= p_{k-1} - p_k \end{aligned}$$

so that finding the quantities p_k will solve our original problem of determining the quantities $P(A_k)$.

The general approach will be to derive a recursive equation, of the form

$$p_{k+1} = ap_k + bp_{k-1}$$

and then to "solve" this equation, in a manner similar to that of finding solutions to differential equations.

Using the ubiquitous "break big events down into small events" idea, define the quantities

$$q_k = P(B_k \text{ and } \textit{no car in } S_k)$$

and

$$r_k = P(B_k \text{ and } \textit{car in } S_k)$$

so that

$$p_k = q_k + r_k \qquad \textbf{(2.3.A1)}$$

Consider $p_{k+1} = P(B_{k+1})$. How can the event B_{k+1} occur; that is, how can we have no chance to cross the road in the first $k+1$ slots? This can happen in one of two ways:

a. We had no chance to cross during the first k slots, and the kth slot contained no car.

b. We had no chance to cross during the first k slots, and the kth slot contained a car.

If Case (a) occurs, B_{k+1} will occur if S_{k+1} contains a car, which has probability 0.3. If Case (b) occurs, then B_{k+1} will automatically occur and will have probability 1.0. Thus

$$p_{k+1} = 0.3q_k + r_k \qquad \textbf{(2.3.A2)}$$

Similar analyses (left to the reader) yield the following relations:

$$q_{k+1} = 0.7r_k \qquad \textbf{(2.3.A3)}$$

$$r_{k+1} = 0.3p_k \qquad \textbf{(2.3.A4)}$$

After doing some algebra with Equations (2.3.A1)–(2.3.A4), we find that

$$p_{k+1} = 0.3p_k + 0.21p_{k-1} \qquad (k \geq 3) \qquad \textbf{(2.3.A5)}$$

This last equation is a special case of

$$x_n = a_1x_{n-1} + a_2x_{n-2} + \cdots + a_kx_{n-m} \qquad \textbf{(2.3.A6)}$$

The general solution of this equation is

$$x_n = c_1r^{n_1} + \cdots + c_mr_m^n$$

where $r_1, \ldots, r_m$ are the roots r of the equation

$$r^n = a_1r^{n-1} + a_2r^{n-2} + \cdots + a_kr^{n-m}$$

(The reader may notice a parallel here with the solution of linear differential equations having constant coefficients.)

Thus in our case [Equation (2.3.A5)], we must find the roots r of

$$r^2 = 0.3r + 0.21$$

which are $r_1 = 0.63$ and $r_2 = -0.33$. Thus

$$p_k = c_1 0.63^k + c_2(-0.33)^k \qquad \textbf{(2.3.A7)}$$

for some constants c_1 and c_2.

We are almost done; the only task remaining is to determine c_1 and c_2. To do this, we calculate p_2 and p_3 "by hand," in the *ad hoc* manner we used originally in Example 2.3.6. For example, p_2 is the probability that at least one of the first two slots contains a car, which is $(0.3)^2+2(0.3)(0.7)=0.51$. Similarly, we find that $p_3 = 0.363$.

Substituting these values for $k=2$ and $k=3$ in Equation (2.3.A7), we find that $c_1=1.393$, and $c_2= -0.4$. So (finally!) we have our answer:

$$P(\text{cannot cross the road in first } k \text{ slots}) = 1.393(0.63)^k - 0.4(-0.33)^k \qquad (k \geq 2)$$

2.4 INTRODUCTION TO SIMULATION

In spite of the powerful tools developed in the last two sections, many probability problems are so complicated that obtaining a mathematical solution of any practical value is extremely difficult, if not impossible. For example, we will find in Chapter 4 that n-fold integrals

$$\int_A \cdots \int f(x_1, \ldots, x_n) dx_1 \ldots dx_n \tag{2.4.1}$$

arise frequently in applications. Any reader who has previously struggled in calculus or physics courses with triple integrals over some region of three-dimensional space can imagine the difficulty of evaluating such n-fold integrals, especially in view of the fact that both the function f and the n-dimensional region A tend to be much more complex than those commonly encountered in previous coursework.

Example 2.4.1

Suppose that in a certain telephone switching system there is a 0.05 second response time for each call. Once a call arrives, any subsequent calls arriving before the response period ends will be lost. Suppose we wish to find the probability that out of the next 20 calls to arrive, at least three will be lost. We will find in Chapter 4 that this probability is a 20-fold integral, with the set A in Equation (2.4.1) being exceedingly complicated.

Clearly, there is a need for some sort of approximation technique. Fortunately, in probability contexts there is a very natural method of approximation, a conceptually simple method derived directly from the intuitive definitions in Section 2.1. To illustrate this method, known as **simulation**, consider how we might approach Example 2.3.4 if we did not have the tools developed in the last section.

Example 2.4.2

Let E denote the event {two heads out of five tosses}. From Informal Definition 2.1.1, we know that $P(E)$ is the long-run proportion of the time that E occurs out of infinitely many repetitions of our experiment, or infinitely many sets of five tosses of the coin. This suggests an approximating technique: We could actually perform the experiment a large number of times, say 1000, and record the proportion of times that E occurs. For example, if out of the 1000 sets of five tosses, 348 result in exactly two heads, then we would estimate $P(E)$ to be 0.348. Of course, this would not be the exact answer, since Informal Definition 2.1.1 technically requires infinitely many repetitions of the experiment, but it should be a fairly accurate approximation.

This method, in the form just presented, is also impractical; no one would want to toss a coin 5000 times. However, the method becomes quite practical if we use a computer to simulate the tossing of a coin. For example, consider the following program:

Program 2.4.1

```
1     program Toss5(input,output);
2
3       var CountE,RepNumber,CountHeads,Toss: integer;
4            Heads: boolean;
5
6     begin
7       CountE:= 0;  {CountE = number of occurrences
8                of E}
9       for RepNumber:= 1 to 1000 do
10        begin
11        CountHeads:= 0;  {initialize count of number
12               of Heads}
13        for Toss:= 1 to 5 do
14          begin
15          simulate tossing the coin;  {this line of
16               English will be replaced by
17               appropriate Pascal code later}
```

```
18                  if Heads then CountHeads:= CountHeads + 1
19                  end;
20                  if CountHeads = 2 then CountE:= CountE + 1
21                  end;
22              writeln('the estimated value of P(E) is ',
23                          CountE/1000)
24          end.
```

The above program is not yet complete; there is still English in Line 15, which simulates tossing a coin. This line will be replaced by Pascal code below, but the important point to note now is that the above program simulates what we were doing by hand—it goes through 1000 sets (Lines 9–21) of five tosses (Lines 13–19) of a coin and counts the number of sets which result in exactly two heads (Line 20). The desired probability is then computed as CountE/1000: the proportion of the 1000 sets which produced two heads.

Replacing the English part of the program turns out to be fairly simple. First we need a **random number generator**, a function subprogram which produces values chosen at random from the numbers between 0 and 1. To be more specific, suppose we have a Pascal function with heading

```
function random(var Seed: integer) : real
```

which, upon being called repeatedly, will produce numbers as described above. Many implementations of languages such as Pascal and FORTRAN include random number generators as built-in functions, so for now we will assume that our implementation of Pascal has a built-in function of the type we are discussing here (construction of such functions will be discussed in Section 3.4).

Let us see how the function works. Consider the following program, which makes a large number of calls to this function. We have chosen 1000 as a "large number" here, just as we chose 1000 in Program 2.4.1, as representing the performance of the experiment 1000 times. Let us see how the function behaves.

Program 2.4.2

```
program Example(input,output);

  var Count,I,Seed: integer;
     X,Proportion: real;

begin
  Count:= 0;
  for I:= 1 to 1000 do
```

```
      begin
      X:= random(Seed);
      if I <= 5 then writeln(X:7:4);
      if (X >= 0.20) and (X <= 0.35) then
         Count:= Count + 1
      end;
   Proportion:= Count/1000;
   writeln(Proportion:7:4)
end.
```

The first thing to note is that each of the 1000 calls to the function random results in a different value of X (you will find out why in Section 3.4). To emphasize this, the program prints out the first five of these values, which turn out to be 0.8541, 0.7192, 0.3762, 0.6122, and 0.0937.

Next, the program has been written to at least partially verify the assertion that the function random chooses values at random from the interval (0,1). The verification consists of checking what proportion of values generated by random fall into the subinterval (0.20,0.35). If the function really does choose numbers in (0,1) at random, then the outputted value of Proportion should be near 0.15, since 15% of the numbers in (0,1) are in (0.20,0.35). The outputted value is 0.1454. (The discrepancy between this and 0.15 is due to the fact that only 1000 numbers were generated, rather than the ideal of infinitely many numbers.) Thus, the function is working as desired (a complete verification would check all subintervals, and also an independence property that we will discuss in Chapter 3).

Now simulation of coin tossing is easy:

```
function Hds(var Seed: integer): boolean;
  var X: real;
begin
  X:= random(Seed);
  if X < 0.5 then Hds:= true
  else Hds:= false
end;
```

This is just what we want: Hds will be "true" half the time, and "false" the other half, just as a real coin would come up heads half the time.

For those readers who like short programs, a more compact form is the following:

```
function Hds(var Seed: integer): boolean;
   begin
     Hds:= (random(Seed) < 0.5)
   end;
```

Now to use this in Program 2.4.1, simply replace Line 15 of Program 2.4.1 by

```
Heads:= Hds(Seed);
```

Results will depend somewhat on the details of the function random. In this text's version (see Section 3.4), the result was that $P(E)$ was estimated to be 0.310, as compared to the exact answer, 0.3125.

Program 2.4.1 follows what is actually a very general outline, given in the following program to find the probability of some event A arising in an experiment.

Program 2.4.3: General Form of Simulation Programs to Find *P(A)*

```
 1    program Genform(input,output);
 2      var {declarations go here}
 3    begin
 4      CountA = 0;
 5      for RepNum:= 1 to TotReps do
 6        begin
 7        simulate performing the experiment;
 8        if A Occurs then
 9           CountA:= CountA + 1
10        end;
11      ProbA:= CountA/TotReps;
12      writeln('P(A) = ',ProbA)
13   end.
```

(As in Program 2.4.1, italics have been used to indicate English *pseudocode*. Here Line 7 is to be replaced by Pascal code specific to the experiment involved.)

The program simulates repeatedly performing the experiment TotReps times. The number of occurrences, CountA, of the event A among the TotReps repetitions of the experiment is recorded. Then our estimate of the probability of that event is simply the proportion of time the event occurred during the simulation run, which is CountA divided by TotReps.

Of course, the larger TotReps is, the better. The value 1000 was used in the coin-tossing program above. In general, one should set TotReps to be as large as possible, subject to the time we are willing to wait for the output. Considerations for improving the speed of a simulation program will be discussed in Section 3.4; these may allow the use of larger values for TotReps. In Section 6.5, we will discuss the degree of accuracy achievable for a given value of TotReps and briefly mention some ways to achieve better accuracy for the same value of TotReps.

It is worth noting the correspondences between Program 2.4.3 and Program 2.4.1:

Program 2.4.3	Program 2.4.1
Line 5	Line 9
Line 7	Lines 13–19
Lines 8–9	Line 20
Line 12	Lines 22–23

Instead of $P(A)$, suppose we are interested in $P(A|B)$. Recalling that $P(A|B)$ is the long-run proportion of the time that A occurs *among those times at which B occurs,* it is clear how to modify Program 2.4.3 to find $P(A|B)$.

Program 2.4.4: General Form of Simulation Programs to Find *P(A|B)*

```
1   program GenForm(input,output);
2     var declarations go here
3   begin
4     CountAB:= 0;  {counts the number of times A
5           occurs among those times B occurs}
6     CountB:= 0;  {counts the number of time B
7           occurs}
8     for RepNum:= 1 to TotReps do
9       begin
10      simulate performing the experiment;
11      if B Occurs then
12        begin
13        CountB:= CountB + 1;
14        if A Occurs then
15          CountAB:= CountAB + 1
16        end
17      end;
18    ProbOfAGivB:= CountAB/CountB;
19    writeln('P(A|B) = ',ProbOfAGivB)
20  end.
```

The program below uses simulation to solve Example 2.3.1.

Program 2.4.5

```
program Example231(input,output);

  var CountB,CountAB,RepNum,Urn2Blue,Seed: integer;
    FirstIsBlue,SecondIsBlue: boolean;

  function IntRandom(M: integer;
              var Seed: integer): integer;
    begin

    {chooses an integer at random from the integers
     1,2,...,M}
    IntRandom := trunc(M*random(Seed) + 1) {assumes that
                       'random' is never equal to 1.0}
    end;

    function BlueCNumBlue,NumMarb: integer;
              var Seed: integer): boolean;
    begin
      {simulates the choice of a marble at random from
       a set of NumMarb marbles, NumBlue of which are blue;
       returns a Boolean value, indicating whether the
       chosen marble turned out to be blue}
       Blue := (IntRandom(NumMarb,Seed) <= NumBlue)
    end;

  begin
    {below, A will refer to the event that the first
     marble chosen is blue, and B will denote the event
     that the second marble chosen is blue}
     CountB := 0;  {CountB records the number of
                      occurrences of B}
     CountAB := 0;  {CountAB records the number of
                      occurrences of A, among those
                      repetitions in which B occurs}
     for RepNum := 1 to 1000 do
       begin
       {perform experiment and record results}
       {choose from Urn 1}
       FirstIsBlue := Blue(4,6,Seed);
       if FirstIsBlue then
          Urn2Blue := 4  {Urn2Blue is number of blue
                          marbles in Urn 2}
       else
```

```
        Urn2Blue  := 3;
      {now choose from Urn 2}
      SecondIsBlue  := Blue(Urn2Blue,9,Seed);
      {now update the counts}
      if SecondIsBlue then
         begin
         CountB  := CountB + 1;
         if FirstIsBlue then
            CountAB:= CountAB + 1
         end
   end;

   writeln('estimate of P(B) = ',CountB/1000);
   writeln('estimate of P(A|B) = ',CountAB/CountB)
end.
```

Running this program produced estimates of 0.379 for $P(B)$ and 0.752 for $P(A|B)$, compared to the exact answers 0.407 and 0.727.

Of course, the examples here had already been solved mathematically and were chosen only to illustrate simulation methods. It is always desirable to find an mathematical solution if possible, but there are many cases in which this is not possible; simulation is then an indispensable practical tool.

Example 2.4.3

Suppose we roll a fair die until we get three consecutive rolls showing the same number, three 4s in a row, for example. What is the probability that it takes at least 50 rolls to achieve this?

In spite of the simplicity of the statement of this problem, this problem is not at all simple to solve mathematically. In fact, it is quite a difficult problem, due to the fact that when one tries to "break big events into small events," it is unclear how to find a good way to enumerate all possible "small events." To fully appreciate the difficulties here, the reader is urged to try to find a mathematical solution (e.g., using the approach in the Appendix of Section 2.3). However, a simulation solution is quite easy.

Program 2.4.6

```
program Three(input,output);

   var Seed,Count,RepNum,TotReps,NumRolls: integer;
```

```
    function RollDie(var Seed:integer): integer;
    begin
       {simulates one roll of the die}
       RollDie := trunc(6*random(Seed) + 1)
    end;

    function RollTil3(var Seed: integer): integer;
       var RollsSoFar: integer;
          Done: boolean;
          Roll: array[1..50] of 1..6;
             {using an array is a little wasteful
              here, but it makes the program simpler}
    begin
       {simulates the experiment of rolling a die until
                getting three identical numbers in a row}

       {function result reports the number of
       rolls needed, with 50 denoting >= 50}
       Roll[1] := RollDie(Seed); {simulate the first two rolls}
       Roll[2] := RollDie(Seed);
       Done := false;
       RollsSoFar := 2;
       repeat
          RollsSoFar := RollsSoFar + 1;
          Roll[RollsSoFar] := RollDie(Seed);
          if (Roll[RollsSoFar] = Roll[RollsSoFar - 1]) and
             (Roll[RollsSoFar] = Roll[RollsSoFar - 2]) then
                Done := true
       until Done or (RollsSoFar = 50);
       RollTil3 := RollsSoFar
    end;

begin
    writeln('enter TotReps'); readln(TotReps);
    Seed := 9999;
    Count := 0;
    for RepNum := 1 to TotReps do
       begin
       NumRolls := RollTil3(Seed);
       if (NumRolls = 50) then Count := Count + 1
       end;
    writeln('approx. prob. = ',Count/TotReps)
end.
```

Running this program with TotReps = 1000 produced a probability estimate of 0.331.

Example 2.4.3 illustrates a very important point: The calculations of some probabilities turn out to be intractable mathematically, but have quite simple simulation solutions. In fact, this situation is encountered quite frequently in applications. Thus, simulation is a tool of great practical value. (This is analogous to the field of differential equations, in which textbook problems always have neat, closed-form solutions, but a great many real-world problems can be solved only by using numerical approximations.) Simulation is especially useful for problems involving some kind of operation through time, as in the cases of both Example 2.4.3 and Example 2.4.1.

The reader may then ask whether one should bother with mathematical solutions at all. The answer is definitely "yes," for a number of reasons:

- For simple problems, a mathematical solution is obtained much more quickly than by writing, debugging, and running a simulation program. Programming always takes more time than one anticipates. This is particularly true in the case of simulation of discrete-event systems, which are studied in Chapter 8. In fact, this kind of simulation is one of the most challenging areas of software engineering, since it usually requires that the programmer keep track of many interacting events simultaneously. Thus, the convenience of a mathematical solution is very appealing.
- Some of the more complicated simulation programs run for hours, or even days. In these cases, a mathematical solution may be much preferable, even if one has to make simplifying assumptions to obtain it.
- A mathematical solution is also much more convenient in problems for which we are interested in the effects of varying the values of several system parameters. For example, recall the application mentioned in Chapter 1 concerning a computer timeshare system. Here we can vary the number of terminals, the memory capacity, the disk capacity, and so forth. The goal is to determine the resulting effects on the mean user response time of the system. If we can obtain a closed-form expression for mean response time here, our investigation will be much easier than it would be using simulation, since the latter would require running the program a large number of times, once for each combination of parameter values.
- Mathematical solutions often make possible qualitative insights that could well be missed if simulation were used. For example, in Chapter 8 we will discuss what are known as **queueing problems**, which concern the behavior of waiting lines. (The lines do not necessarily need to consist of people; they might consist of phone conversations waiting to be routed through a communications channel.) One well-known mathematical formula in queueing theory gives the mean queue wait time, in terms of characteristics of the time of the service for which the items are waiting. The

formula shows that the *variability* of this service time is just as important a factor as the mean. This *qualitative* information is important to those who analyze problems of this type.

- As mentioned above, the discrete-event type of simulation is prone to error and difficult to write. Due to the complexity of such applications, it may be difficult to confidently state that the program is producing correct answers. Mathematical analysis can play a vital validation role here: We first run the program in a setting for which mathematical analysis is practical and if the answers check, we can then run the program in the settings for which mathematical analysis is not possible.

Thus, one should view mathematical and simulation methods as twin tools for solving probability problems; neither tool is sufficient by itself, but together they form a very powerful combination.

2.5 FORMAL CONCEPTS OF PROBABILITY

In the preceding sections, we have presented a highly informal version of the basic concepts of probability. The fundamental idea here has been that probabilities are viewed as long-run frequencies of events among infinitely many repetitions of some "experiment." This is the way probability is viewed both by nonspecialists and by those who apply probability methods in the real world; it is also the fundamental basis for computer simulation methodology, as mentioned previously. However, many books define things much more formally, in order to develop rigorous mathematical theory. Although we will not take such an approach here, we will briefly discuss the typical structure used.

First, an **event** is simply defined as a subset of the sample space. This makes sense in the light of Example 2.1.2 and Figure 2.1: For example, let A be the event that the sum of the two dice is 6, and let B be the event that at least one of the two dice is a 4. Then

$$A=\{(1,5), (2,4), (3,3), (4,2), (5,1)\}$$

and

$$B=\{(1,4), (2,4), \ldots, (6,4), (4,1), (4,2), \ldots, (4,6)\}$$

Figure 2.1 will also facilitate the introduction of the notation commonly encountered in more formal probability analysis. $P(A$ and $B)$ and $P(A$ or $B)$ are written as $P(A \cap B)$ and $P(A \cup B)$, respectively. To see why this is true, let events A and B be as in the last paragraph. Then as we have defined $P(A$ and $B)$ previously, it is the long-run proportion of the rolls of the pair of dice which result in *both* A and B occurring (i.e., the sum of the dice is 6 *and* one of the dice is a 4). This description amounts to saying that we are interested in long-run proportion of the rolls which result in either (2,4) or (4,2). But these two pairs are precisely the set $A \cap B$, so you can see the motivation behind the notation $P(A \cap B)$. Of course, $P(A \cup B)$ can be explained similarly.

Finally, the idea of probability itself is defined differently. The notation $P(A)$ is viewed in terms of mathematical functions; P is the name of a function whose domain is the collection of all subsets of the sample space S. The function P is assumed to have the following properties:

a. $0 \leq P(A)$ for any subset A of S.

b. For any collection of disjoint subsets $A_1, A_2, \ldots$ we have

$$P(\bigcup_{i=1}^{\infty} A_i) = \sum_{i=1}^{\infty} P(A_i)$$

c. $P(S) = 1$.

These assumptions make intuitive sense. What is perhaps surprising is that this minimal set of assumptions actually can be used to formally derive all of our intuitive notions of probability. For example, Equation (2.1.1) can actually be proved using only (a) and (b) above, as a theorem known as the Strong Law of Large Numbers.

FURTHER READING

Richard Barlow and Frank Proschan, *Statistical Theory of Reliability and Life Testing: Probability Models*, Holt, Reinhart and Winston, 1975.

C.L. Liu, *Introduction to Combinatorial Mathematics*, McGraw-Hill, 1968.

Sheldon Ross, *A First Course in Probability* (2nd edition), Macmillan, 1984. A deeper, more theoretical treatment of probability modeling. An entire chapter is devoted to combinatorics.

Kishor Trivedi, *Probability and Statistics, with Reliability, Queueing and Computer Science Applications*, Prentice-Hall, 1982. Many interesting applications.

S. Gill Williamson, *Combinatorics for Computer Science*, Computer Science Press, 1985.

EXERCISES

Note: Most exercises have either an M or S designation, or both. An M indicates that a mathematical solution is required, while an S indicates that the problem is to be done using simulation. Many problems have both designations, and are to be done both ways;

this helps develop insight, and the two answers serve as checks on each other.

In each chapter, the first few exercises are extensions of examples from the text, followed by new problems.

2.1 (M, S) In the setting of Example 2.2.1, find P (exactly one king and exactly one heart).

2.2 (M, S) In the setting of Example 2.2.6, suppose you bought five tickets, and your friend bought two. Adapt the analysis in that example to find the probability that you and your friend win exactly one prize each.

2.3 (M, S) In the setting of Examples 2.3.1 and 2.3.2, suppose we know that at least one of the marbles chosen was blue. What is the probability that the first one drawn was blue? What is the probability that both marbles drawn were blue?

2.4 In the setting of Example 2.3.6 concerning highway crossing time, suppose you and a friend are driving together, in separate cars. You are in the lead car, with your friend following behind. After you cross successfully, you wait for your friend to get across the road.

a. (M, S) What is the probability that you wait at most 6 seconds for your friend?

b. (S) What is the probability that your friend gets to the other side of the road within 24 seconds of the time you first reach the highway?

2.5 (M) Review Example 2.3.8 regarding Triple Modular Redundancy (TMR), and consider the following. It was mentioned that, for example, if we observe $X=1$, $Y=1$, and $Z=0$, then probably Z was in error, and the true stored bit was a 1. Find the probability that this assumption is true,

$$P(T=1|X=1,Y=1,Z=0)$$

where T is the true bit value. (You may be surprised at the answer, in light of the 1% error rate we would get without using TMR.) Do this computation again for Quadruple Modular Redundancy (i.e., for the case in which you observe $W=1$, $X=1$, $Y=1$, and $Z=0$, where W is the value observed for a fourth stored copy of the data).

2.6 Consider Example 2.1.1. A coin is chosen from the box, and that coin is tossed twice. Let H_1 and H_2 represent the events that the first and second tosses result in heads, respectively.

a. (M) Find the exact values of $P(H_1)$, $P(H_2)$, and $P(H_1 \text{ and } H_2)$, and note that $P(H_1 \text{ and } H_2) \neq P(H_1)P(H_2)$. (Note: If the nonindependence of H_1 and H_2 sounds strange to you, consider what would happen if the coin chosen from the box were to be tossed 10,000 times instead of just twice. If you knew the results of the first 9,999 tosses, you would have a good idea as to which coin had been chosen because you would know whether its probability of heads is 0.9 or 0.1. *Given this information*, you would then have an excellent chance of correctly guessing the outcome of the 10,000th toss. This implies nonindependence of the tosses; if tosses were independent, early tosses should provide no information helpful to guessing the outcomes of later tosses. The tosses are *conditionally* independent, given knowledge of which coin was chosen, but not unconditionally independent.)

b. What difficulty would we have if we were to try to use simulation methods to determine whether H_1 and H_2 are independent events in (a)?

2.7 (M, S) Four men and four women will be seated at random in eight seats arranged in a row. Find the probability that men and women sit in alternate seats.

2.8 (M, S) A theater has r rows of s seats each. Three friends arrive at the theater, and are told that all but l of the rs tickets have been already sold. Those $rs - l$ ticketholders have already entered the theater and chosen seats (with choices assumed to be random, i.e., with each seat assumed to have the same probability of being chosen). What is the probability that the three friends will be able to sit together? Find the answer for general r, s, and l, and then check it for a specific numerical case of your choice, using simulation.

2.9 (M, S) Suppose I toss a fair coin twice, without your seeing the outcomes. If I tell you that at least one of the two tosses resulted in a head, what is the probability that the other toss resulted in a head?

2.10 (M, S) Suppose we roll a fair die repeatedly until we obtain a 2. Find P(exactly 4 rolls are needed). Suppose we roll until we obtain two 2s. Find P(exactly 5 rolls are needed).

2.11 (M, S) At a certain widget factory, there are two inspectors. Inspector A is faster than Inspector B, so 60% of all widgets are inspected by A. However, A makes more errors than B. No faulty widget is supposed to leave the factory, but 10% of all widgets leaving the factory after inspection by A are faulty; for B the rate is only 4%. If the widget we buy turns out to be faulty, what is the probability that it was inspected by A?

2.12 (M, S) An urn contains 4 blue and 2 white marbles. We draw marbles from the urn without replacement, until we obtain a white one. Let N be the number of drawings needed. Find $P(N=i)$, $i=1, 2, 3, 4, 5$.

2.13 (M, S) Suppose an urn contains 3 blue marbles and 1 yellow marble. Each time we draw a marble from the urn, we replace it with a marble of the opposite color. We continue to draw marbles from the urn until exactly 2 of the marbles in the urn are blue; let N be the number of draws needed. Find $P(N=i)$, for several values of i of your choice.

2.14 An office has 4 phone lines. Each is busy about 10% of the time. Assume that the lines act independently.

a. (M, S) What proportion of the time are all 4 busy?

b. (M, S) What percentage of the time is it the case that there are 3 busy lines?

c. (S) Suppose that after a call is completed at a phone, the time until the start of the next busy period for that phone is either 1, 2, or 3 time units, with probability 1/3 each. Also, a phone call lasts 1, 2, 3, or 4 time units, with probability 1/4 each. Suppose that at Time 0, calls have just completed at each of the 4 phones. What is the probability that there are at least 2 busy phones at Time 20? Comment on the feasibility of doing this problem mathematically.

2.15 (M, S) Six applicants for a job have been deemed qualified by the company, and ranked in priority order. Three positions are open. Past experience has shown that 20% of those applicants who are offered this kind of position do not accept the offer.

a. What is the probability that the sixth-ranked applicant will be offered the job?

b. What is the probability that both the sixth-ranked and fourth-ranked applicants will be offered positions?

2.16 (M, S) Suppose we keep rolling a fair die until we get a total of at least 10 dots. Find the probability that when we stop, the total is *exactly* 10.

2.17 (M) A very useful relation is **Bonferroni's inequality**.

a. Prove that for any two events A_1 and A_2, we have

$$P(A_1 \text{ or } A_2) \leq P(A_1) + P(A_2)$$

b. Using the previous inequality and mathematical induction, show that for any n events $A_1, \ldots, A_n$, we have

$$P(A_1 \text{ or } \cdots \text{ or } A_n) \leq \sum_{i=1}^{n} P(A_i)$$

c. Suppose a computer message consists of 100 bits. Each bit has a probability 0.001 of being in error, and the bits act independently. First find the exact probability that there is at least one erroneous bit (use Rules 2 and 4). Then, using (b), find an upper bound for this probability. Compare this upper bound to the exact answer.

2.18 (M) A dispatcher at a warehouse has two delivery trucks available and 20 different sites to which deliveries will be made. The dispatcher will assign each truck to 10 sites. She will also specify the order in which the sites are to be visited. All of this information will go on one scheduling sheet, listing site, truck, and delivery order for that site. In how many different ways can this sheet be filled out?

2.19 A building has six floors and is served by two freight elevators, A and B. The elevators initially are both at Floor 1. The elevators carry orders of freight only, one order at a time. When a freight loader summons an elevator, the elevator closer to the summoner's floor is the one that moves; in the case of a tie, Elevator A moves. After an elevator's load reaches its destination floor, the elevator remains there until it is summoned (rather than, say, automatically returning to Floor 1). Of all summons, 50% are made from Floor 1, and 10% each from the other five floors. Of summons made from Floor i, the destination is Floor j with probability 0.20 ($j = 1, 2, 3, 4, 5, 6$; $j \neq i$).

a. (M, S) After the second summons, what is the probability that Elevator B is on Floor 3?

b. (S) Elevator A will make slightly more moves than B. What percentage of all moves, in the long run, will be made by A?

2.20 (M, S) As indicated in Example 2.3.8, it is common for bits of data, stored or transmitted, to be in error. One method for dealing with this involves adding a **parity** bit to the data.

For example, consider textual data (e.g., the computer files that store the manuscript for this book, or a corporate report transmitted across the continent through a computer network). A common system for storing characters (letters, punctuation marks, etc.) uses 7 bits per character.

For example, the letter C is stored as 1000011. A parity bit could be added to each stored character, say at the left end, resulting in a total of 8 bits of storage for each character. We will adopt "even" parity in this example. This means that the parity bit will be chosen so that the total number of 1 bits in a character is even. For example, the letter C, stored as 1000011, has three 1s (the first, sixth, and seventh bits). Since three is an odd number, we set the parity bit to be 1; the 8-bit storage for C will then be 11000011 (the leftmost 1 is the parity bit), which has four 1s, an even number as desired.

On the other hand, the letter A has the code 1000001, which already has an even number of 1s, so its parity bit would be 0. Thus, its 8-bit storage would be 01000001.

The point is that when a character is examined, we first count the number of 1 bits. If this number turns out to be odd, we know that there is at least one bit in error.

Note however that if two bits are in error, there will be some "canceling," so the total number of 1s will still be even, and we will not notice the error. We will notice any odd number of bit errors, but not any even number.

Consider the letter C. Suppose each bit has probability 0.01 of being in error (note that this includes the parity bit), and that the bits are independent.

a. What is the probability that an error of some type occurs?

b. What is the probability that, when the character is examined and the 1 bits counted, the count will be odd, indicating an error?

c. Suppose the number of 1 bits in the character turns out to be odd. What is the probability that the character itself (the original 7-bit pattern, without parity bit) is correct? In other words, what is the probability that the parity bit is the source of the incorrectness in the 1s count?

2.21 (M) Imagine a set of n objects, of which one is "special." Suppose we wish to choose a subset of size k of the n objects. We could do this in two ways:

a. Choose all k for our subset from among the "nonspecial" objects.

b. Choose our subset so that it includes the "special" object and $k-1$ others.

By thinking in this way, express a, b, c, and d as functions of n and k, in the following relation:

$$\binom{n}{k} = \binom{a}{b} + \binom{c}{d}$$

Verify this relation, by using the definition of

$$\binom{x}{y}$$

as

$$\frac{x!}{y!(x-y)!}$$

2.22 (M) Consider a system of streets in the form of a grid of 2 rows of n columns. (This problem could also be described in terms of a computer communications network.) There are $2n$ intersections, with coordinates (i, j) $(i=1, 2;\ j=1, 2, \ldots n)$. The streets connect these points, and run horizontally [i.e., from (i, j) to $(i, j \pm 1)$] and vertically [from (i,j) to $(3-i,j)$]. We wish to travel from the point (1,1) to $(1, n)$.

a. How many paths are possible? (Hint: Break this down according to the total number of turns made.)

b. Suppose, unknown to the drivers, the block between (2, k) and (2, k+1) is closed, for some k satisfying $1 < k < n - 1$. Suppose also that drivers choose paths between (1,1) and (1, n) at random, before starting out. What proportion of drivers will have to alter their plans?

2.23 (S) Suppose you toss a coin repeatedly, winning one dollar each time you toss a head, and losing one dollar for each tail. Suppose you start with s dollars, and will play until you either reach a goal of g dollars, or go broke (using more extensive analysis, it can be shown that one of these will occur with probability 1.0; i.e., your fortune cannot wander forever). Answer the following questions, for values of s, g, and t of your choice. (Caution: for some values, the simulation will have an extremely long run time.)

a. Find the probability that you ever reach g dollars.

b. Find the probability that you make fewer than t tosses in all.

c. Using the approach in the Appendix of this chapter, do a mathematical analysis of (a) and (b).

2.24 (S) The 18th-century mathematician Buffon suggested the following technique for finding the value of π. You take a large piece of paper ruled with parallel lines 2 inches apart, and you toss a 1 inch needle at random on the paper, say 1000 times. You then estimate the value of π to be the reciprocal of the proportion of the time your tossed needle intersected one of the ruled lines. Write a simulation program (this is easier than actually doing needle-tossing) to try this, and verify that the technique actually works. (Note: You will be using the value of π in your program. This is not "cheating," since we are merely simulating the needle-tossing process, which does *not* require knowledge of the value of π. The simulation is done because we are lazy!)

2.25 (S) Consider the following card game, played with a standard 52-card deck. There are n players, sitting in a circle. Let P_1 denote the dealer, and then let $P_2, \ldots, P_n$ denote the other players, in clockwise order. At the beginning of the game, five cards are dealt to each player, so that there are still $52 - 5n$ cards remaining in the deck. P_1 takes the top card from the deck, and thus will now have six cards. He or she then gives one of those six cards to P_2. P_2 now has six cards, and gives one of them to P_3. Each time P_1 gets a turn, he or she draws another card from those remaining in the deck. This continues until some player has four cards of one kind in his or her hand, or until the deck is exhausted.

It is clear that all n players do *not* have an equal chance of winning. Investigate how much disparity in winning probabilities there is among the n players, say for $n = 4$. Assume that each player adopts the following strategy: As soon as you get two cards of a certain kind, discard any card given to you which is not of that kind; before that time, choose your discard at random from among your five cards (in the case of the dealer, replace "given to you" by "you take from the top of the deck").

2.26 (M) In order to make some machines less prone to breakdown, the concept of **redundancy** is sometimes used. This refers to the idea of intentionally including superfluous parts in the design of a machine in order to increase the probability that the machine works. We will investigate an example of that idea here.

Suppose that the basic machine M consists of two components, A and B, and that both of them must work in order for the machine to work. We can introduce **redundancy at the component level** as follows: For each component, add a backup component, A' for A and B' for B. Also add some switching circuits S_A and S_B, which will automatically switch operation to the backup part in case the primary part fails. For example, if A fails, S_A should sense that failure and automatically switch operation from A to A'; then M would run with A' and B, instead of A and B. The word *should* is used here, since the switch itself is also subject to failure.

Another approach is that of **redundancy at the system level**. Here we have a whole separate machine M' acting as backup, with a switch S_M. If M fails for any reason, whether because of A or B, then S_M automatically switches operation from M to M'.

Let p be the probability that a component is working, assumed in this example to have the same value for A, B, A' and B', and let q be the probability that a switch works.

a. Suppose $q=1$. Show that component redundancy gives greater reliability than system redundancy. Give an intuitive explanation for this.

b. If q is smaller (possibly much smaller) than 1, is it possible for system redundancy to be better than component redundancy? If so, give specific numerical values for p and q which would make this occur. If you believe that this is not possible, prove your impossibility claim.

2.27 (S) A particle moves in the XY plane, traveling only on the integer lattice points

$$\{(i,j): 0 \le j \le i,\ 0 \le i \le 10\}$$

Each jump to a new lattice point is made by choosing one of the vertical or horizontal immediate neighbors of the current point at random. For example, if the particle is currently at (3,1), then it jumps to one of (3,2), (3,0), (2,1) or (4,1) with probability 1/4 each; if the current position is (1,1), a jump is made to one of (1,0) or (2,1) with probability 1/2 each. The particle starts at (10,0). Find the probability that the particle reaches the diagonal line

$$\{(0,0),(1,1),(2,2),\ldots,(10,10)\}$$

within 15 steps.

CHAPTER 3

Random Variables and Distributions

3.1 DISCRETE AND CONTINUOUS RANDOM VARIABLES

A **random variable** is simply some numerical outcome of our experiment. For the sake of completeness, we mention that in the formal description introduced in Section 2.5, a random variable would be defined as a function whose domain is the sample space, and whose range is the set of real numbers.

Example 3.1.1

Suppose our experiment is to roll two dice. Let X denote the number on the first die, and Y the number on the second die. Then X and Y are random variables, as are $S=X+Y$ and $Z=XY$.

The word *random* is used to indicate that a random variable is the result of a random experiment, and a random variable will thus take on different values from one repetition of the experiment to the next. For example, the random variable S takes on values from 2 to 12. The first time we perform the dice-rolling experiment, S might turn out to be 7, while the next time it might be 2, and so on.

The random variables in Example 3.1.1 take on integer values only, but there are other kinds of random variables too:

Example 3.1.2

Suppose we throw a dart at the interval [0,1], in a completely random manner, that is, in such a way that all points in [0,1] are equally likely. Let R be the point in [0,1] which is hit. Then R could be 0.2, 0.549, 0.8333, or any of the numbers in the whole continuum between 0 and 1.

There is a fundamental difference between the two types of random variables introduced above. To illustrate this, let's look a little closer at R. Consider $P(R=0.2)$. We will determine this probability indirectly. First, note that since all points in [0,1] are equally likely to be hit, the probability of the dart's hitting inside any subinterval should be equal to the length of the subinterval; that is, if

$$0 \leq a \leq b \leq 1$$

we have

$$P(a < R < b) = b - a$$

For example, $P(0.3 < R < 0.7)$ should be 0.4, since 40% of all the numbers in [0,1] are in the range 0.3 to 0.7.

Thus $P(0.2 - c < R < 0.2 + c)$ should be 2c. But the probability $P(R=0.2)$ should be smaller than this, since the singleton set $\{0.2\}$ is much smaller than the set of all points in $(0.2 - c, 0.2 + c)$. Thus

$$P(R=0.2) < P(0.2 - c < R < 0.2 + c)$$

so that we have $P(R=0.2) < 2c$. This will be true for any c ($c=0.1$, 0.01, 0.001, etc.). Thus, we have found that $P(R=0.2)$ is less than 0.2, 0.02, 0.002, and so on; in other words, $P(R=0.2)$ is smaller than any positive number, which forces us to conclude that it must be equal to zero.

The same reasoning implies that $P(R=t)=0$ for any t in [0,1]. At first, this is a little disturbing, but only if one thinks of a zero probability as necessarily indicating impossibility. It is better to recall that the intuitive meaning of probability is the long-run proportion of the time that the given event occurs, in infinitely many repetitions. Under this interpretation, the statement that $P(R=0.2)=0.0$ merely says that, while R might occasionally take on the value 0.2, this happens so rarely that the proportion of repetitions for which it occurs eventually shrinks to zero as we continue to repeat the experiment. This is not really surprising; R comes *near* 0.2 fairly often, but is almost never *exactly* equal to 0.200000. In fact, another point of view is that the random variable R as described above is only an idealization anyway. Our dart has nonzero width, and our measuring instruments have only finite precision. Thus the notion that R is *exactly* equal to 0.200000 does not really make physical sense. This is not a problem—

our model that R can be anywhere in [0,1] with equal chance is still realistic—but you can see that we should not be bothered by the seeming anomaly $P(R=0.2)=0.0$.

Thus, there is a fundamental difference between the types of random variables discussed above. For example, compare X in Example 3.1.1 to R in Example 3.1.2. X does take on some values with nonzero probability; in fact, those nonzero probabilities sum to 1.0: $P(X=1)+P(X=2)+\cdots+P(X=6)=1/6+1/6+\cdots+1/6=1.0$. On the other hand, R puts zero probability on any value. This difference leads to the following important definitions.

Definition 3.1.1

a. If a random variable V has

$$\sum_t P(V=t)=1.0 \tag{3.1.1}$$

then we say that V is **discrete**.

b. If a random variable V has

$$P(V=t)=0.0 \tag{3.1.2}$$

for all t, we say that V is **continuous**.

The preceding definition is rather abstract. However, the distinction between discrete and continuous random variables is quite important, so you should spend some time thinking about it, beginning by reviewing the definition in light of the examples presented so far. It will also help to keep in mind that discrete random variables are typically integer-valued, and continuous ones typically take on values throughout some interval (a,b), possibly with $a=-\infty$ and/or $b=\infty$. You will find that you will gradually become very comfortable with these concepts.

Definition 3.1.1 is not exhaustive; other kinds of random variables exist (see, for example, Exercise 3.32). However, in most practical applications, we work exclusively with the two kinds defined above.

In the rest of this section, we will deal with the idea of the **distribution** of a random variable. This is a central concept, which must be mastered before one can do probabilistic and statistical applications. Fortunately, it is a very simple concept; it merely formalizes the notion that not all values taken on by a random variable occur with equal frequency. For example, consider the dice experiment described at the beginning of this section.

There the random variable S takes on the values 2, 3, . . . , 12, but not with equal frequencies; for example, $P(S=2)=1/36$ while $P(S=7)=1/6$.

The **distribution** of a random variable is a complete description of the relative frequencies of occurrence of the different values taken on by that random variable. This description could have various forms. For example, for discrete random variables, we could simply provide a list of all the probabilities of individual values. For example:

$$P(S=2) = \frac{1}{36}$$

$$P(S=3) = \frac{1}{18}$$

$$\vdots$$

$$P(S=12) = \frac{1}{36} \quad \textbf{(3.1.3)}$$

This list of probabilities is formally defined as the **probability mass function** of S.

Definition 3.1.2

> If V is a discrete random variable, the **probability mass function** of V is denoted by p_V and defined by
>
> $$p_V(t) = P(V=t) \quad \textbf{(3.1.4)}$$
>
> for any real t.

For example, from Equation (3.1.3) we see that $p_S(3) = 1/18$. The function p_S is graphed in Figure 3.1. Note that the purpose of the subscript on the function p is to indicate which random variable is being discussed; thus, we use the subscript S here.

The probability mass function, then, serves as a form for describing the distribution of a discrete random variable. However, for continuous random variables it is not possible to use exactly this form of description, since as we have seen, all probabilities of individual values are zero. Thus, we need some other way to graphically describe the frequency variation of a continuous random variable.

The concept of a **density function** serves this purpose. If we have a continuous random variable V, its density function, denoted by f_V, serves as the analog of the function p_V we would have used if V had been discrete. To

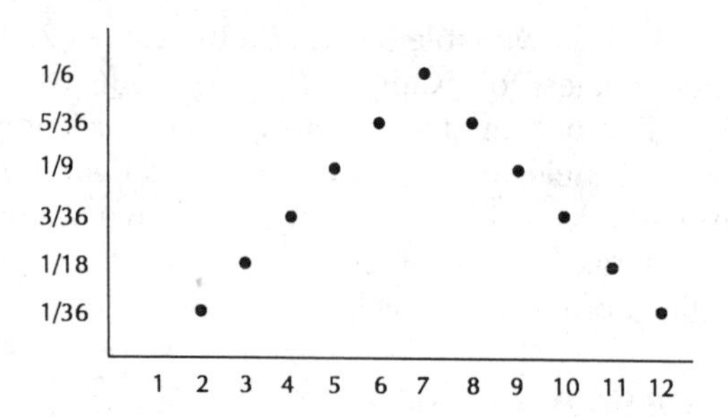

Figure 3.1 Probability mass function for the sum of two dice.

introduce this concept, look again at Figure 3.1, which plots the probability mass function of the discrete random variable S. The height of the graph for $p_S(3)$ is 1/18, while for $p_S(7)$ it is 1/6. However, the first aspect that a viewer of the graph notices about the heights of $p_S(3)$ and $p_S(7)$ is not the *absolute* values of the two heights, but rather their *relative* values. In other words, what the viewer pays most attention to is that the graph is much higher at 7 than at 3. This difference in heights at 3 and 7 tells us that over many repetitions of our experiment, the event $S=7$ occurs much more often than $S=3$. Since the vertical axis of the graph is marked numerically, the viewer can determine that S takes on the value 7 three times as often as the value 3. *However, even if the vertical axis were not marked*, the viewer could immediately tell that the relative frequency of occurrence of $S=7$ is several times that of $S=3$. The exact value of the ratio may not even be important to the viewer, although of course he or she could use a ruler to find it.

Thus, for a discrete random variable V, a large part of the usefulness of the graph of p_V is that it gives the viewer an immediate visual idea of which values of V occur more often than other values. For a continuous random variable V, the density function f_V serves an analogous purpose. The only difference is that since $P(V=t)=0$ for all t, we do not graph the relative frequencies of individual values of V, but rather the relative frequencies of *regions* in the horizontal axis. For example, if $f_V(6.4)$ is much larger than $f_V(10.7)$, this tells us that over many repetitions of the experiment, V will take on values *near* 6.4 much more often than values *near* 10.7—in fact $f_V(6.4)/f_V(10.7)$ times more often.

The graph for R in Example 3.1.2 should thus be that of Figure 3.2. The graph should have constant height between 0 and 1, reflecting the fact that R takes on values in all equal-size regions between 0 and 1 with equal frequency. For example, R takes on values near 0.2 with exactly the same frequency as it does near 0.7. Similarly, the graph should have zero height outside of [0,1], since R never is outside this range.

The only remaining question is then what the constant height between 0 and 1 should be. This question is settled by adopting the convention that the total area under any such curve should be equal to 1.0, in analogy with

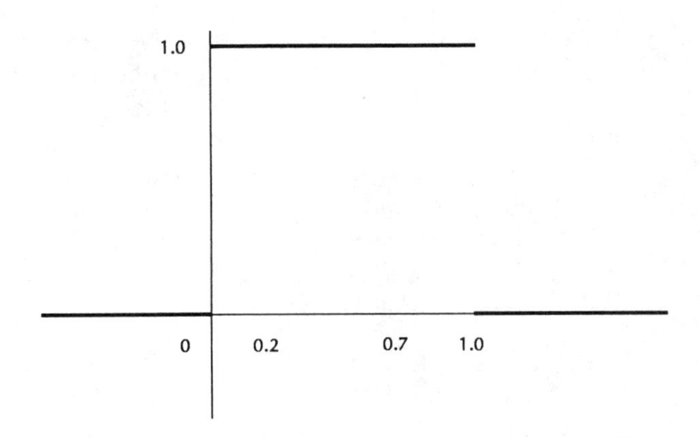

Figure 3.2 Graph for R in Example 3.1.2.

Equation (3.1.1). In the case of the density of R in Figure 3.2, the height of the nonzero portion of the curve would then also be equal to 1.0.

Recalling from calculus that integrals are analogs of sums, we are ready to present the formal definition of a density function.

Definition 3.1.3

A **density function** is any function satisfying the following two conditions:

$$g(t) \geq 0 \qquad (-\infty < t < \infty) \tag{3.1.5a}$$

$$\int_{-\infty}^{\infty} g(t)\, dt = 1 \tag{3.1.5b}$$

The concept of a density is initially difficult for many students of probability modeling. The best way to become comfortable with it is to constantly keep in mind specific examples, such as Example 3.1.2 and the following.

Example 3.1.3

Suppose we have a large bag filled with straight pieces of wire, of lengths ranging from 2 to 6 inches, with an equal number of pieces for each length between 2 and 6. Assume that there are so many pieces of wire in the bag that we may treat the total number of pieces as infinite. Our experiment is to reach into the bag and take out a piece of wire. Let L be the length of the wire drawn from the bag. It is reasonable that the chance of a particular wire being chosen is proportional to its length,(e.g., wires of length 5.2 are twice as likely to be chosen as wires of length 2.6). In the field of statistical sampling theory, this is called length-biased sampling, reflecting the fact that the sampling process is biased in favor of choosing the longer wires. This then implies that $f_L(t)$ should be proportional to t, $f_L(t) = ct$ for some c. The requirement that the total area under the curve f_L be 1.0 then allows us to determine c:

$$1.0 = \int_2^6 ct\,dt = 16c \tag{3.1.6}$$

so $c = 1/16$. Thus, the density of L is equal to $t/16$ for t in (2, 6), and is equal to 0 outside that range. This density is graphed in Figure 3.3.

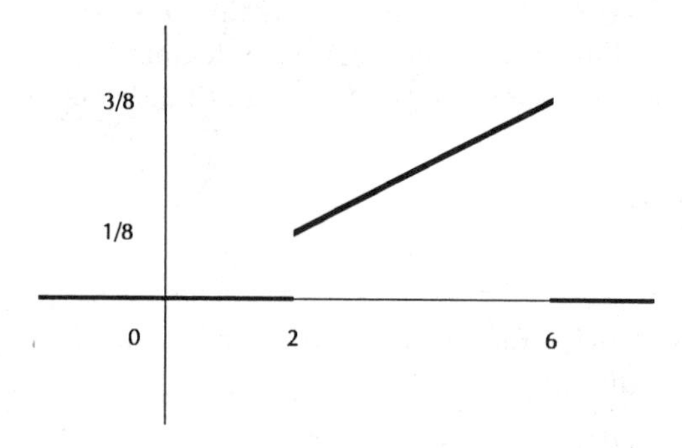

Figure 3.3 Density function for Example 3.1.3.

Density functions can be used to calculate probabilities for continuous random variables. Consider the last example; suppose we wish to find the probability of choosing a wire between 4.4 and 5.6 inches in length, that is, $P(4.4 < L < 5.6)$. For discrete random variables, probabilities of this type turn out to be sums involving the probability mass function; for example, in Example 3.1.1 we would have for the probability of rolling a sum between 3 and 7

$$\begin{aligned} P(3 < S < 7) &= P(S = 4) + P(S = 5) + P(S = 6) \\ &= \sum_{i=4}^{6} p_S(i) \\ &= \frac{1}{3} \end{aligned} \tag{3.1.7}$$

For continuous random variables, the analog of summing a probability mass function turns out to be integrating a density function. (While we will not discuss technical reasons for this here, the reader can get at least some understanding by recalling the calculus definition of an integral in terms of a limit of sums.)

Finding Probabilities for Continuous Random Variables

If V is a continuous random variable, then for any real numbers a and b we have

$$P(a < V < b) = \int_a^b f_V(t)\, dt \quad \textbf{(3.1.8)}$$

Both here and throughout the book, the reader should keep in mind that variables of integration are merely placeholders. There is no special significance to the use of the letter t in (3.1.8). We of course obtain the same numeric result if we replace t and dt everywhere by s and ds:

$$P(a < V < b) = \int_a^b f_V(s)\, ds$$

It is worth pointing out that Equation (3.1.8) is consistent with Equation (3.1.2), which said that continuous random variables put probability zero on an individual point, since

$$P(V = t) = \int_t^t f_V(s)\, ds = 0.0 \quad \textbf{(3.1.9)}$$

Also, the same reasoning shows that in calculating probabilities such as in Equation (3.1.8), it does not matter whether endpoints are included or excluded; for example, $P(a \le V \le b) = P(a < V < b)$.

Example 3.1.4

We can now answer the question raised in Example 3.1.3: What is the probability that the wire we choose happens to be between 4.4 and 5.6 inches in length? From the preceding discussion, we now know that this probability is

$$\int_{4.4}^{5.6} \frac{t}{16}\, dt = 0.375 \quad \textbf{(3.1.10)}$$

Although distributions are usually described using probability mass or density functions, an alternate method is sometimes useful.

Definition 3.1.4

For any random variable V, the **cumulative distribution function** (c.d.f.) of V is denoted F_V, and is defined by

$$F_V(t) = P(V \leq t) \tag{3.1.11}$$

for any real number t.

Example 3.1.5

Consider the random variable S in Example 3.1.1. Suppose that we need to find $F_S(3.158)$.

$$\begin{aligned} F_S(3.158) &= P(S \leq 3.158) \\ &= P(S = 2 \text{ or } S = 3) \\ &= \frac{1}{36} + \frac{1}{18} \\ &= \frac{1}{12} \end{aligned} \tag{3.1.12}$$

so $F_S(3.158) = 1/12$.

Example 3.1.6

For the random variable L in Example 3.1.3, the value of the distribution function at (say) 2.5 is

$$\begin{aligned} F_L(2.5) &= P(L \leq 2.5) \\ &= P(2.0 \leq L \leq 2.5) \\ &= \int_{2.0}^{2.5} \frac{t}{16} dt \\ &= 0.07 \end{aligned} \tag{3.1.13}$$

The cumulative distribution functions in the last two examples are graphed in Figures 3.4 and 3.5. Notice the crucial distinction—the graph of the c.d.f. for the discrete random variable S has jumps, whereas the graph for L, a continuous random variable, is continuous.

Note that since the cumulative distribution function of a random variable is obtained by integrating the density (as in the last example), the

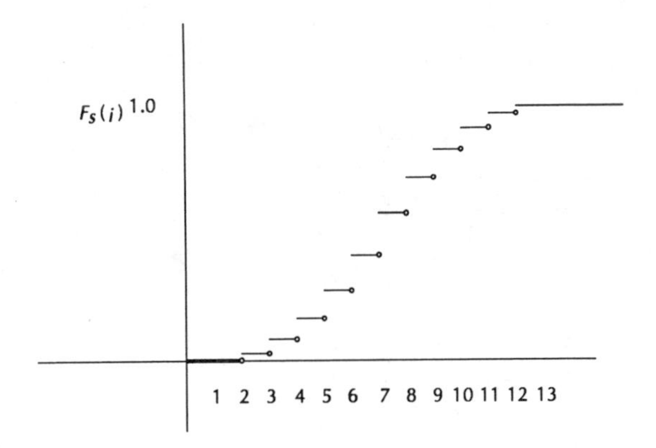

Figure 3.4 c.d.f. for function in Figure 3.1.

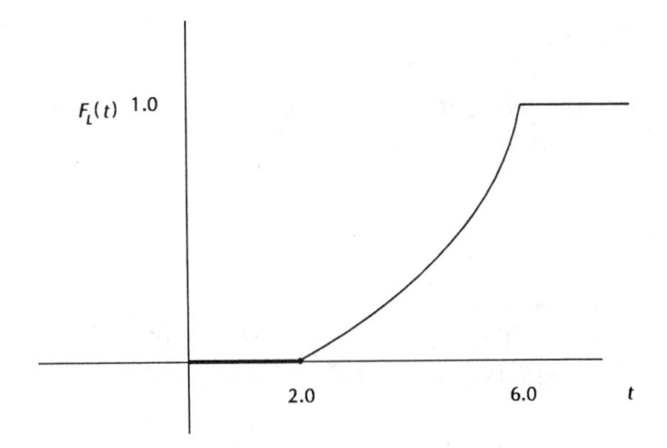

Figure 3.5 c.d.f. for function in Figure 3.3.

density can in turn be obtained by differentiating the cumulative distribution function. Thus, each of these two items can be obtained from the other. An important consequence of this is that, given a random variable X with known density, it is easy to find the density of some new random variable Y which is defined as a function of X, say $Y=q(X)$. This notion is made concrete in the following example.

Example 3.1.7

Suppose X has density value 0.5 within the interval $(-1,1)$, and 0 outside the interval. If we define a new random variable Y by the equation $Y=X^2$, then we can derive the density of Y as follows: Since Y is obviously never negative, its density has the value 0 for negative arguments. Now for $0 \leq t \leq 1$, we have

$$
\begin{aligned}
f_Y(t) &= \frac{d}{dt} F_Y(t) \\
&= \frac{d}{dt} P(Y \leq t) \\
&= \frac{d}{dt} P(X^2 \leq t) \\
&= \frac{d}{dt} P(-t^{1/2} \leq X \leq t^{1/2}) \\
&= \frac{d}{dt} \int_{-t^{1/2}}^{t^{1/2}} 0.5 \, ds \\
&= \frac{d}{dt} t^{1/2} \\
&= \frac{1}{2} t^{-1/2}
\end{aligned}
$$

Following is another example, which combines several of the ideas we have covered so far.

Example 3.1.8

Suppose a system contains two identical devices, both of which must work for the system as a whole to work. Let us suppose that device lifetime L is continuous, and has density

$$
f_L(t) = \frac{t}{3750} \tag{3.1.14}
$$

for t between 50 and 100 hours, and 0 for all other values of t.

Let us find the density of S, which is the lifetime of the system as a whole. Note that $S = \min(L_1, L_2)$, where L_1 and L_2 are the lifetimes of the two devices in the system. Note that each L_i has the density given in Equation (3.1.14). Note too that the density of S will have the same range as that of L, 50 to 100, though its density will be different from that of L.

Proceeding as in the last example, we have

$$
\begin{aligned}
f_S(t) &= \frac{d}{dt} F_S(t) \\
&= \frac{d}{dt} P(S \leq t) \\
&= \frac{d}{dt} P(\min(L_1, L_2) \leq t)
\end{aligned}
$$

$$= \frac{d}{dt}[1 - P(\min(L_1, L_2) > t)]$$

$$= -\frac{d}{dt}P(\min(L_1, L_2) > t) \qquad \textbf{(3.1.15)}$$

Now the event $\min(L_1, L_2) > t$ will occur if and only if $L_1 > t$ and $L_2 > t$. As before (e.g., Example 2.3.7), we will assume independence of the two components. Then

$$\frac{d}{dt}P(\min(L_1, L_2) > t) = \frac{d}{dt}P(L_1 > t \text{ and } L_2 > t)$$

$$= \frac{d}{dt}[P(L_1 > t)P(L_2 > t)] \qquad \textbf{(3.1.16)}$$

Since each L_i has the density given in Equation (3.1.14), we have

$$P(L_i > t) = \int_t^{100} \frac{u}{3750}du = \frac{1}{3750}\left(5000 - \frac{t^2}{2}\right) \qquad \textbf{(3.1.17)}$$

Thus, combining the three equations, we find that $f_S(t)$ is -1 times the derivative of the square of the expression in Equation (3.1.17), so

$$f_S(t) = \frac{t}{3750^2}(10{,}000 - t^2) \qquad \textbf{(3.1.18)}$$

for $50 < t < 100$.

The densities of L and S are graphed in Figure 3.6. Note the difference: f_L is constantly increasing; components with lifetimes toward the high end of its range (100) are more prevalent (e.g., if you go to the manufacturer of the components and test a large number of them, more of them will turn out to be near 100 than 50). But f_S is quite different. Its values are highest near the low end of its range (50). Thus,

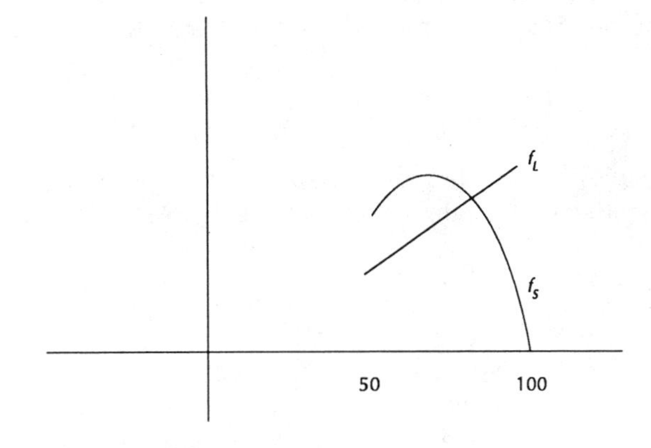

Figure 3.6 Two density functions for Example 3.1.8.

if you test many systems and record the *system* lifetime in each case, you will find that the recorded values will be near 50 more often than near 100.

Now that we have the density of S, we can use it to find probabilities. For example, the probability that the system lasts between 60 and 70 hours is

$$P(60 < S < 70) = \int_{60}^{70} \frac{t}{3750^2}(10,000 - t^2)dt = 0.266$$

Keep in mind what this means: If we take a large number of these components, and form many copies of the system, 26.6% of the copies will have lifetimes between 60 and 70 hours.

We can also find the mean of S (i.e., the mean lifetime of the system) using the techniques in the next section.

3.2 MEASURES OF LOCATION AND DISPERSION

Although the distribution of a random variable is described completely by its probability or density function, we are often interested in certain summary measures. For example, suppose we are comparing several distributions. It is convenient (although in some cases perhaps an oversimplification) to base our comparison on one or two summary values for each distribution. In this section we discuss two broad classes of summary measures, called **measures of location** and **measures of dispersion**.

Measures of location roughly indicate the "center" of a distribution. You are probably somewhat familiar with the two most common measures, the **mean** (or **expected value**) and the **median**.

Informally speaking, the mean of a random variable V is the long-run average value of V, in infinitely many repetitions of the experiment:

$$\lim_{n\to\infty} \frac{V_1 + V_2 + \cdots + V_n}{n} \qquad \textbf{(3.2.1)}$$

where V_i is the value that V happens to take on in the ith repetition of the experiment. This long-run average is denoted by $E(V)$, or simply EV.

Example 3.2.1

Suppose we play a game in which we roll a fair die. If the die comes up 1 or 2, we win four dollars; otherwise we win only one dollar. Let W be the amount we win. Then W is a discrete random variable, with probability mass function

$$p_W(1) = \frac{2}{3}$$

$$p_W(4) = \frac{1}{3} \tag{3.2.2}$$

In this context, EW is the long-run average winning per game, over infinitely many plays of the game. To find EW, consider Equation (3.2.1), and let N_1 and N_4 be the number of times $W = 1$ and $W = 4$, respectively, among the first n plays of the game. Then

$$W_1 + \cdots + W_n = N_1 + 4N_4 \tag{3.2.3}$$

so

$$\begin{aligned} \lim_{n\to\infty} \frac{W_1 + \cdots + W_n}{n} &= 1 \lim_{n\to\infty} \frac{N_1}{n} + 4 \lim_{n\to\infty} \frac{N_4}{n} \\ &= 1\left(\frac{2}{3}\right) + 4\left(\frac{1}{3}\right) \\ &= 2 \end{aligned} \tag{3.2.4}$$

Thus $EW = 2$; our average winnings after a large number of plays of the game (technically, infinitely many plays) will be \$2.00 per play. Note that two dollars is thus also the "fair" fee which might be charged us to play the game; with this fee, neither we nor the casino would make any profit in the long run. (Of course, the casino would charge a bit more than \$2.00, due to overhead and profit.)

Note that Equation (3.2.4) has the form $p_W(1) + 4p_W(4)$. The reasoning which led to that form generalizes into the following formula.

Expected Value of a Discrete Random Variable

If V is a discrete random variable, its expected value is given by

$$EV = \sum_s s p_V(s) \tag{3.2.5}$$

where the sum extends over all values of s taken on by V, e.g., 1 and 4 in the last example. Note again that the variable s is merely a place-holder here; for example, Equation (3.2.5) would be numerically unchanged if we were to replace s everywhere by t.

The formula for continuous random variables is analogous to Equation (3.2.5), with summation replaced by integration, and a density function instead of a probability mass function.

Expected Value of a Continuous Random Variable

The expected value of a continuous random variable V is

$$EV = \int_{-\infty}^{\infty} s f_V(s)\, ds \tag{3.2.6}$$

Example 3.2.2

The expected value of the random variable L in Example 3.1.3 is

$$EL = \int_2^6 s \frac{s}{16} ds = 4.33 \tag{3.2.7}$$

Thus, if we perform the experiment in Example 3.1.3 repeatedly, the long-run average value of L that we observe will be 4.33. By the way, note that EL did *not* turn out to be halfway between 2 and 6, but instead had a value to the right of the halfway point. This should not be surprising, since the density of L, $f_L(t) = t/16$, indicates that L takes on its larger values (near 6) more often than its smaller values (near 2).

Example 3.2.3

Recall Example 3.1.8, in which we had a two-component system, which fails if either component fails. We found in that example that the density of the system lifetime S is

$$\frac{t}{3750^2}(10000 - t^2)$$

for t between 50 and 100, 0 elsewhere. Then from Equation (3.2.6), we have that the mean system lifetime S is

$$ES = \int_{50}^{100} t \frac{t}{3750^2}(10000 - t^2)\, dt = 69.63$$

Thus, if we take a large number of these components, and form many copies of the two-component system, the average lifetime of all these systems will be about 69.63.

By comparison, let us find the mean lifetime of individual components. In Example 3.1.8, the density of component lifetime L was given to be

$$\frac{t}{3750}$$

for t between 50 and 100 hours, 0 elsewhere. Thus, the mean component lifetime is

$$EL = \int_{50}^{100} \frac{t^2}{3750} dt = 77.78$$

This is larger than the mean system lifetime, 69.63, which is what we would expect: The two-component system fails if *either* of its two components fails; that is, its lifetime depends on the weaker of the two components it happens to contain. Thus the mean system lifetime should be smaller than the mean lifetime of individual components.

It should be mentioned here that the term "expected value" is rather misleading. We do not necessarily "expect" the random variable to take on that value with high probability. In fact, $P(V = EV)$ is equal to zero in both of the examples above. Instead, the proper interpretation of EV is that of a long-average, as described above. (A physical interpretation as a "center of gravity" is also possible, but has little meaning in probability contexts.)

It should also be mentioned that, mathematically speaking, the expected values of some random variables do not exist. An example is the **Cauchy density**:

$$f(t) = \frac{1}{\pi(1 + t^2)} \qquad -\infty < t < \infty$$

The integral in Equation (3.2.6) does not exist for this density, even though on an intuitive basis it is clear that 0 should be considered the "center" of this distribution. In fact, the **median** of this distribution, which is another measure of location defined below, does exist and equals 0. The median can be thought of as the halfway point in the frequency variation of a random variable.

Definition 3.2.1

If for a given random variable V there is a number c which has the property that

$$P(V < c) = P(V > c) = 0.5 \qquad \textbf{(3.2.8)}$$

then we say that c is a **median** of V.

Since Equation (3.2.8) implies that $P(V=c)=0$, the median as defined here is usually a meaningful concept only for continuous random variables.

Example 3.2.4

Returning to Example 3.1.3, the median of L can be found by using the definition and back-solving:

$$\begin{aligned} 0.5 &= P(L < c) \\ &= \int_2^c \frac{1}{16} s\, ds \\ &= \frac{1}{16}\left(\frac{1}{2}c^2 - 2\right) \end{aligned}$$

so $c = \sqrt{20}$, or about 4.47. Thus, if we perform this experiment repeatedly, L will be less than 4.47 half of the time, and greater than 4.47 the other half of the time.

Recall that measures of location are supposed to give us a quick idea of the "center" or "general location" of the distribution of a random variable. What, then, is the best such measure? Of course, there is no single answer to this question. In some contexts, the expected value is highly appropriate. For example, recall the discussion following Equation (3.2.4), concerning the value of a "fair" fee for the game. Here the expected value is quite meaningful as a measure of location, at least from the point of view of the casino.

Suppose the casino has expenses amounting to \$0.10 per game, and suppose they charge \$$x$ as the entry fee for the game. Consider what happens after n plays of the game. The casino has collected entry fees of nx dollars, and has incurred expenses of

$$n(0.10) + W_1 + \cdots + W_n$$

as in Equation (3.2.3). Assuming the casino stays in business, n grows larger and larger each day, so that by Equation (3.2.4) the approximation

$$\frac{W_1 + \cdots + W_n}{n} \approx 2$$

becomes more and more accurate.

Thus, after some time, the net profit of the casino is approximately

$$nx - [n(0.10) + 2n] \tag{3.2.9}$$

so that x must be at least \$2.10 for the casino to break even. This result should not be very surprising to the reader, but it does point out the fact that the mean is a useful measure of location in some contexts. We could *not* have made a similar analysis if we had used the median as our measure of location; if the median winnings for some game is $4.25 and the expenses are $0.35, it is not necessarily true that the casino can break even by setting the entry fee at $4.60.

The same principles imply that expected value is a very useful concept in the insurance industry and many other similar contexts. Here is an application that arises in a computer science context.

Example 3.2.5

One of the problems that arise with computer databases is that of storage capacity. For example, the government's Social Security Administration needs to keep records for the entire employment history, benefit information, and other relevant data, for each citizen in the country. This is a truly huge volume of data. It would be nice to have a way to compactify data and thus reduce storage requirements somewhat.

One way this can be done is through **Huffman coding**. While a complete explanation of this method would be out of place in this text, we will give an example here and refer the reader to computer science texts for details on how to use the method in general.

We take as our example the following simple case: Suppose a certain database consists only of various combinations of the letters A, B, C, and D. Suppose further that after some investigation we find that A is the most commonly appearing letter in the database records, with 70% of the recorded letters being A. Also suppose the percentages for B, C, and D are 20%, 5%, and 5%, respectively.

Recall from Chapter 2 (Example 2.3.8) that data in computers, in this case a disk, are stored as 0s and 1s. For nonnumerical data such as letters, codes must be agreed upon. In the current example, we could use the code 00 for A, 01 for B, 10 for C, and 11 for D. In this way, each letter in the database will take up two bits of storage space. Let us call this storage scheme described above as the fixed-size method, since each letter needs a fixed number of bits of storage (2).

However, an alternate storage scheme based on Huffman coding would be to have a single bit 0 represent an A, the pair of bits 10 represent a B, and the triples 110 and 111, represent C and D, respectively.

Suppose our database will store n letters, and let B_i denote the number of bits needed to store the ith letter, $i = 1, 2, \ldots, n$. In the fixed-size method, $B_i = 2$ for all i, while for Huffman storage, B_i can be either 1, 2, or 3, with probabilities 0.70, 0.20, and 0.10, respectively.

The mean number of bits for a letter in the fixed-size method is 2, since the number of bits is nonrandom. For the Huffman method the mean is obtained by using Equation (3.2.5), since the number of bits is a *discrete* random variable:

$$1(0.7) + 2(0.2) + 3(0.10) = 1.4$$

It can be proved that Huffman coding is optimal; that is, it minimizes the mean number of bits of storage for letters. Thus, the value 1.4 here is the best possible.

For storing the n letters in a database, the number of bits of storage needed will be

$$B_1 + \cdots + B_n$$

For the fixed-size method, this expression will be exactly equal to $2n$, so a total of $2n$ bits of storage will be needed. On the other hand, if the Huffman method is used, we have, by the reasoning which led to Equation (3.2.9),

$$\frac{B_1 + \cdots + B_n}{n} \approx 1.4$$

so

$$B_1 + \cdots + B_n \approx 1.4n$$

Thus the Huffman code (the variable-size method) reduces total storage requirements by a proportion of approximately $(2n - 1.4n)/2n$, or 30%.

Huffman coding can be used in a similar way in the telecommunications field. In this case, instead of saving storage space, we save transmission time, which in tremendously high-volume communications traffic is quite significant.

The last two examples show that the relation of the expected value to a total can be very advantageous, and thus the expected value is often a good choice for a measure of location. However, in some settings it may have some undesirable properties. In particular, it is sometimes too sensitive to extreme values in the distribution. For example, consider the distribution of income in the population of some town. If there a few people in the town who have extraordinarily high incomes, with a large gap separating them from the incomes of the other people, these high values may inflate the mean so much as to make it a misleading measure of the center of the income distribution. In such a situation, many analysts prefer to use the median.

In addition to using a measure of location to summarize a distribution, a measure of dispersion is needed as well. Suppose that in a study of income in a certain town, we find that the median income is \$15,000. This says that half of the income values are below this amount, and half are above. However, it does not give us any information concerning the variation in income from one person to the next. Are most of the income values concentrated closely around \$15,000, or is there a wide variation?

The role of a measure of dispersion, then, is to give some indication of range or variability of a distribution. Note that the overall range itself (e.g., [2,6] in Example 3.1.3) is not really a very good measure. To see this, the density of L in Example 3.1.3 is shown in Figure 3.7, and another density, say for some random variable M, is shown superimposed. The densities indicate that both random variables have range (2,6), but that M tends to be near 4 more often than does L; conversely, M takes on the extreme values, near 2 and 6, more rarely than does L. Intuitively, M has less variability than L, and yet this fact would not be apparent if pure range—(2,6) for each variable—were used as our measure of variability.

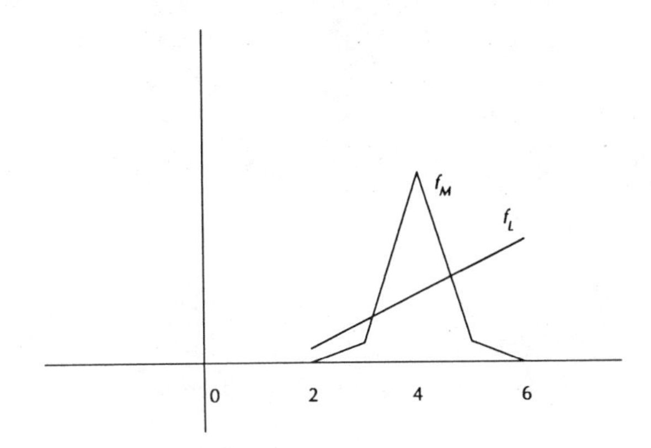

Figure 3.7 Density functions for two random variables.

There are many possible measures of dispersion, usually chosen according to the measure of location used. Since expected value is the most commonly used measure of location, we will concentrate here on developing a corresponding measure of dispersion (a measure suitable for the median is presented in Exercise 3.33).

One natural measure of variability would be $E(|X-\mu|)$, the average distance from X to $\mu = EX$. However, this measure was rejected by the early developers of statistics, mainly due to limitations in the state of mathematics at the time. Instead, the quantity

$$\sigma = \{E[(X-\mu)^2]\}^{1/2}$$

was chosen as the measure of dispersion. This quantity, which is called the **standard deviation** of X, is the square root of the average squared distance from X to μ. This is not the same as the average distance from X to μ as considered above, but the two measures are somewhat comparable.

The standard deviation is by far the most commonly used measure of dispersion. Often we do not bother with the square root, simply reporting the value of the **variance** $E[(X-\mu)^2]$, denoted Var(X).

We have so far not indicated how quantities such as $E(|X-\mu|)$ and $E[(X-\mu)^2]$, or more generally $E[g(X)]$ for any function g, are calculated. Since X is a random variable, $Y=g(X)$ is a random variable too. Thus, to find $E[g(X)]$, one approach would be to first find the probability mass function or density function of Y from that of X (the technique for doing this was given at the end of Section 3.1). Then we could find EY directly, from applying Equations (3.2.5) or (3.2.6) to Y. However, it can be proved that such a procedure would be equivalent to the following much easier method.

Finding the Expected Value of a Function of a Random Variable

a. If X is a discrete random variable, then

$$E[g(X)] = \sum_t g(t)\, p_X(t) \tag{3.2.10}$$

b. If X is a continuous random variable, then

$$E[g(X)] = \int_{-\infty}^{\infty} g(t)\, f_X(t)\, dt \tag{3.2.11}$$

Example 3.2.6

Suppose we roll a single fair die and let X be the resulting number. Let us find the variance and standard deviation of X. We need to find μ first. Since X is discrete, we use Equation (3.2.5):

$$\mu = EX = 1\left(\frac{1}{6}\right) + \cdots + 6\left(\frac{1}{6}\right) = 3.5 \tag{3.2.12}$$

Then from Equation (3.2.10), with $g(t)=(t-3.5)^2$, we have

$$\text{Var}(X) = (1-3.5)^2\left(\frac{1}{6}\right) + \cdots + (6-3.5)^2\left(\frac{1}{6}\right) = 2.92 \tag{3.2.13}$$

The standard deviation is then $\sqrt{2.92} = 1.71$.

Example 3.2.7

To find the variance of wire length L in Example 3.1.3, first recall that the mean of L was found to be 4.33. Thus from Equation (3.2.11) we have

$$\text{Var}(L)=\int_2^6 (t-4.33)^2\left(\frac{1}{16}t\right) dt=1.25$$

The standard deviation is then 1.12.

Example 3.2.8

Suppose a public garage charges parking fees of \$1.50 for the first hour or fraction of an hour and \$1.00 per hour subsequently. If the time parked has the density

$$0.6e^{-0.6t} \qquad (t > 0)$$

then we can determine the mean revenue by using Equation (3.2.11) with $g(t)=1.5$ for $0 \le t \le 1$, and equal to $1.5+(t-1) = t + 0.5$ for $t > 1$. The expected value of the fee will be

$$\int_0^\infty g(t)0.6e^{-0.6t}\, dt = \int_0^1 (1.5)\, 0.6e^{-0.6t}\, dt + \int_1^\infty (t + 0.5)\, 0.6e^{-0.6t}\, dt \qquad \textbf{(3.2.14)}$$

which has the value 2.38.

Expected value and variance are quite important concepts in probability and statistics. The following properties are used regularly, and thus should be understood thoroughly and memorized.

Properties of Expected Value and Variance

For any constants a and b we have

a. $E(aX+b)=aEX+b$ **(3.2.15)**

b. $\text{Var}(X)=E(X^2)-(EX)^2$ **(3.2.16)**

c. $\text{Var}(aX+b)=a^2\text{Var}(X)$ **(3.2.17)**

We now offer proofs of these properties. For the discrete case of **a**,

$$\begin{aligned} E(aX + b) &= \sum (at + b)p_X(t) \\ &= a\sum tp_X(t) + b\sum p_X(t) \\ &= aEX + b(1) = aEX + b \end{aligned}$$

The continuous case is similar. For **b**, we let $\mu = EX$. Then

$$\begin{aligned} \mathrm{Var}(X) &= E[(X-\mu)^2] \\ &= E(X^2 - 2\mu X + \mu^2) \\ &= E(X^2) - 2\mu EX + \mu^2 \\ &= E(X^2) - (EX)^2 \end{aligned}$$

using an argument similar to that in the proof of **a**. For **c**, see Exercise 3.14.

Example 3.2.9

Equation (3.2.16) provides a very convenient way to find variances. To illustrate its use, let X be the number resulting when we roll a certain weighted die. The weights in the die have been chosen such that X has probability 1/4 each of taking on the values 3 and 4, with numbers 1, 2, 5, and 6 having probability 1/8 each. We can use property (b) above as a shortcut method for finding $\mathrm{Var}(X)$. Due to the symmetry of the distribution of X around 3.5, we do not need to bother with computing EX; it is clear that $EX = 3.5$, just as for a fair die. Thus

$$\begin{aligned} \mathrm{Var}(X) &= E(X^2) - 3.5^2 \\ &= 1^2\left(\frac{1}{8}\right) + 2^2\left(\frac{1}{8}\right) + 3^2\left(\frac{1}{4}\right) + 4^2\left(\frac{1}{4}\right) \\ &\quad + 5^2\left(\frac{1}{8}\right) + 6^2\left(\frac{1}{8}\right) - 12.25 \end{aligned}$$

which has the value 2.25. Thus, the standard deviation of X is $\sqrt{2.25} = 1.5$.

This last example gives us another opportunity to illustrate the role of variance in describing a distribution. Suppose Casinos A and B each offer a game involving rolling a die, with Casino A using an ordinary die and Casino B a weighted die as in Example 3.2.9. Suppose the player wins one dollar per dot showing on the die (e.g., two dollars if he or she rolls a 2). Ignoring expenses, each should charge $3.50 as an entry fee for the game. A player in either casino will exactly break even over the long run; however, note that the standard deviation of the weighted die—found to be 1.5 in the last example—is less than that of the ordinary die—found to be 1.71 in Example 3.2.6. The implication of this difference in standard deviations is that a player at Casino B will experience less *fluctuation* in net winnings over time.

Example 3.2.10

Suppose the wires in Example 3.1.3 weigh 20 grams per inch. We can use Equations (3.2.15) and (3.2.17) to find the mean and variance of W, the weight of the wire chosen. Since $W = 20\ L$, we simply use $a = 20$ and $b = 0$ in Equations (3.2.15) and (3.2.17).

$$EW = 20(EL) = 20(4.33) = 86.67$$

$$\text{Var}(W) = 20^2\text{Var}(L) = 400(1.25) = 500$$

It was stressed in Chapters 1 and 2 that in spite of the impressive wealth of mathematical tools that have been developed for applications of probabilistic modeling, many problems arise in practice that are too complicated to be solved mathematically; in these cases, computer simulation is an extremely valuable tool. The same is true for expected value and variance. We first try to obtain a mathematical solution to the problem at hand, but if we fail in this effort, we can get an approximate solution via simulation.

Note that the more complex a problem is, the less likely it is that a mathematical solution can be found, even by experts who specialize in mathematical analysis of probability models. Simulation allows *any* problem to be solved by a nonexpert, without the need to be able to invent some clever mathematical trick. As discussed at the end of Section 2.4, a mathematical solution is preferable if one can be found, but simulation provides a vitally important alternative in the numerous cases in which a mathematical approach cannot be found.

The general structure of a simulation program to find a mean of a random variable V follows directly from the long-run average interpretation of expected value, in Equation (3.2.1). We simulate repeating the experiment a large number of times, and take as our approximation for EV the average value of V during these repetitions. Following is the general form of such a simulation program.

Program 3.2.1: General Form of Simulation Programs to Find *EV*

```
program GenForm(input,output);
   var  {declarations go here}
```

```
begin
   SumV:= 0;
   for Rep:= 1 to TotReps do
      begin
      simulate performing the experiment;
      SumV:= SumV + V
      end;
   EV:= SumV/TotReps;
   writeln('EV  =  ',EV)
end.
```

Of course, the larger the value of TotReps used, the more accurate the output. This program structure is illustrated in the following two examples.

Example 3.2.11

A bag contains 20 marbles, 16 blue and 4 yellow. Suppose we choose a marble at random from the bag and replace it with a marble of the opposite color. We continue this process until exactly half of the marbles in the bag are blue; let T be the number of times we perform the process. Let us find $E(T)$ and $\mathrm{Var}(T)$.

Program 3.2.2

```
program Marble(input,output);
   const NMarb = 20;
         InitBlue = 16;

   var Seed,NBlue,Rep,TotReps,T: integer;
       SumT,SumT2,ET,VarT: real;

begin
   SumT := 0;
   SumT2 := 0;
   writeln('enter number of repetitions for the
                          simulation');
   readln(TotReps);
   Seed := 9999;
   for Rep := 1 to TotReps do
      begin
      NBlue := InitBlue;
      T := 0;
```

```
      repeat
         T := T + 1;
         if random(Seed) < NBlue/NMarb then
            NBlue := NBlue - 1
         else
            NBlue := NBlue + 1
      until NBlue = NMarb div 2;
      SumT := SumT + T;
      SumT2 := SumT2 + sqr(T)
      end;
   ET := SumT/TotReps;
   VarT := SumT2/TotReps - sqr(ET); {using Equation (3.2.16)}
   writeln('the mean and variance of T');
   writeln('are approximately',ET:8:2,VarT:8:2)
end.
```

We ran this program with TotReps = 100, and found ET and Var(T) to be approximately 17.50 and 128.91, respectively.

Note the standard deviation of T in this last example is then 11.35. This is a rather large value, in view of the fact that it is almost as large as $ET = 17.50$. This indicates that T can take on much higher values than 17.50.

Example 3.2.12

In Example 2.3.6 we were concerned with the length of time needed for a car to cross a busy highway. Let us find the expected value of this time.

In reading the program below, recall that the probability that a time slot is open (i.e., that no car passes during that time slot) is 0.7, and that we need two consecutive open slots to be able to cross the highway.

Program 3.2.3

```
program CrossHwy(input,output);

   var SlotsNeeded,SumT,T,Rep,TotReps,Seed: integer;
       LastSlotOpen,CurrentSlotOpen: boolean;
```

```
begin
   writeln('enter TotReps,Seed');
   readln(TotReps,Seed);
   SumT := 0;
   for Rep := 1 to TotReps do
      begin
      LastSlotOpen := (random(Seed) > 0.3);
      CurrentSlotOpen := (random(Seed) > 0.3);
      if LastSlotOpen and CurrentSlotOpen then
         SlotsNeeded := 2
      else
         repeat
            LastSlotOpen := CurrentSlotOpen;
            CurrentSlotOpen := (random(Seed) > 0.3);
            SlotsNeeded := SlotsNeeded + 1
         until LastSlotOpen and CurrentSlotOpen;
      T := 2*SlotsNeeded;
      SumT := SumT + T
      end;
   writeln('ET = ',SumT/TotReps:8:2) end.
```

Running this program with TotReps = 500 and Seed = 9999, the author obtained $ET = 9.94$, i.e. on the average, then, over many trips it will take about 10 seconds to cross the highway.

3.3 PARAMETRIC FAMILIES OF DISTRIBUTIONS

Consider the function $g(x) = cx^2$. For any given value of c, the graph of g is a parabola centered around the y axis. Each value of c corresponds to a different parabola, thus producing a **parametric family** of curves, shown for a few values of c in Figure 3.8. The quantity c here is called a **parameter** for this family of curves.

In applications of probability modeling, we often encounter a similar situation in a **parametric family of distributions**. To see this, recall Example 2.3.4, in which we found the probability of getting two heads in five tosses of a fair coin. The answer turned out to be $10(0.5)^2\ (0.5)^3$, reflecting the following reasoning:

a. Each of the 2 heads has probability 0.5 each, contributing the factor $(0.5)^2$.

b. The remaining $5 - 2 = 3$ tosses must be a tail, with probability $1 - 0.5 = 0.5$ each, contributing the factor $(0.5)^3$.

c. There are $\binom{5}{2} = 10$ possible orders for the 2 heads and 3 tails (HHTTT, HTHTT, TTHHT, etc.).

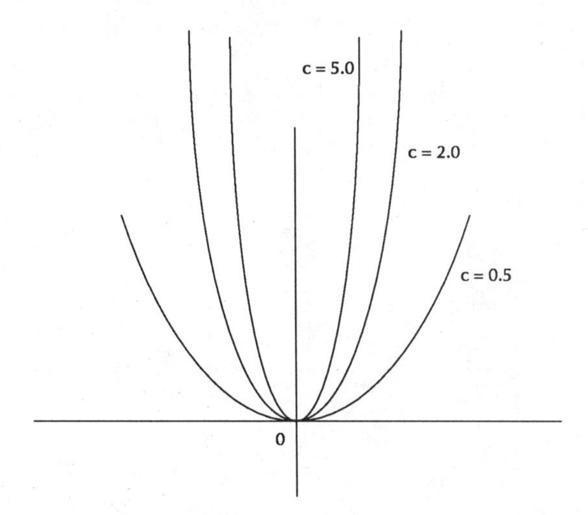

Figure 3.8 The parametric family of curves defined by $g(x) = cx^2$.

This reasoning is easily generalized: If we have a coin whose probability of heads is p, then the probability of getting k heads in n tosses of the coin is

$$\binom{n}{k} p^k (1-p)^{n-k}$$

Note that if we let X denote the number of heads obtained, then this last expression is the probability mass function of X, $p_X(k)$. Thus, we have discovered a parametric family of probability mass functions:

$$p_X(k) = \binom{n}{k} p^k (1-p)^{n-k} \tag{3.3.1}$$

For each fixed value of n and p, this function of k is a probability mass function; thus n and p are the parameters for this family, just as c was the parameter for the family $g(x) = cx^2$. k is the analog of x here, and n and p jointly are analogs of c. To complete our definition of the family, we should also state the ranges for the parameters n and p, although in this case they are clear: n can be any positive integer, and p any number in the interval [0,1].

Similarly, we very often use families of densities. For example, suppose an operations research engineer is designing a special telephone system. She has available old data on the lengths of calls and is using this data in her design. Suppose a frequency plot of the data has roughly the form of one of the curves in the exponential family graphed in Figure 3.9, something like ae^{-bt}. Since a density must have total area 1.0, we must have $a=b$ (Exercise 3.19). Thus the curve has the form be^{-bt}. Thus, we have again encountered a *family* of functions, one function for each value of b. The value of b in the engineer's case would be estimated from the data. (The use of data in choosing a parametric family, and then estimating the resulting parameter or parameters, is treated in Chapter 6.)

Figure 3.9 An exponential family.

In this section we will introduce several of the most widely used families of distributions, with others being covered in the exercises.

The Family of Binomial Distributions

This is a family of discrete distributions, with probability functions given by

$$p_X(k) = \binom{n}{k} p^k (1-p)^{n-k} \qquad \textbf{(3.3.2)}$$

This of course is the family described in the coin example above. However, it is widely applicable to a variety of situations. The probability mass function (3.3.2) is denoted $B(n, p)$.

The binomial distribution is appropriate in contexts which satisfy the following conditions:

a. The experiment consists of n identical stages, often called **trials**.

b. The trials are independent.

c. Each trial has only two possible outcomes, which, for the sake of identification only, are named "success" and "failure."

d. The probability of success is the same in each trial.

e. We are interested in X, the total number of successes in the n trials.

Then X has the probability mass function (3.3.2).

Example 3.3.1

To illustrate these conditions, consider again Example 2.3.4. A trial consists of tossing the coin once, and there are 5 trials. Success can be taken as getting a head, with p equal to 0.5. Then X has a $B(5,0.5)$ distribution, where X is the number of heads in the five tosses. We are interested in $P(X=2)$, which we can evaluate using Equation (3.3.2) with $n=5$, $p=0.5$ and $k=2$. Alternately, success could be defined as getting a tail, also with p equal to 0.5, and we would then be interested in $P(X=3)$.

Example 3.3.2

A certain **fault-tolerant** machine has ten identical components. It is designed to work properly as long as at least four of the components are good. The components may be assumed to act independently, and each has a 0.04 chance of failing. We can find the probability that the machine works by using a binomial distribution: We take $n=10$ and $p=0.96$, with a trial being identified with a component, and success being defined as the component working. Then our desired probability may be expressed as $P(X \geq 4)$, where X is the number of working components. This probability is $P(X=4)+ \cdots + P(X=10)$, which is then found, using Equation (3.3.2), to be equal to 1.0000 (to four decimal places). Thus the machine is quite reliable.

The expected value of a binomial random variable is found using Equation (3.2.5):

$$EX=\sum_{k=0}^{n} k\binom{n}{k}p^k(1-p)^{n-k} \tag{3.3.3}$$

This expression turns out to have the value np, and the variance of X is found to be $np(1-p)$ (see Exercise 3.16 and also Example 4.2.5).

In the context of the binomial distribution, we have a fixed number of trials (n), and a random number of successes (X). A **geometric distribution** arises in the opposite context: The number of trials, N, is random, and the number of successes is fixed at 1.

For example, suppose we roll a fair die until the first time a 4 appears. Then a *trial* consists of a roll of the die, and we can let *success* be getting a 4. Let N be the number of trials needed. Then N, the number of trials is variable, while 1, the number of successes, is fixed.

The probability that N is (say) 8 is

P(the first number is not a 4 and
the second number is not a 4 and
. . .
the seventh number is not a 4 and
the eighth number *is* a 4)

which, because of the independence of each trial, is $(1 - 1/6)^7(1/6)$.

The Family of Geometric Distributions

In general, if the probability of success is p, and N is the number of trials need for the first success, then

$$p_N(k) = (1-p)^{k-1}p \tag{3.3.4}$$

Our notation for this distribution will be Geom(p). Again, the notation shows that the parameter for this family is p, so for each value of p, there will be a different probability mass function (3.3.4).

The expected value of N is

$$EN = \sum_{k=1}^{\infty} k(1-p)^{k-1}p \tag{3.3.5}$$

To derive the numerical value in Equation (3.3.5), let $q = 1 - p$, and recall the formula for a geometric series:

$$\sum_{m=0}^{\infty} u^m = \frac{1}{1-u}$$

Then

$$EN = p\sum_{k=1}^{\infty} k\, q^{k-1}$$

$$= p\,\frac{d}{dq}\sum_{k=1}^{\infty} q^k$$

$$= p \frac{d}{dq}\left[\frac{1}{1-q} - 1\right]$$

$$= p \frac{1}{(1-q)^2}$$

$$= \frac{1}{p}$$

Similarly, we can derive $\text{Var}(N) = (1-p)/p^2$.

Example 3.3.3

One of the toll booths at a certain highway toll station is automatic, accepting exact change only. Five percent of all motorists using this booth are unaware of this requirement, and thus require time-consuming intervention from the station personnel. Suppose we have just finished going through this intervention process for a car, and let N denote the number of cars that will pass through this booth until the next one requiring intervention arrives.

Assuming the cars to be independent of each other, this is an example of a geometric distribution. Each car corresponds to one trial. Let us refer to a car's need for intervention as success (recall that we use *success* and *failure* merely as names for the two outcomes of a trial; the words have no positive or negative connotations). Then $p = 0.05$. Then, for example, $P(N = 10)$ is equal to

$$p(1-p)^9 = 0.0315 \qquad \textbf{(3.3.6)}$$

We now consider the family of negative binomial distributions. In spite of the fact that the name of this family contains the word *binomial,* it is related more closely to the geometric family than to the binomial family. Recall that a geometric random variable arises as the count of the number of trials needed to get the first success. The negative binomial family is an extension of this concept: We count the number of trials needed to get the rth success.

More formally, consider a sequence of independent success/failure trials as in the binomial and geometric cases, and let N be the number of trials needed to get r successes. Let p denote the probability of success on a single trial. Let us find the probability function of N; $p_N(k) = P(N = k)$, $k = r, r+1, r+2, \ldots$.

How can the event $N = k$ occur? We would have to have exactly $r-1$ successes in the first $k-1$ trials, and then a success on the kth trial. But the probability of getting $r-1$ successes in the first $k-1$ trials is easy to get, because this event fits the situation defined for the binomial family:

$$P(r-1 \text{ successes in the first } k-1 \text{ trials}) = \binom{k-1}{r-1} p^{r-1}(1-p)^{(k-1)-(r-1)}$$

Thus, since the first $k-1$ trials are independent of the kth trial, we have the following.

The Family of Negative Binomial Distributions

$$p_N(k) = P(N=k) = \binom{k-1}{r-1} p^{r-1}(1-p)^{k-r} p = \binom{k-1}{r-1} p^r (1-p)^{k-r}$$

for (k=r, r+1, r+2, . . .). We will denote this family of distributions by $NB(p, r)$. There is a different distribution for each pair of values for p and r.

Example 3.3.4

Suppose a candy bar company has a promotional contest, in which 10% of the candy bars have a coupon inserted into the packaging. You get a prize if you collect 5 coupons. What is the probability that you will need to buy more than 30 candy bars to get the 5 coupons?

Let N be the number of candy bars you will need. N has a negative binomial distribution with $p=0.10$ and $r=5$. Then

$$P(N > 30) = 1 - P(N \leq 30) = 1 - \sum_{k=5}^{30} \binom{k-1}{4} 0.1^5 0.9^{k-5} = 0.82$$

Thus, there is an 82% chance that 30 candy bars will not be enough.

The Family of Poisson Distributions

As with the last two families of distributions, the Poisson family is also discrete. The form of the probability mass function for a

Poisson-distributed random variable V is

$$p_V(k) = \frac{e^{-b}b^k}{k!} \tag{3.3.7}$$

where k ranges through all nonnegative integers. The probability mass function (3.3.7) is denoted $P(b)$. The parameter b can take on any positive real value.

The Poisson model has empirically been found to accurately describe "counting" random variables in a large number of applications. Typically, the random variable of interest is a count of the number of events occurring within a given interval of time, such as

a. The number of radioactive particles emitted from a substance in a one hour interval.

b. The number of accidents in a factory in a one week time period.

c. The number of cars passing a certain point on a highway in a one minute period.

d. The number of requests for an input/output device in a computer system in a 10 second interval.

Processes of this kind are known as **Poisson processes**. In these cases, b is proportional to the length l of the time interval involved (i.e., $b = wl$ for some constant of proportionality w).

The Poisson distribution is somewhat remarkable in that its mean and variance turn out to be equal, both having the value b (Exercise 3.17). The parameter $w = b/l$ in this discussion may be considered an *intensity parameter*, representing the mean number of events occurring per unit time.

In addition to *temporal* settings such as those just listed, the Poisson model is also often useful in the analysis of *spatial* counts. For example, epidemiologists have found that the distribution of the number of incidents of a certain disease in a given region can often be described well by Equation (3.3.7), for an appropriate value of b determined by observed data for that disease. Here b is proportional to the area a of the region (i.e., $b = wa$ for some constant w, with w now representing mean number of incidents per unit area).

Example 3.3.5

Modern technology allows a large and complex set of electronic circuitry to be manufactured as one unit, called a **chip**. This capability in fact is the major impetus for the electronics revolution that has been taking place since about 1970.

Unfortunately, the manufacture of chips sometimes produces flaws. Suppose these flaws have a spatial Poisson distribution, with a mean of $w=0.002$ flaws per square millimeter. Consider chips of size 5 mm by 5 mm, or 25 mm^2. What proportion of such chips will have at least one flaw?

Let N be the number of flaws on a chip of this size. Then the probability mass function of N will be

$$p_N(k) = \frac{e^{-b} b^k}{k!}$$

where $b=25(0.002)=0.050$ and k is any nonnegative integer. Then

$$P(N \geq 1) = 1 - P(N=0) = 1 - \frac{e^{-0.050} 0.050^0}{0!} = 0.0488 \qquad \textbf{(3.3.8)}$$

Therefore, 4.88% of the chips will be flawed.

Another use for the Poisson distribution is as an approximation to the binomial distribution (Exercise 3.35).

Having introduced some families of probability mass functions, we now turn to some families of density functions.

The Family of Uniform Distributions

This family is a generalization of the distribution in Example 3.1.2. The idea is that a random variable having distribution in this family arises by choosing a number from some bounded interval $[a, b]$ in such a way that all numbers are equally likely to be chosen.

As with the reasoning in Example 3.1.2, we see that the density of such a random variable, say V, must be constant within the interval $[a, b]$, signifying that V takes on all values in that region with equal frequency, and 0 outside the interval (i.e., V never takes on values outside the interval).

Since any density must integrate to 1 [Equation (3.1.5b)], the constant height of f_V in $[a, b]$ must be $1/(b-a)$. Thus

$$f_V(t) = \frac{1}{b-a} \qquad \textbf{(3.3.9)}$$

for t in [a,b], and 0 elsewhere. The density is denoted $U(a, b)$. The parameters a and b can take on any real values, subject of course to $a < b$.

One application of the uniform model is in the analysis of roundoff errors in numerical computation. For example, suppose we believe a certain type of computation to be accurate to within 0.1. We may model the error to be uniformly distributed on the interval $[-0.1, 0.1]$.

Another example involves the design process for a disk-based output system in a computer.

Example 3.3.6

Recall Example 2.3.9, in which computer disk systems were introduced. Let us continue to assume the characteristics of the disk drive in that example: The read/write head moves from one track to another at the rate of 3.2 milliseconds per track, the drive is rotating at a speed of 30 ms for a complete rotation, and the actual data transfer (read or write) takes 1.2 ms.

One of the disk actions mentioned was the **rotational delay**, during which the read/write head is waiting for the proper sector in the disk to rotate around to the position underneath the head so that the sector can be read or written. Let us denote this time by R. It is reasonable to assume that R could take on any value in the interval $[0,30]$ with equal likelihood: R has a $U(0,30)$ distribution.

Here is an example of how the knowledge of the distribution of R can be used: Suppose the read/write head is currently at Track 25. What is the probability that the next disk access (read or write) will take more than 160 ms?

Let I be the track number of the next requested access, and T the total access time. How can the event $T > 160$ occur? First note that to move to Track I, the seek time S will be $|I - 25|\ (3.2)$. Now since $T = S + R + 1.2 = |I - 25|(3.2) + R + 1.2$, we have (using the idea in Chapter 2 of breaking big events down into small events)

$$\begin{aligned} P(T > 160) &= \sum_{k=0}^{75} P(I = k \text{ and } T > 160) \\ &= \sum_{k=0}^{75} P(I = k)P(T > 160 | I = k) \\ &= \sum_{k=0}^{75} P(I = k)P[R > 160 - |k - 25|(3.2) - 1.2] \end{aligned}$$

Now assuming as before that all tracks are accessed with equal frequency, we have $P(I = k) = \frac{1}{76}$ for all k. To evaluate

$$P[R > 160 - |k - 25|(3.2) - 1.2]$$

we will use the fact that R has a $U(0,30)$ distribution, but we need to consider cases.

For example, look at the case $k=50$. Here

$$P[R > 160 - |k-25|(3.2) - 1.2] = P(R > 78.8) = 0$$

since the largest value R can take on is 30. Similarly,

$$P[R > 160 - |k-25|(3.2) - 1.2] = 0$$

for $k \leq 65$. For $k=66$,

$$P[R > 160 - |k-25|(3.2) - 1.2] = P(R > 27.6) = \int_{27.6}^{30} \frac{1}{30-0}\,dt = 0.08$$

Similarly, for $k=67$ through 74, we have

$$P[R > 160 - |k-25|(3.2) - 1.2] = P(R > 238.8 - 3.2k) = \int_{238.8-3.2k}^{30} \frac{1}{30}\,dt = \frac{30 - 238.8 - 3.2k}{30}$$

For the case $k=75$, the reader should check that $P[R > 160 - |k-25|(3.2) - 1.2] = 1$.

Finally putting all of this together, we have

$$P(T > 160) = \sum_{k=66}^{74} \frac{1}{76}\,\frac{3.2k - 208.8}{30} + \frac{1}{76}(1) = 0.07$$

Before going on to the next family of distributions, we note also that the $U(0,1)$ distribution is a fundamental building block in the simulation of random variables having various distributions. This fact will be illustrated extensively in Section 3.4.

The Family of Normal Distributions

This family of densities has the form

$$\frac{1}{\sqrt{2\pi}b} e^{-(t-c)^2/2b^2} \qquad \textbf{(3.3.10)}$$

so that for each value of b and c, we have the famous "bell shape" of Figure 3.10. Since the curve is symmetric around $t=c$, the mean for a random variable having the density (3.3.10) must be c, since the values larger and smaller than c will eventually "cancel out" in the integral used to find the mean, Equation (3.2.6). (Actually, this will happen only if the mean exists, as it does here.) Integration by parts shows that the variance is b^2. The density is denoted by $N(c, b^2)$. The parameter c can take on any real value, and b can take on any positive real value.

The normal (or **Gaussian**) distributions are very prominent in statistical applications, largely due to the *Central Limit Theorem*. This famous theorem, which is discussed in more detail in Section 4.4, says that under fairly general conditions, the sum of a large number of random variables has an approximately normal distribution, whatever the distributions of the summands.

Many empirically observed random variables seem to have a distribution that can be described well by the normal model (3.3.10). For example, consider the total monthly sales of a company to its 50 major customers. Since this is a sum, its distribution is probably close to normal. Of course, as in most applications of probability models, the normal model is only an approximation: For example, most quantities in practice have lower and upper bounds (e.g., the sales total above is nonnegative) while exact normal random variables extend from $-\infty$ to ∞. Thus, the distribution of total sales cannot be *exactly* normal. However, the normal model is a very useful approximation.

As with all continuous random variables, probabilities for normally distributed random variables take the form of an integral of the density. However, in the normal case a certain difficulty arises, as shown in the following example.

Figure 3.10 A normal distribution.

Example 3.3.7

Suppose tires of a certain type have lifetime L, normally distributed with mean 40,000 miles and standard deviation 6,000, and that we wish to find the proportion of tires that have lifetime under 35,500 miles, or $P(L < 35{,}500)$.

Proceeding as usual, this probability is equal to

$$\int_{-\infty}^{35{,}500} \frac{1}{\sqrt{2\pi}\,6000} e^{-(t-40{,}000)^2/2(6000)^2}\, dt \tag{3.3.11}$$

A problem then arises in that none of the usual methods of integration (e.g, integration by parts) will work in calculating this integral. In fact, it can be proved that *no* method will work! Thus, the value of the integral must be found approximately, using techniques in the field of numerical analysis (we will finish this example later).

Since the normal distribution is so important, numerical methods have been used to create tables of probabilities such as in the previous example. At first, it might seem that for each pair of values b and c, a separate tabulation of the values $P(X < t)$ would be needed. However, it turns out (fortunately) that only one tabulation is needed—that for the case $b=1$ and $c=0$. The reason for this stems from the following special property.

The Linearity Property of the Normal Family

> Suppose X has a $N(c, b^2)$ distribution. Given any two constants d_1 and d_2, with d_1 nonzero, define the random variable $Y = d_1 X + d_2$. Then Y too has a normal distribution, with mean $d_1 c + d_2$, and variance $d_1^2 b^2$.

The proof of this property is outlined in Exercise 3.18. Note the mean and variance claimed for Y are consistent with Equations (3.2.15) and (3.2.17), so the values for EY and $\text{Var}(Y)$ indicated above do not really say anything new. Instead, the remarkable part is that Y inherits the "normal-ness" of X; that is, Y too has a member of the family of bell-shaped curves for its density.

The Linearity Property implies that only one table is needed for the normal family of distributions, that for the normal distribution having $c=0$ and $b=1$, which is called the **standard normal distribution**. To see that this is the only distribution that need be tabulated, let us return to Example 3.3.7. Define a new random variable $Z = (L - 40000)/6000 = (1/6000)\ L - 20/3$. Then, using the Linearity Property with $X=L$, $Y=Z$, $c=40{,}000$, $b=6000$, $d_1 = 1/6000$ and $d_2 = -20/3$, we see that Z has a $N(0,1)$ distribution.

Thus, since our original problem $P(L < 35{,}500)$ can be re-expressed as

$$\begin{aligned} P(L < 35,500) &= P(L - 40,000 < -4500) \\ &= P\left(\frac{L - 40,000}{6000} < -0.75\right) \\ &= P(Z < -0.75) \end{aligned} \qquad \textbf{(3.3.12)}$$

we see that a table for the probabilities $P(Z < t)$ will provide the answer to our question in Example 3.3.7. Specifically, look at Table 2 in Appendix B. The proper row is found by looking for -0.7 along the left-hand margin, and the column is indicated by the heading 0.05 along the top margin. The table entry for this row and column combination is 0.2266, which means that $P(Z < -0.75) = 0.2266$. Thus, only about 23% of the tires will last less than 35,500 miles.

Of course, the same trick works for any normally distributed random variable X. If X has a $N(c, b^2)$ distribution, then we can define a new random variable $Z = (X - c)/b$, which will have the $N(0,1)$ distribution. Then our original probability statement concerning X can be transformed to a probability that can be found from the $N(0,1)$ in Table 2 in Appendix B.

Example 3.3.8

Suppose flashlight batteries have lifetimes L, normally distributed with mean 50 hours and standard deviation 15 hours. Our flashlight requires one battery. If we put in a new battery before we go on a camping trip, what is the probability that the battery will last more than 80 hours? This question is easily answered.

$$\begin{aligned} P(L > 80) &= P(Z > 2) \\ &= 1 - P(Z \leq 2) \\ &= 1 - P(Z < 2) \\ &= 1 - 0.9772 \\ &= 0.0228 \end{aligned} \qquad \textbf{(3.3.13)}$$

(Note that $P(Z \leq 2)$ and $P(Z < 2)$ are equal, since Z is a continuous random variable.) Thus, we should take extra batteries if we expect to use the flashlight more than 80 hours (the question of *how many* extra is investigated in Exercise 4.31).

Another use of the normal distribution is as an approximation to the binomial distribution, complementing the Poisson approximation mentioned earlier. This procedure is discussed in Section 4.4.

The Family of Exponential Distributions

This family of densities, a few of whose members were graphed in Figure 3.9, has the form

$$be^{-bt} \tag{3.3.14}$$

for positive t, and takes the value 0 for nonpositive t. The mean and variance of a random variable X having this density are easily found to be $1/b$ and $1/b^2$ (Exercise 3.19). The notation for this family is Expon(b), with the parameter b allowed to be any positive real value. Note that if X has an exponential density, then X takes on smaller values more often than larger values.

The density (3.3.14) is often used in modeling duration, such as lifetimes or waiting times. For example, it has been found to be a good model for the lifetimes of airplane air conditioners (see the analysis of such a data set in Chapter 6). Also, density (3.3.14) can be shown to be the density of waiting times between successive counts in a Poisson process (see Exercise 3.43).

The exponential densities have a curious property, as illustrated in the following example.

Example 3.3.9

A group of people arrived at a bus stop 15 minutes ago, just after a bus departed; they are waiting for the next bus. One person in the group mentions that in his experience, the average waiting time between successive buses is about 10 minutes, so there probably will be a bus arriving very soon. Another person in the group just happened to have been keeping complete records in the past, and has observed that the waiting times seemed to have the density (3.3.14), with a value 0.1 for b, consistent with the 10 minute mean observed by the first person.

Let us investigate the first person's belief that the next bus is likely to arrive soon since they had already waited 15 minutes. For example, we can find the probability that the bus will arrive in the next 3 minutes: Let W be the final waiting time, or the actual time the group will have waited when the bus finally does arrive. We already know that in this instance $W > 15$, so the probability in question is conditional; that is, we want to find

$$P(W < 18 | W > 15)$$

In our calculation of this probability, we might as well be a bit more general, finding

$$P(W < t+3|W > t)$$

Recalling Rule 3 of Section 2.3, we have

$$\begin{aligned} P(W < t + 3|W > t) &= \frac{P(W < t + 3 \text{ and } W > t)}{P(W > t)} \\ &= \frac{P(t < W < t + 3)}{P(W > t)} \\ &= \frac{\int_t^{t+3} 0.1e^{0.1s} ds}{\int_t^{\infty} 0.1e^{0.1s} ds} \\ &= \frac{e^{-0.1t} - e^{-0.1(t+3)}}{e^{-0.1t}} \\ &= 0.26 \end{aligned} \tag{3.3.15}$$

Thus, the probability of the bus arriving within the next 3 minutes, given that the group has been waiting 15 minutes already, is only 0.26, so the first person's guess was wrong. Furthermore, the fact that the t dropped out in calculation (3.3.15) shows that the probability of arrival within the next 3 minutes would have been 0.26 regardless of the time at which the group had asked the question, whether they had been waiting 15 minutes or 5 minutes or any other time. For example, immediately upon arrival of the group (i.e., $t=0$) the probability that the next bus would come within 3 minutes was also 0.26.

It has been found that density (3.3.14) is a good model for the lifetime of some electronic components. Suppose that such a component has mean lifetime 10 months. Then an analysis similar to that above shows that a 15-month-old component has no more (and no less) chance of failing within the next 3 months than that of a *new* part failing within its first 3 months. This special property is often called "memorylessness," indicating the 15-month-old component doesn't remember those 15 months of wear.

Of course, just as the lifetimes of some kinds of components have exponential distributions, some others do not. For a type that does not, we must search for another parametric family that fits the lifetime data we have gathered for that component type.

The exponential distribution is also frequently used as a model for service times for items using a machine or some other "server" (possibly human). It is interesting to note that in some cases for which the exponential is *not* appropriate, we can actually take advantage of this fact.

Example 3.3.10

In some computer timesharing systems, there are two or mores queues of different priorities. Each queue is of the **Round Robin** type: The queue is circular, with jobs taking turns, running for a short period of time each turn; a job will thus usually take several turns before it finishes. We wish to give short jobs higher priority, but we do not know the run times R of jobs beforehand. A newly arriving job is at first assigned to the queue of highest priority. If the job is not done after taking several turns in this queue, it is transferred to a lower-priority queue, on the assumption that it will be a long-running job. This would not be possible if the run times had an exponential distribution; in such a case, no matter how many turns a job has taken, the chance of its finishing within the next few turns would always be the same. Of course, for this scheme to be successful, it is not sufficient that we know the distribution is nonexponential; the distribution must have the property that the longer a job has been running, the less likely it is to finish soon; that is, $P(R < t + s | R > t)$ should be a decreasing function of t.

The Family of Gamma Distributions

This family of densities has the form

$$c\, b^r\, t^{r-1}\, e^{-bt} \qquad \textbf{(3.3.16)}$$

for positive t, and zero otherwise. Here the parameters b and r must both be positive real numbers. The quantity c is not a parameter, but rather a value that makes the density integrate to 1.0, as required for any density [Equation (3.1.5b)]. Setting

$$1.0 = \int_0^\infty c\, b^r t^{r-1} e^{-bt}\, dt$$

and making the change of variable $s = bt$, we find that

$$c = \left(\int_0^\infty s^{r-1}\, e^{-s}\, ds \right)^{-1}$$

This integral, which arises in many areas of applied mathematics, is a function of r known as the **gamma function**, denoted $\Gamma(r)$. The notation for this distribution is $\gamma(r, b)$.

Note that the exponential distribution is a special case of the gamma family, in which $r=1$. As mentioned above, random variables measuring lifetime, waiting time, or some other duration are sometimes modeled with an exponential density, but often this is too restrictive, so that other values of r are used instead. The shape for $r=2$ and $b=1$, shown in Figure 3.11, is typical. Note that in contrast to an exponential distribution, for which the most frequently occurring value (**mode**) is 0, the most frequently occurring value for a random variable having the density in Figure 3.11 is 1.

Figure 3.11 A gamma distribution ($r = 2$, $b = 1$).

The cases in which r is an integer in density (3.3.16) form an important subfamily of the gamma distributions, called the **Erlang** distributions. It will be shown in Example 4.3.8 that if the random variables Y_i are independent (a notion similar to that of Definition 2.3.2, to be formally defined in Section 4.2), and exponentially distributed with parameter b, and that if we set

$$Y=Y_1+\cdots+Y_r$$

then Y has the density (3.3.16). This fact will be used in the next section.

Another important subfamily of the gamma class is the set of **chi-square** distributions. Here r is of the form d/2 for a positive integer d, and $b=1/2$. This is then a one-parameter family, parameterized by d. The parameter d is called the **degrees of freedom**. It can be shown that if $Y = Y_1^2 + \cdots + Y_d^2$, where the random variables Y_i are independent and have $N(0, 1)$ distributions, then Y has a chi-square distribution with d degrees of freedom. This distribution is tabulated in Table 4 in Appendix B.

Example 3.3.11

A machine is supposed to insert a pin at a certain position P on a flat surface. However, there is some error in the position at which the machine actually makes the insertion. Let X denote the horizontal error and Y the vertical error. Suppose X and Y are independent random variables with distribution $N(0, 3^2)$. Let W be the distance from the actual insertion point to the intended point. Let us find $P(W > 7.5)$.

From the Linearity Property of the Normal Family, we know that $X/3$ and $Y/3$ have $N(0,1)$ distributions. Thus, from the discussion, $Q = W^2/9$ will have a chi-square distribution with 2 degrees of freedom. Thus, Table 4 in Appendix B tells us that

$$P(W > 7.5) = P(Q > 6.25) \approx 0.05$$

There are other families of distributions related to the chi- square. For example, a **Student-*t* distribution** with d degrees of freedom is defined to be that of the random variable

$$\frac{Z}{\sqrt{Y/d}} \tag{3.3.17}$$

where Z has a $N(0,1)$ distribution, Y has a chi-square distribution with d degrees of freedom, and Z and Y are independent random variables. The Student-t distributions thus form a one-parameter family. Also, an **F distribution** is defined to be that of the random variable

$$\frac{W_1/d_1}{W_2/d_2} \tag{3.3.18}$$

where W_1 and W_2 are independent chi-square random variables, with degrees of freedom d_1 and d_2, respectively. Thus, the F distributions form a two-parameter family. The Student-t and F distributions will be discussed again in Chapters 6 and 7.

The Family of Weibull Distributions

This is a two-parameter family with density

$$\gamma\beta t^{\beta-1}e^{-\gamma t^{\beta}} \qquad (t > 0) \tag{3.3.19}$$

It has been found very useful in reliability modeling, with component lifetime density often well described by a member of this family.

3.4 SIMULATION OF RANDOM VARIABLES HAVING GIVEN DISTRIBUTIONS

We have seen that simulation is a very powerful problem-solving tool. In this section we continue our development of simulation methodology, presenting methods for simulating random variables having specified probability or density functions. Thus, if we are studying a problem in which a random variable has, say, an exponential distribution, then we will be able to study the problem through simulation.

In our previous simulation work (Section 2.4 and the last part of Section 3.2), we assumed that our computer system has an **intrinsic** (i.e., built-in) function called "random," which simulates a $U(0, 1)$ random variable. We will now show how this is done.

The function below is called a **linear congruential generator**.

Program 3.4.1

```
function random(var Seed: integer): real;
   const C = 25173;
         D = 13849;
         M = 32768;
begin
    Seed := (C*Seed + D) mod M;
   random := Seed/M
end;
```

Several remarks are in order:

a. The choices of C, D, and M depend essentially on two factors: the size of the maximum allowable integer for the computer and language being used and accuracy considerations, which are discussed below.

b. This function is only an approximation; it does not truly simulate $U(0, 1)$. For example, the function cannot take on *all* values in $[0, 1]$; it can only take on values in the set

$$\left\{\frac{0}{M}, \frac{1}{M}, \ldots, \frac{M-1}{M}\right\} \tag{3.4.1}$$

Of course, the finite word size of a computer implies that a similar problem would occur with other random number generation schemes.

c. In order to be acceptable, the values of the function must satisfy two conditions. The first condition is the actual uniformity of the function values in [0,1]; for example, the value of Q resulting from

```
Count := 0;
for I := 1 to 1000 do
   begin
   X := random(Seed);
   if (X > 0.25) and (X < 0.40) then
      Count := Count + 1
   end;
Q := Count/1000
```

should be approximately 0.15. Similarly, the value of Q should be approximately $b - a$, if 0.25 and 0.40 are replaced by a and b above; this must be true for all values of a and b between 0 and 1 ($a < b$).

The second condition is that successive calls to the function should simulate independence; for example, the value of Q resulting from

```
Count := 0;
for i := 1 to 1000 do
   begin
   X := random(Seed); Y := random(Seed);
   if (X < 0.4) and (Y < 0.3) then
      Count := Count + 1
   end;
Q := Count/1000
```

should be approximately 0.12; similar results should hold for n-tuples of X values (see Exercise 3.20).

d. We would at least hope that our U(0,1) generator hits all numbers in the set (3.4.1). Fortunately, this can be assured by choosing C, D, and M to satisfy the following conditions (see the Knuth reference at the end of this chapter):

(i) D and M are relatively prime.

(ii) C − 1 is divisible by every prime factor of M.

(iii) If M is divisible by 4, then so is C − 1.

e. The above version of "random" may have to be modified a little if your compiler aborts a job which has produced multiplicative or additive overflow. Also, you should check to make sure that your Pascal *mod* function produces nonnegative values only.

f. Most Pascal compilers have their own built-in functions like "random." If yours does, then you need not bother with writing your own, although you should still learn about how such functions are constructed.

Functions such as "random" are called **random number generators** (because they necessarily have some imperfections, as discussed above, they are often described as **pseudorandom**). We will use "random" as a foundation for developing random number generators for other distributions. These generators can then be used in simulation studies.

Example 3.4.1

Recall Example 2.4.1. Suppose interarrival times between calls have an exponential distribution with mean 0.25 second. If we had a function Expon similar to "random" above, but generating a random variable having the exponential density [Equation (3.3.14)], rather than U(0,1) we could find *P*(at least 3 lost calls out of 20) by running Program 3.4.2.

The program section beginning with the line

```
TimeNow := 0.0
```

and ending with the line

```
NextCallArrive := TimeNow + Expon(B,Seed)
```

simulates the experiment, that is, it corresponds to Line 7 of Program 2.4.3. We need keep track of the system only at those times at which changes occur, which are (a) when a new call arrives (NextCallArrive) and (b) when the system becomes free after handling a call arrival (TimeNextFree). The simulated current time (TimeNow) is updated for each new call. For most readers, this program will be considerably more difficult to follow than those we have encountered up to this point, as well as harder than those we will see in the next few chapters. Read it slowly, possibly several times. "Walk through" the execution of the "for" loop, for the first few values of I, making up your own values for Expon (which will be different for each successive call to Expon). Soon you will find yourself agreeing that it does mimic the operation of the telephone system as described.

Program 3.4.2

```
program Lost(input,output);

   const ResponseTime = 0.05;
         B = 4.0;  {recall that EX = 1/B for expon.
                              random vars.}

   var NLost,I,Rep,Count,Seed: integer;
         NextCallArrive,TimeNextFree,
            TimeNow,Prob: real;

   function Expon(B: real; var Seed: integer): real;
   begin
      {function body goes here}
   end;

begin
   Seed := 9999;
   Count := 0;
   for Rep := 1 to 1000 do
      begin
      TimeNow := 0.0;
      TimeNextFree := 0.0;
      NextCallArrive := Expon(B,Seed);  {time Call #1
                              will arrive}
      NLost := 0;
      for I := 1 to 20 do
         begin
         TimeNow := NextCallArrive;  {Call #i has arrived}
         if TimeNow >= TimeNextFree then  {Call #i is
                              accepted, resulting in a system
                              lockup for the next 0.05 seconds}
            TimeNextFree := TimeNow + ResponseTime
         else
            NLost := NLost + 1;
         NextCallArrive := TimeNow + Expon(B,Seed)  {time
                              when Call #i+1 will arrive}
         end;
      if NLost >= 3 then Count := Count + 1
      end;
   Prob := Count/1000.0;
   writeln('P(at least 3 lost) = ',Prob:6:3)
end.
```

We will now discuss methods for writing Expon and other nonuniform random number generators. A number of methods have been developed. The most widely used "general" method (in the sense of being applicable to a large class of distributions) is based on the inverse of the cumulative distribution function and is sometimes called the "inverse c.d.f." method. Suppose we need a random number generator for a continuous random variable having cumulative distribution function H and density h. We will assume here that the graph of h is in one piece (i.e., not like that in Figure 3.12; otherwise we would run into some difficulties that would obscure the principles here). Let $G = H^{-1}$, where the -1 does not mean reciprocal, but rather functional inverse; if the function w is defined by $w(x) = x^2$, then w^{-1} is defined by $w^{-1}(y) = \sqrt{y}$, and if $v(x) = \sin x$, then $v^{-1}(y) = \arcsin y$.

Then in order to generate a random variable X having distribution H, we can simply use the function "random" to generate a $U(0,1)$ random variable U and then set $X = G(U)$. To see that this works, observe that

$$\begin{aligned} F_X(t) &= P(X \leq t) \\ &= P(G(U) \leq t) \\ &= P(U \leq G^{-1}(t)) \\ &= P(U \leq H(t)) \\ &= H(t) \end{aligned} \tag{3.4.2}$$

where the last equality is due to the fact that since U has a U(0,1) distribution, $P(U \leq s) = s$ for any s in [0,1]. Thus $F_X = H$, as desired.

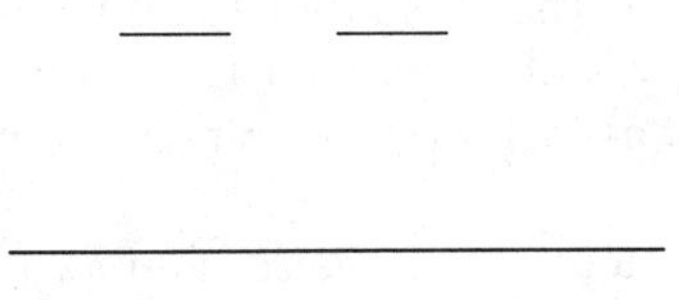

Figure 3.12 A two-piece density function.

The method we have just discussed, which is called the **inverse transformation method**, can be used to write the function Expon used in Program 3.4.2. For example, suppose $b = 1$ in the formula for the exponential density family [Equation (3.3.14)]. Then for $t > 0$ we have

$$H(t) = \int_0^t e^{-s}\, ds = 1 - e^{-t} \tag{3.4.3}$$

We can find the inverse function of H by setting $v = H(t) = 1 - \exp(-t)$, and solving for t in terms of v. We get $t = -\ln(1 - v)$, or $G(v) = -\ln(1 - v)$. Thus, the Pascal statement

```
W := -ln(-random(Seed))
```

will generate a random variable having the desired density. The student should verify that this really works by executing the above statement (say) 1000 times and drawing a chart of the relative frequencies of the values of w generated. The chart should look approximately like one of the curves in Figure 3.9 (as usual, we would need an infinite number of values of w in order to exactly reproduce that figure).

We can go through the same procedure to find the function G for general b, and thus write the random number generator Expon, for any value of b in Equation (3.3.14).

Program 3.4.3

```
function Expon(B: real; var Seed: integer): real;
   var U: real;
begin
   U := random(Seed);
   if U < 0.0001 then U := 0.0001;
   Expon := -ln(U)/B
end;
```

On some systems, the ln function does not accept numbers that are too close to 0, so U needs to be truncated at some point, such as the 0.0001 value used here. The reason for using ln(U) rather than ln(1 – U) is that if T has a $U(0,1)$ distribution, then 1 – T also has this distribution; thus ln(U) is just as good as ln(1 – U), and since it requires less computation, we use this form.

Unfortunately, the *inverse transformation* method used above is impractical to implement for some distributions because of difficulties in getting a closed-form expression for G. In such cases we usually look for a method which depends on some special property of the distribution in question.

Consider the problem of writing a random number generator for the family of Erlang distributions, discussed in Section 3.3. Finding a closed-form expression for H is difficult here, although it can be done by repeated integration by parts (the fact that r is an integer in the Erlang density function—Equation (3.3.16)] makes this possible). Finding an expression for G from H is worse; in fact, it is impossible.

However, we do know that an Erlang random variable has the same distribution as that of a sum of r independent exponentially distributed random variables, each having parameter b (this was mentioned without proof in Section 3.3 and will be proved in Example 4.3.8). This special property of the Erlang family, then, makes it easy to write a random number generator for the family:

Program 3.4.4

```
function Erlang(B: real; R: integer;
    var Seed: integer): real;
     var Sum: real;
         I: integer;

begin
    Sum := 0;
    for I := 1 to R do
        Sum := Sum + Expon(B,Seed);
    Erlang := Sum
end;
```

This idea of capitalizing on some special property of a distribution in order to write a random number generator for the distribution is quite common, as illustrated in the following examples. Constructing a generator for the binomial family is easy, done by simply simulating the setting in which it occurs, as in the following program.

Program 3.4.5

```
function Binom(N: integer; P: var Seed: integer): integer;
    var Trial,NSuccess: integer;
begin
    NSuccess := 0;
    for Trial := 1 to N do
        if random(Seed) < P then
            NSuccess := NSuccess + 1;
    Binom := NSuccess
end;
```

A generator for the family of geometric distributions is similarly easy:

Program 3.4.6

```
function Geom(P: real; var Seed: integer): integer;
    var Trial: integer;
```

```
begin
   Trial := 0;
   repeat
      Trial := Trial + 1
   until random(Seed) < P;
   Geom := Trial
end;
```

While the inverse transformation method can be used for writing a generator for the $U(a,b)$ family, it is easier just to utilize the fact that an expanded and shifted version of a random variable which takes on values uniformly in (0,1) will be uniformly distributed on the expanded and shifted version of (0,1):

Program 3.4.7

```
function Unif(A,B: real; var Seed: integer): real;
begin
   Unif := (B - A)*random(Seed) + A
end;
```

Random number generators for the Poisson and normal families also depend on special properties of those families, although these properties are less obvious than those used above. The random number generator for the Poisson family is based on the following fact, proved in Exercise 3.43: $W_1, W_2, W_3, \ldots$ are independent random variables with an exponential distribution, then the random variable N has a Poisson distribution, where N is defined to be the largest number k such that

$$W_1 + W_2 + \cdots + W_k < 1.0$$

This fact then leads to the following random number generator.

Program 3.4.8

```
function Poisson(B: real; var Seed: integer): integer;
   var Count: integer;
       Sum: real;
begin
   Sum := 0;
   Count := 0;
   repeat
```

```
         Count := Count + 1;
         Sum := Sum + Expon(B,Seed)
      until Sum >= 1.0;
      Poisson := Count - 1
   end;
```

Generating a distribution in the normal family is especially difficult using the inverse method; it is impossible to obtain a closed-form expression even for the function H. Thus, we again use a special property of distribution. Box and Muller have shown that if V and W are independent U(0,1) random variables, and we set

$$X = \cos(2\pi V)\sqrt{-2\ln(W)}$$

and

$$Y = \sin(2\pi V)\sqrt{-2\ln(W)}$$

then X and Y will be independent $N(0,1)$ random variables. Then the linearity property of the normal family can be used to transform X and Y to the distribution $N(c, b^2)$ as in the following program.

Program 3.4.9

```
procedure Normal(B,C: real; var X,Y: real; var Seed:
   integer);
   var T1,T2: real;
begin
   T1 := 6.28*random(Seed);
   T2 := sqrt(-2.0*ln(random(Seed)));
   X := cos(T1)*T2; X := B * X + C;
   Y := sin(T1)*T2; Y := B * Y + C
end;
```

Note that this method actually gives us *two* normal random variables, X and Y, rather than one, as with the other generators presented above. For example, if we wish to generate 20 normal random variables, we need call this procedure only 10 times. Another comment that should be made here is that the computation of transcendental functions, such as square root, natural logarithm, sine, and cosine functions, take up a substantial amount of computer time. Thus, variations of the above procedure, which require less computation of this kind, are often used.

This concern about computation time involved in a random number generator is worth discussing a bit more, in a general context, rather than just in the context of the normal distribution generator above. Floating-point (type "real" in Pascal) computation is always slower than the corresponding fixed-point (type "integer" in Pascal) action (e.g., a floating-point addition is slower than a fixed-point addition). Furthermore, if the computer does not include floating-point operations in its basic instruction set, as is the case with most microcomputers, the difference between floating-point and fixed-point computation times may be tremendous since the floating-point operations will have to be implemented in software. This will be especially true for computation of transcendental functions, since they involve many floating-point calculations.

Even on a system with good floating-point facilities, a real-world simulation program may use very large amounts of computer time (e.g., several hours). Thus, the slowness of floating-point computation in some computers is clearly a very serious practical problem for the simulation analyst, especially in light of the prevalence of microcomputers. We will briefly discuss two solutions:

a. Most microcomputers, especially those of the "PC" type, can accomodate the addition of a floating-point "chip" (e.g., Intel 8087). This chip augments the basic instruction set of the computer with many floating-point operations and can speed up floating-point computation by an order of magnitude. Of course, the user should make sure that the compiler he or she is using has been designed to make use of the chip.

b. Some "compilers" are actually **interpreters**. The distinction between these is beyond the scope of this book, but it suffices to say that a program will run many times slower when translated by an interpreter. (This is true for all computation, not just the floating-point portions.) It is thus very important to make sure that the translator being used is a true compiler, not an interpreter.

There are numerous advanced methods for random number generation. The interested reader can consult the references below, but we will present one example here, the **rejection method**: Suppose we wish to simulate a random variable X which has density h and cumulative distribution function H. Suppose h has maximum value c, and $h(t)$ is nonzero only for $a < t < b$. (If these bounds do not exist, then h can be truncated, and an approximate generator can be obtained.) As noted several times before, H^{-1} may be difficult to compute, so that the inverse transformation method is infeasible. By contrast, in the rejection method, we need not compute either H or its inverse; only h is used, as follows. We continue to generate variables U_1 and U_2 which are uniformly distributed on (a,b) and (0,c), respectively, until $U_2 < h(U_1)$. X is then U_1. This algorithm is shown in program form in the following.

Program 3.4.10

```
function X(A,B,C: real; var Seed: integer): real;
   var U1,U2: real;
begin
   repeat
      U1 := A + (B − A)*random(Seed);
      U2 := C*random(Seed)
   until U2 < h(U1);
   X := U1
end;
```

The proof that this program produces the desired result (i.e., that it will generate a random variable X having density h) will be presented in Example 4.2.8.

Note, however, that this algorithm is not necessarily a perfect solution to our problems. For example, the expected number of iterations of the "repeat" loop above is $c(b - a)$ (Exercise 3.23). This number will be large if $c(b - a)$ is much larger than 1, that is, if the density curve h does not come close to "filling up" the box bounded by the points $(a,0)$, (a, c), $(b,0)$ and (b,c).

FURTHER READING

Paul Bratley, Bennett Fox, and Linus Schrage, *A Guide to Simulation*, Springer-Verlag, 1983.

John Chambers, *Computational Methods for Data Analysis*, Wiley, 1977.

Donald Knuth, *The Art of Computer Programming* (Vol. 2, *Seminumerical Algorithms*), Addison-Wesley, 1981.

William Kennedy and James Gentle, *Statistical Computing*, Marcel Dekker, 1980.

Sheldon Ross, *A First Course in Probability*, Macmillan, 1976. A deeper, more theoretical treatment of probability modeling.

Reuven Rubinstein, *Simulation and the Monte Carlo Method*, Wiley, 1981.

Kishor Trivedi, *Probability and Statistics, with Reliability, Queueing and Computer Science Applications*, Prentice-Hall, 1982. Many interesting applications.

EXERCISES

3.1 (M) In Example 3.1.1, find the probability mass function p_Z of Z.

3.2 Consider Example 3.1.3, in which we have a bag filled with wires of various lengths.

a. (S) Using the inverse transformation method from Section 3.4, write a Pascal function to simulate the random variable L.

b. (M, S) Find $P(L > 5)$.

c. (S) Verify the findings in Example 3.2.2, 3.2.4, and 3.2.7, that $EL = 4.33$, median$(L) = 4.47$ and Var$(L) = 1.25$.

d. (M, S) Find F_L (4.0).

e. Suppose we draw 10 wires at random from the bag, and let N denote the number of these which are less than 4 inches long. To what parametric family that we have discussed does the distribution of N belong? Specify the parameter values.

3.3 (M, S) Recall Example 2.3.7, concerning a system consisting of a component and a spare part. Let L be the lifetime in months of the two-component system. Find p_L, EL and Var(L).

3.4 (M, S) Consider Example 3.2.5, involving Huffman coding for data compression. Find the variance of the number of bits in a letter stored in the database.

3.5 (M, S) Consider Example 3.3.5, involving flaws in electronic chips.

a. What proportion of chips will have more than one flaw?

b. Suppose we double the area of the electronic chips. What proportion of chips will have at least one flaw? Compare to the result for standard-size chips.

3.6 (M) Consider Example 3.3.11, regarding insertion of a pin by machine. Find the density of W, using the information known about the distribution of Q. (Review Examples 3.1.7 and 3.1.8 before you start.)

3.7 (S) Recall Example 3.3.6, regarding the computer disk drive. Use simulation to find $P(T > w)$, for various values of w.

3.8 (M, S) In the setting concerning tires in Example 3.3.7, suppose the tires are guaranteed for 52,000 miles.

a. Find the proportion of tires that fail the guarantee.

b. Suppose the guarantee has a prorated feature: If a tire fails after x miles ($x < 52{,}000$), then the manufacturer will refund a proportion $52000 - x/52000$ of the original purchase price ($50). What is the mean cost to the manufacturer for maintaining this policy? (*Caution*: Keep in mind that the manufacturer pays nothing if the failure occurs after 52,000 miles. It would be helpful to review Example 3.2.8 before proceeding.)

3.9 (M) Review Example 3.3.9, about the arrival patterns of buses. If we have exponentially distributed waiting times between successive buses, the quantity $q(t) = P(W < t + 3|W > t)$ is constant in t (i.e., the graph of q is a horizontal line). Draw this graph for the case in which interarrival times have a $U(0,20)$ distribution, and for the case in which these times have the density $t/200$, $0 < t < 20$.

3.10 (S) Verify the entry 0.612 in the $N(0,1)$ table.

3.11 (M) Suppose X has a $U(0,1)$ distribution and $Y = X^3$. Use the method in Examples 3.1.7 and 3.1.8 to find the density of Y, and then calculate EY from that density. Then calculate EY using Equation (3.2.11), verifying that equation.

3.12 (M) Show that for $g(t) = at + b$, $E[g(X)] = aEX + b$. Do this separately for the discrete and continuous cases.

3.13 (M) Let μ and ν denote the mean and median of a random variable, respectively. Give an example of a density function for which $\mu < \nu$, and give one for which $\mu > \nu$.

3.14 (M) Derive Equation (3.2.17): Show that

$$\mathrm{Var}(aX+b) = a^2\mathrm{Var}(X)$$

Try to use previously proved properties of mean and variance, in Equations (3.2.15) and (3.2.16), to make your work easier. Give an intuitive interpretation of what Equation (3.2.17) says in the case $a=1$.

3.15 (M) For the density $a\exp(-bt)$, $(t > 0)$, show that we must have $a=b$.

3.16 (M) Show that EX in Equation (3.3.1) is np. *Hint:* Set up the proper summation expression, and then do some algebraic manipulations, including the Binomial Formula learned in high school algebra classes:

$$(a+b)^r = \sum_{i=0}^{r} \binom{r}{i} a^i b^{r-i}$$

3.17 (M) Show that EX in Equation (3.3.7) is equal to b. *Hint:* Set up the proper summation expression, and then do some algebraic manipulations aimed at using the Taylor series expansion for e^u:

$$e^u = \sum_{i=0}^{\infty} \frac{u^i}{i!}$$

3.18 (M) Derive the Linearity Property of the Normal Family. *Hint*: Do separate derivations for d_1 positive and negative. Review Examples 3.1.7 and 3.1.8 before you start.

3.19 (M) Show that the mean and variance for the density be^{-bt} are $1/b$ and $1/b^2$.

3.20 (S) Write a program to verify that the function "random" in Section 3.4 does satisfy the "independence of triples" condition mentioned in that section.

3.21 Write a program that will generate a $N(0,1)$ table like that of Table 2 in Appendix B, approximating areas under the $N(0,1)$ density function by either rectangles or trapezoids.

3.22 (M) Find the mean and variance of the distribution uniform on [a, b].

3.23 (M) Show that the expected value of the number of iterations in Program 3.4.10 is $c(b-a)$.

3.24 (M) This problem will derive some relationships between EX and F_X.

a. Show that if X is a nonnegative, continuous random variable, then

$$EX = \int_0^\infty P(X > t)\, dt$$

Hint: Express $P(X > t)$ as an integral and reverse the order of integration in the resulting double integral.

b. Show that if X is a nonnegative, integer-valued random variable, then

$$EX = \sum_{n=1}^{\infty} P(X \geq n)$$

3.25 (M) Consider the set of all distributions totally contained within the interval $[a,b]$, that is, arising from random variables ranging in that interval. Which such distribution, not necessarily from a parametric family, has the greatest variance?

3.26 (M, S) Suppose a gambling casino has the following game: The player rolls a die until a face having at least 5 dots appears. The player wins $1 for each roll.

a. How much should the casino charge as an entry fee in this game, excluding overhead and profit?

b. Do part (a) again for the following variation of the game: The player continues to roll the die until he or she accumulates a *total* of at least 5 dots.

3.27 (M, S) A store restocks a certain item daily and needs to decide how many items per day it will order. This value, denoted by s, will be fixed in a long-term contract with the wholesaler. The store makes a profit of $2 on each unit sold and loses $1 on each unit left unsold at the end of a day. The demand D varies from day to day, with $P(D=n)$ being proportional to $10-n$, $(n=0, 1, 2, \ldots, 9)$. Find the optimal value of s.

3.28 The **hypergeometric distribution** arises as the number of Type 1 items obtained, when choosing n items at random from a set of r items of Type 1 and s items of Type 2.

a. Relate this distribution to Example 2.2.1, in which we found the probability of getting exactly two hearts in a 5-card hand drawn from a standard 52-card deck. Let X denote an appropriate random variable for this experiment, and relate the probability found in that example in terms of p_X. Specify the values of n, r, and s that are relevant to this example.

b. (M) Write down the general probability mass function for this distribution.

c. (S) Write a general Pascal function that will simulate a random variable having this distribution, taking r and s as function arguments.

d. (M, S) Using the general tools developed in (b) and (c), find the answer for the special case described in (a).

3.29 (M, S) Suppose battery lifetime for a certain brand has an exponential distribution with mean 10 hours.

a. If the batteries are guaranteed for 30 hours, what percentage will fail to meet the guarantee?

b. Suppose we have five 1-cell flashlights. At a certain time, we turn on all five at the same time, and let them burn continuously. What is the probability that exactly three are still burning 40 hours later?

c. Suppose you have a 5-cell flashlight that fails if any individual cell fails. Find the probability that the flashlight is still working after 20 hours of use.

d. Find the mean life of the 5-cell system in (c). *Hint*: First find the cumulative distribution function of the system lifetime.

3.30 (S) Assume that a dam starts out half full, and its daily inflow, $I_1, I_2, \ldots$ is a sequence of independent U(0,50) variables. Evaporation is 6 units per day, and total capacity of the dam is 400 units. Let D be the number of days until the dam reaches its overflow point. Find $E(D)$ and $P(D > k)$ for several values of k.

3.31 (M, S) A civil engineer is stationed on a certain road to do a traffic survey. She wants to take data on 25 trucks. Ten percent of the vehicles passing are trucks. What is the probability that she will have to wait for more than 200 vehicles to pass before she has the data she wants?

3.32 Let T be the time elapsed (in minutes) between a person's arrival at an elevator waiting area and the subsequent arrival of the elevator. Suppose that T has density function $2e^{-2t}$ for $t > 0$, but that we adopt the policy of waiting at most 0.8 minutes before we give up and walk instead of waiting further. Let W be the amount of time we stay in the waiting area (i.e., the time we wait until either the elevator arrives or we give up and take the stairs instead). Suppose it takes 0.25 minute to ride the elevator to our destination, compared to 0.50 to take the stairs. Let Y be the total time elapsed, between the time we arrive at the elevator waiting area to the time we reach our desination.

a. (M) Show that W is neither discrete nor continuous, and graph F_W.

b. (M, S) Find EW and Var(W). *Hint*: Review Example 3.2.8. To do something similar here, write W as $g(T)$ for some appropriate function g.

c. (M, S) Find EY and Var(Y).

3.33 (M, S) In Section 3.2, there was a short discussion comparing the relative merits of the mean and median as measures of location. It was pointed out that one problem with the mean is its sensitivity to extreme values. The standard deviation suffers similarly, as a measure of dispersion. Just as the median provides an alternative to mean for measuring location, the **interquartile range** γ provides a similar alternative to the standard deviation as a measure of dispersion. Let X denote the random variable whose distribution is being described. γ is defined in terms of the **quantiles** of X, where the qth quantile of X is the value t for which $P(X \leq t) = q$. (For example, since the median is the "halfway" point for X, it is the 0.5th quantile of X.) γ is then defined as the distance between the "1/4-way" and "3/4-way" points (i.e., $\gamma =$(0.75th quantile) − (0.25th quantile). Find the median and interquartile range for an exponential random variable whose mean is 2.0. *Hint* for the simulation part of the problem: Generate (say) 1000 values of X, and sort them from smallest to largest. The median will then be (approximately) the 500th in this sorted list.

3.34 (M, S) Consider Example 3.2.11, with b blue marbles and $(n - b)$ yellow ones. $E(T)$ can be found in an *indirect* way, as follows. Let $g(k)$ denote the value of $E(T)$ for the situation in which we start out with $b=k$ blue marbles in the bag ($k=0, 1, \ldots, n$); for example, the simulation program in Example 3.2.11 finds $g(16)$ for the case in which there are 20 marbles altogether. We will be able to find many $g(k)$ values all at once. To do this, first find values of c and d such that

$$g(k)=c[g(k-1)+1]+d[g(k+1)+1]$$

and explain why such an equation is true. Note that this is actually a *system* of equations for various values of k. Now to solve them, let's turn to a specific example, that of $n=6$. First set up the above equation for $k=4$ and $k=5$, and a modified one for $k=6$. Then note that $g(3)$ is by definition equal to 0. You now have four equations in four unknowns, which you can solve. Check one of your answers by modifying and running the program in Example 3.2.11.

3.35 (M) One can show that if X is a $B(n, p)$ random variable with np small (say, less than 10), then the approximate distribution of X is Poisson with mean np. Try this approximation on the following problem: Suppose 5% of some population of people have a special blood type. If 15 people are selected at random from that population, what is the probability that exactly one of them has the special blood type? Find the exact answer and the approximate answer, and compare.

3.36 (M) In many computer memory systems, memory is divided into units called **pages**. Pages at any given time are marked either "empty" or "in use," with a list of these conditions maintained by the computer's operating system (OS) indicating the status of each page. When a new user program is loaded into memory, the OS needs to find enough empty pages to accommodate it. Suppose the probability of a page being empty is p, a user program requires u pages, and the pages are independent. Consider a system in which the OS checks its page list sequentially (i.e., it checks Page 0 for being empty, then Page 1, then Page 2, etc., until it finds a total of u empty pages). Let X be the number of pages that need to be checked. To what parametric family discussed in this chapter does the distribution of X belong?

3.37 (S) You are driving down a long street (for simplicity, assumed to be infinitely long), looking for a parking place. The parking places are numbered in increments of 1, so that there is one space for each integer (including the nonpositive integers). You start at Space -20, heading toward Space 0, which is your destination; of course, you hope to find an open space as close to 0 as possible. You can see two spaces ahead at a time, so that when you are at Space i, you know whether Spaces j ($j \leq i + 2$) are open or not, but do not know about the ones further ahead. However, you do know that in general a space is occupied with probability 0.9 and open with probability 0.1, and that the spaces are independent. You have decided on the following class of strategies: You will pass up Spaces -20 through $-c$, even if open, and then stop the first time you encounter an empty space after $-c$; at that time, you take the space closest to 0 among the three spaces you can see at that time. What is the optimal value for you to choose for c, from the point of view of the expected value of your final distance to 0?

3.38 Recall the TMR concept introduced in Example 2.3.8. Suppose three identical devices A, B, and C are used to make a TMR system having a 1-bit output. The output is determined by "voting" among the three devices, in the spirit of TMR. Let V denote the voting circuitry; note that it too can fail. Suppose each of A, B, and C has an exponential lifetime with mean μ_{dvc}, and that V has an exponential lifetime with mean μ_{vtr}. Note that the system becomes unreliable either if more than one of A, B, and C fail or if V fails.

a. (M) At time r, an inspector arrives and finds that all components are working. Let $r+W$ denote the first time at which the system becomes unreliable (i.e., W is the remaining reliable time of the system after the inspector arrives). Find the density of W. [*Hint*: Review Example 3.1.7 first, and note from that example that we can find the density of W if we find $P(W \leq t)$. Then note that we can find this latter probability if we find $P(W > t)$.] Would this problem be workable as stated if the distributions involved had not been exponential?

b. (S) Suppose that when a device (A, B, or C) fails, a repair begins immediately, and that repair time has a $U(10,15)$ density. However, if V fails, it is not repairable, so the system goes down. Suppose $\mu_{\text{dvc}} = 150$ and $\mu_{\text{vtr}} = 300$. Find the probability that the system is up at n hours, for $n = 100, 200, 300, \ldots, 1500$.

3.39 (M, S) As we know from Example 2.3.8, computers store information in binary digits, or bits. About half of all bits are 0s, and half are 1s. As in that example, let us refer to individual bits within a number according to the power of 2 to which that bit corresponds. For example, consider the number 1101 (13 in base-10). Then the 2^3 bit is 1, the 2^2 bit is 1, the 2^1 bit is 0, and the 2^0 bit is 1.

In the design of computer hardware to do addition, one concern is the existence of carries. For example, in adding 0101 and 0100 $(5+4)$ there is a carry out of the third bit from the right (i.e., the summing of the 2^2 bits produces a carry into the 2^3 bit position).

Furthermore, a carry can propagate. For example, in adding 10101 and 10111 (21 and 23), the carry out of the 2^0 bit position will in turn cause a carry out of the 2^1 bit position, which in turn will cause a third carry, out of the 2^2 bit position. That will be a total of 3 consecutive carries (there will also be a carry out of the 2^4 bit position, but that one will not be consecutive with the others).

Let n be the number of bits in a number, with n being fairly large (in some supercomputers, the size of a number may be 60 bits or more). Find the expected value of the number of consecutive bit positions in which a carry occurs, if the 2^0 bits of the two addends are both 1s. You may assume consecutive bits to be probabilistically independent.

3.40 (M) Show that if X is a nonnegative random variable, then the **Markov Inequality**

$$P(X \geq c) \leq \frac{EX}{c}$$

holds for any $c > 0$. (*Hint* for continuous case: Start by writing down the integral for EX, and break it into two parts, integrating from 0 to c and from c to ∞.) Then use this to show the **Chebychev Inequality**: For any random variable Y having finite variance,

$$P(|Y - EY| \geq d) \leq \frac{\text{Var}(Y)}{d^2}$$

Illustrate both of these inequalities for the U(0, 1) distribution and for the exponential distribution with mean 1.0, by choosing several values of d and calculating both sides of the inequalities.

3.41 Suppose buses arrive at times $T_1 < T_2 < T_3 < \cdots$, where the interarrival times $I_1 = T_1, I_2 = T_2 - T_1, I_3 = T_3 - T_2, \ldots$ are independent random variables, each having an exponential density with mean 1.0. We arrive at time 54.2. Let W be the amount of time we must wait for the next arriving bus. Suppose we are interested in the value of EW. Since the average time between successive buses is $EI_i = 1.0$, and since our arrival will be somewhere between two successive bus arrivals, we might guess that EW should be less than 1.0, maybe even 0.5; the latter value would seem to correspond to our arrival occurring "on the average" halfway between two successive bus arrivals.

a. (M, S) Cite material from the text that suggests that our conjecture is wrong, and use that material to determine the actual value of EW. Check by using simulation.

b. (S) Find EW for the case in which each I_i has a U(0,1) distribution.

c. (S) Consider the interval (A, B), where A is the arrival time of the last bus to arrive before time 54.2, and B is the arrival time of the first bus to arrive after time 54.2. For both the exponential and U(0,1) cases, find the mean length of the interval and compare to EW.

3.42 (S) (*Note*: This problem is a simplified version of one that arises in the area of computer database design.) Suppose we have n one-dimensional objects which we wish to put in bins of length 1.0. The length of an object L has density $f_L(t) = 2(1 - t)$ for t in (0,1), zero elsewhere. We have n objects, with independent lengths $L_1, \ldots, L_n$. The n objects must be packed into consecutive bins in the order of their indices. This means that Object 1 is packed first, then Object 2, and so on. Also, an object cannot be split between 2 bins.

This last condition implies that there may be a fair amount of wasted space in the bins used. Let W represent the total amount of wasted space, and B be the number of bins used. To clarify conditions and notation, consider the following example:

Suppose $n=3$, and $L_1 = 0.62$, $L_2 = 0.25$, and $L_3 = 0.44$. Then Objects 1 and 2 go into Bin 1, and Object 3 goes into Bin 2. There is $1.0 - (0.62+0.25)=0.13$ space wasted in Bin 1, and $1.0 - 0.44=0.56$ space wasted in Bin 2. $W=0.13+0.56=0.69$, and $B=2$.

Assume $n=10$.

a. Find EB, $\mathrm{Var}(B)$, and $P(B \leq t)$ for several values of t.

b. Find $E(W/B)$, the expected proportion of empty space.

3.43 (M) Consider the setting of Exercise 3.41. It turns out that the independent, exponential intervals imply that the *count* of the number of events (in this case number of bus arrivals) occurring before any fixed time t has a Poisson distribution with mean $1.0t$:

$$P(N(t)=k)=\frac{e^{-t}t^k}{k!}$$

Prove this, by using the description of the Erlang distribution in Section 3.3. *Hint*: First find the cumulative distribution function F_V for an Erlang random variable V having $b=1$. Do so by integrating by parts.

CHAPTER 4

The Multivariate Environment

4.1 DISTRIBUTIONS OF RANDOM VECTORS

The machinery developed in the last chapter to describe the variation of random quantities is not quite complete. In that chapter we discussed the variation of random quantities *individually*; here we treat the problem of describing *concurrent* variation.

Example 4.1.1

Suppose we have four fair coins, denoted by A, B, C, and D. We toss the four coins, and note whether each one comes up heads. Let X be the total number of heads obtained from Coins A and B, Y the total from B and C, and Z the total from C and D. Then each of X, Y, and Z has a $B(2,0.5)$ distribution. However, this latter statement does not completely describe the situation, since such a statement fails to mention that there is some degree of dependence between X and Y, and between Y and Z, but not between X and Z. For example, both X and Y include Coin B in their counts, so that if we know that X is 2, then we know that Y is at least 1; thus X and Y are not independent, in some sense. On the other hand, X and Z do act independently. (We will formally define the notion of independence later, relying for the time being on intuitive descriptions only.)

To solve the description problem which arises in the example, we need to extend the concepts of probability and density functions developed in the last chapter. We will treat the discrete case first. Suppose we have discrete random variables S and T. We found above that the functions p_S and p_T carry insufficient information; instead, we need to have a description of how S and T interact; that is, we need to know the values of $P(S=a$ and $T=b)$, over the range of values of a and b. We thus have the following definition.

Definition 4.1.1

For any pair of discrete random variables S and T, the **bivariate probability mass function** $p_{S,T}$ is defined by

$$p_{S,T}(a,b)=P(S=a \text{ and } T=b) \tag{4.1.1}$$

The **multivariate probability mass function**, for k discrete random variables rather than just two, is defined similarly. For $k=3$,

$$p_{S,T,U}(a,b,c)=P(S=a \text{ and } T=b \text{ and } U=c)$$

Example 4.1.2

Consider the variables S and Z in Example 3.1.1. We can compute, for example, $p_{S,Z}(6,8)$ as

$$P(S=6 \text{ and } Z=8)=P(\text{we roll either } (2,4) \text{ or } (4,2))=\frac{2}{36}=\frac{1}{18}$$

Example 4.1.3

Let us find $p_{X,Y}$ in Example 4.1.1.

$$\begin{aligned} p_{X,Y}(a,b) &= P(X = a \text{ and } Y = b) \\ &= P(X = a)P(Y = b|X = a) \end{aligned}$$

Since X has a binomial distribution, regardless of its relation to Y, $P(X=a)$ is easily evaluated using the probability mass function for the binomial distribution, Equation (3.3.2). After algebraic simplification, we find its value to be $\binom{2}{a}/4$. However, $P(Y=b|X=a)$ is more complicated; it must be handled separately for different values of a. First, consider $a=0$. Here we know that Coin B was not heads, so $P(Y = 0|X = 0)$ and $P(Y = 1|X = 0)$ boil down to $P(C$ was tails) and

P(*C* was heads), which are each equal to 1/2. Calculation of $P(Y = b|X = 2)$ is similar, with $P(Y = b|X = 1)$ being a bit more elaborate (Exercise 4.8).

There is an analogous concept for continuous random variables, called the *bivariate density function* (compare to Definition 3.1.3).

Definition 4.1.2

A bivariate density function is a function $g(u,v)$ such that

a. $g(u, v)$ is nonnegative for all values of u and v.
b. $\int_{-\infty}^{\infty}\int_{-\infty}^{\infty} g(u, v)du\,dv = 1.0$.

Multivariate density functions, for k continuous random variables rather than two, are defined similarly. (Some sets of two or more continuous random variables do not have densities. However, these occur only rarely in practice.)

If X and Y are continuous random variables, their density function is denoted by $f_{X,Y}$. As in the one-variable case, $f_{X,Y}(u, v)$ is not equal to $P(X = u$ and $Y = v)$. The latter value is still zero. However, $f_{X,Y}$ does describe the frequency variation of the pair (X,Y), just as was discussed in Section 3.1 for the one-variable case. If for example $f_{X,Y}$ (12.1,6.8) is much smaller than $f_{X,Y}$ (5.0,9.2), then over many repetitions of the experiment, the pair (X,Y) will turn up near (12.1,6.8) much less often than near (5.0,9.2).

Multivariate densities are used to find probabilities involving k continuous random variables.

Finding Probabilities Involving *k* Continuous Random Variables

Let E be any subset of the two-dimensional plane. Then

$$P((X,Y) \text{ is in } E) = \int_E\int f_{X,Y}(u,v)\,du\,dv \qquad \textbf{(4.1.2)}$$

In the general case of k random variables, the desired probability is the k-fold integral of the multivariate density function, over the indicated set. (As before, note that the variables u and v here could be replaced by s and t or any other pair.)

Example 4.1.4

Suppose bus lines A and B intersect at a certain point, allowing transfers between lines. Buses from both lines are supposed to arrive at 3:00, although they are always late to various degrees. Let X denote the tardiness of the Line A bus, and let Y be defined similarly for the bus on Line B; both quantities are measured in hours. Suppose the bivariate density is defined by

$$f_{X,Y}(u,v) = 2 - u - v \qquad \textbf{(4.1.3)}$$

for $0 < u,\ v < 1$, and has the value zero for all other (u,v) pairs. Suppose a passenger on Line B has arranged to meet a friend at this bus stop, and that the friend will take Line A to get there. We wish to find the probability that the person arriving on Line B will have to wait more than 6 minutes for the friend.

Translating this to symbols, the desired probability is $P(X > Y + 0.1)$. Thus the set E in Equation (4.1.2) is

$$\{(u,v) : 0 < u, v < 1 \text{ and } u > v + 0.1\} \qquad \textbf{(4.1.4)}$$

which is graphed as the shaded region in Figure 4.1. By Equation (4.1.2), $P(X > Y + 0.1)$ can then be obtained by integrating Equation (4.1.3) over this set:

$$P(X > Y + 0.1) = \int_{0.1}^{1} \int_{0}^{u-0.1} (2 - u - v)\,dv\,du = 0.405 \qquad \textbf{(4.1.5)}$$

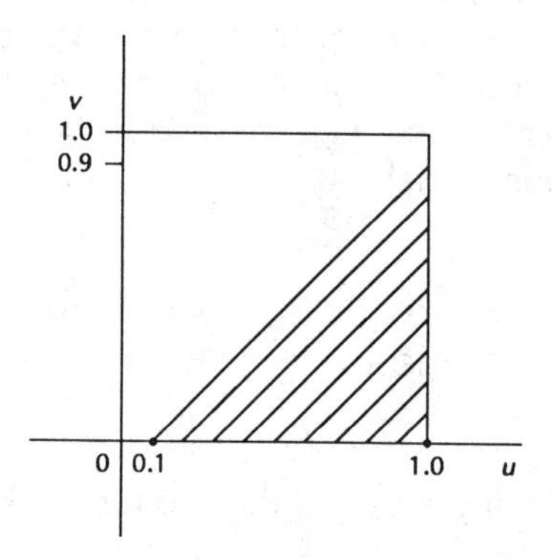

Figure 4.1 Graph of $P(X > Y + 0.1)$.

To review the concepts we have developed so far, it is worthwhile to view this geometrically (Figure 4.2). Since $f_{X,Y}$ is a function of u and v, it has been graphed as a surface lying above the UV plane, with the height of the surface indicated on the W axis. Notice that the portion of the surface above the region near (0,0) is much higher than near (1,1), indicating that

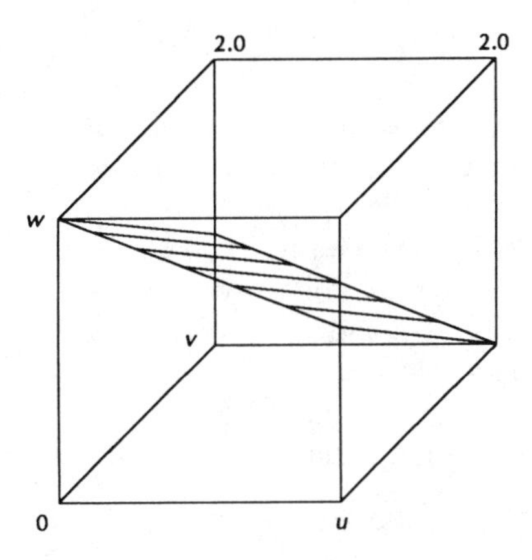

Figure 4.2 $f_{X,Y}$

it is much more common for both buses to be only a little late than it is for both buses to be about an hour late. On the other hand, $f_{X,Y}(0,1)$ and $f_{X,Y}(1,0)$ are intermediate values, indicating that it is somewhat common for one of the buses to be nearly on time but the other very late. Finally, note that Equation (4.1.5) is the *volume* delineated by this surface and the set (4.1.4) in the UV plane. This is the analog of the fact that probabilities involving *one* random variable (say Z) turn out to be *areas* delineated by f_Z and the set under consideration.

In Chapter 3 the concept of the expected value of a function of a random variable, $E[g(X)]$, was presented. This concept generalizes to the multivariate case.

Finding the Expected Value of a Function of Two or More Random Variables

a. If X and Y are discrete random variables, then

$$E[g(X,Y)] = \sum_{u,v} g(u,v)p_{X,Y}(u,v) \qquad \textbf{(4.1.6)}$$

b. If X and Y are continuous random variables having a bivariate density, then

$$E[g(X,Y)] = \int_{-\infty}^{\infty}\int_{-\infty}^{\infty} g(u,v)f_{X,Y}(u,v)\,du\,dv \qquad \textbf{(4.1.7)}$$

c. For the cases of more than two variables, use the multivariate analogs of Equations (4.1.6) and (4.1.7).

Example 4.1.5

A bag contains tags numbered 1 through 10. A game allows us to choose two tags from the bag, without replacement. We win an amount in dollars equal to the product of the numbers on the two tags drawn. We can use Equation (4.1.6) to find the expected value of the number of dollars won as follows: Let X be the number on the first tag drawn, and Y the second. We wish to find $E(XY)$, so $g(u,v) = uv$ in Equation (4.1.6). We will need to know the values of $p_{X,Y}$ for our computation. But this turns out to be simple. For example $p_{X,Y}(7, 2)$ is by definition equal to

$$\begin{aligned} P(X = 7 \text{ and } Y = 2) &= P(X = 7)P(Y = 2 | X = 7) \\ &= \left(\frac{1}{10}\right)\left(\frac{1}{9}\right) \\ &= \frac{1}{90} \end{aligned}$$

Similarly, $p_{X,Y}(u, v)$ is equal to 1/90 for all the other possible values that the pair (X,Y) can take on: (1,2), . . . , (1,10), (2,1), (2,3), . . . , (2,10), . . . , (9,1), . . . , (9,8), (9,10). Then $E(XY)$ is

$$\sum_{u=1}^{10} \sum_{v=1, v \neq u} (uv)\left(\frac{1}{90}\right) = 29.33$$

Example 4.1.6

Recall the material on computer disk drives from Example 2.3.9. Data are stored on concentric tracks on the disk, and the disk is continually spinning. In order to access a given data item, the read/write head must first be moved laterally along the arm to the proper track, and then must wait for the data item to rotate around under it. The head movement along the arm is called a **seek**. There are a number of different policies for scheduling a set of several requests for disk access, most of which are primarily concerned with achieving a low average seek time; the smaller the average seek time, the more requests can be serviced per unit of time.

Let us denote the innermost extreme position of the read/write head as 0, and the outermost position as 1. If the total number of tracks is large, then the position of the read/write head will be well approximated as a continuous random variable. Let X and Y be the track positions of two consective accesses to the disk. Then if the bivariate density of X and Y is known, the mean seek distance will be

$$E[|Y-X|]=\int_0^1\int_0^1 |v-u| f_{X,Y}(u,v)du\,dv \quad \textbf{(4.1.8)}$$

In simple cases $f_{X,Y}$ is known and Equation (4.1.8) can be evaluated easily. For example, suppose $f_{X,Y}(u,v) = 1$ for any (u,v) pair satisfying $0 < u,\ v < 1$, with the value 0 for other pairs. Then by (4.1.8), the mean seek distance is

$$\int_0^1\int_0^1 |v-u| 1\ du\ dv=\frac{1}{3}$$

This is a good example of the conceptual difference between univariate and bivariate density functions. A bivariate density function $f_{X,Y}$ will uniquely determine the univariate densities f_X and f_Y [see Equations (4.1.9) below], but univariate densities will *not* uniquely determine bivariate densities.

To illustrate this, suppose in this disk example that all files are accessed equally often, so that all tracks are accessed equally often (assuming that each file occupies a single sector and is thus entirely contained within one track). This implies that X and Y have $U(0,1)$ distributions: $f_X(t) = 1$ and $f_Y(t) = 1$ for all t in (0,1). However, there are many bivariate densities $f_{X,Y}$ that are consistent with this.

The bivariate density $f_{X,Y}$ will be determined by the nature of our disk access request scheduling policy. For example, one of the more sophisticated disk scheduling policies is called Shortest Seek Time First (SSTF). Whenever a disk access request is completed, the scheduler will scan all the requests waiting to be processed and choose the one whose track is closest to the current track position of the read/write head, so that the request chosen by the scheduler is the one that would result in the shortest seek time. In this case, Y should be fairly highly correlated with X since under this scheduling policy, two consecutive disk accesses will tend to be for tracks near each other. On the other hand, if a first-come, first-served (FCFS) scheduling policy is used, then X and Y might be independent.

It should be noted that, as we have seen many times before, many problems are intractable mathematically, and simulation is thus the tool to use in such cases. For example, for analyzing complicated disk scheduling policies such as SSTF above, simulation tends to be the only usable tool. In Section 4.1.A, we present such a simulation analysis, comparing SSTF to FCFS.

Since multivariate probability and density functions give complete information about the frequency variation of the corresponding random variables, the univariate probability and density functions can be obtained from them. For example, if we know $f_{X,Y}$ we can use it to find the individual (called the **marginal**) distribution of X, f_X. We present the bivariate cases below; the k-variate cases are similar.

Finding Marginal Distributions from Multivariate Distributions

$$p_X(m) = \sum_n p_{X,Y}(m,n) \quad \textbf{(4.1.9a)}$$

$$p_Y(n) = \sum_m p_{X,Y}(m,n) \quad \textbf{(4.1.9b)}$$

$$f_X(s) = \int_{-\infty}^{\infty} f_{X,Y}(s,t)dt \quad \textbf{(4.1.9c)}$$

$$f_Y(t) = \int_{-\infty}^{\infty} f_{X,Y}(s,t)ds \quad \textbf{(4.1.9d)}$$

As a "proof" of Equation (4.1.9a), we present the following example. Let X and Y denote the numbers that appear when we roll two fair dice. Then $p_{X,Y}(m,n) = 1/36$ for $m, n = 1, 2, 3, 4, 5, 6$, and so on.

$$\begin{aligned} p_X(m) &= P(X = m) \\ &= P(X = m \text{ and } Y = 1 \text{ or } \cdots X = m \text{ and } Y = 6) \\ &= \sum_{n=1}^{6} P(X = m \text{ and } Y = n) \\ &= \sum_{n=1}^{6} \frac{1}{36} \\ &= \frac{1}{6} \end{aligned}$$

From this example, the reasoning behind Equations (4.1.9a) and (4.1.9b) is apparent, and the next two equations are intuitive analogs. Here is an example of the use of Equation (4.1.9c).

Example 4.1.7

Suppose (X,Y) has a uniform distribution on the triangle

$$T = \{(s,t) : 0 < t < 1 - s < 1\}$$

(the shaded region in Figure 4.3). The uniformity implies that $f_{X,Y}(s,t)$ has a constant value on that set, say c. Condition (b) of Definition 4.1.2 then implies that $c = 2.0$:

$$f_{X,Y}(s,t) = 2.0$$

for points (s, t) in T, and zero elsewhere (recall finding c in Example 3.1.3). We can now obtain the individual densities of X and Y using Equation (4.1.9b). Before we start, we might ask what we should expect. For example, the range of X is the interval (0, 1); will X have a $U(0,1)$ distribution? The answer is that we expect X not to be uniform on (0,1), but rather to have a density which is higher near 0 than near 1 since more points in T have s values near 0 than near 1. Let us see how true this is. We use Equation (4.1.9c):

$$\begin{aligned} f_X(s) &= \int_0^{1-s} 2\,dt \\ &= 2 - 2s \end{aligned}$$

So our prediction is confirmed: f_X is *not* the $U(0,1)$ density, and $f_X(s)$ is higher for values of s near 0. Similarly, $f_Y(t)$ may be found to be $2 - 2t$.

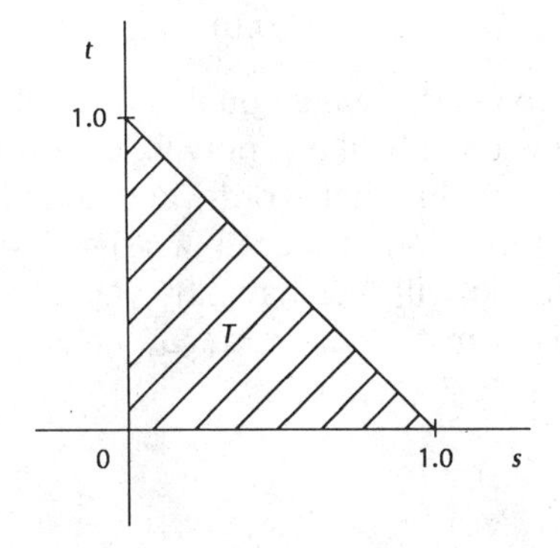

Figure 4.3 Area under consideration in Example 4.1.7.

We can also define *conditional* probability and density functions, in the obvious ways.

Example 4.1.8

Recall the random variables S and Z in Examples 3.1.1 and 4.1.2. We can define the conditional probability function

$$p_{Z|S}(a,b) = P(Z = a|S = b) \tag{4.1.10}$$

This is then equal to

$$\frac{P(S = b \text{ and } Z = a)}{P(S = b)} \tag{4.1.11}$$

Thus, we have the following relation between the several probability mass functions:

$$p_{Z|S} = \frac{p_{S,Z}}{p_S} \tag{4.1.12}$$

For example,

$$p_{Z|S}(8,6) = \frac{p_{S,Z}(8,6)}{p_S(6)} = \frac{1/18}{5/36} = \frac{2}{5}$$

Example 4.1.9

Consider the random variables X and Y in Example 4.1.7. Suppose it is known that $X = 0.4$. Recall the set T from that example, and define the subset

$$L = \{(0.4, t) : 0 < t < 0.6\}$$

All the points in T were equally likely; thus, all the points in the subset L are now equally likely, now that we know that (X,Y) is somewhere in the region L. In other words, intuition would say that the conditional density $f_{Y|X}(z, 0.4)$ of Y given $X = 0.4$ should be uniform in the interval (0, 0.6), and will thus have the value 1/0.6 for all z in that interval.

This is in fact the case. By analogy with Equation (4.1.12), we have

$$f_{Y|X} = \frac{f_{X,Y}}{f_X} \tag{4.1.13}$$

so that

$$f_{Y|X}(z, 0.4) = \frac{f_{X,Y}(z, 0.4)}{f_X(0.4)} = \frac{2.0}{2 - 2(0.4)}$$

which is equal to 1/0.6, the density of $U(0,0.6)$, as our intuition had suggested.

You should keep in mind the regions of definition of the conditional functions in Equations (4.1.12) and (4.1.13). For instance, in the last example, $f_{Y|X}(z, 0.4)$ is zero for values of z outside (0,0.6), and $f_{Y|X}(z, w)$ is not even defined for values of w outside (0,1), since the event $\{X=w\}$ will never occur for such w.

Example 4.1.10

Let X be a random variable representing the lifetime of a light bulb. Suppose that for a certain light socket, whenever the bulb burns out, it is immediately replaced with a new bulb, and when that bulb dies, we again replace it with a new bulb, and so on, with this process continuing indefinitely. Suppose that we walk into the room at a certain time w, long after the process has begun, and we wonder how much longer the current bulb will burn. Let Z be this quantity, called the **residual lifetime** of the bulb. We will derive the density of Z, using the idea of conditional densities.

Let Y be the *total* lifetime of this particular bulb. The first thing to note is that $f_Y \neq f_X$. This may seem strange to you at first, but it is actually quite intuitive: In this situation, we are more likely to encounter a long-lived bulb than a short-lived one (recall Example 3.1.3, in which we chose a piece of wire from a bag, and longer pieces were assumed more likely to be chosen than were shorter ones). For example, suppose that $f_X(10.2) = f_X(28.7)$, that is, among all bulbs manufactured, the ones that last 10.2 hours are equally prevalent as those with lifetime lasting 28.7 hours. Then we would have

$$f_Y(28.7) = \frac{28.7}{10.2} f_Y(10.2)$$

Thus, among lifetimes that appear with equal frequency in shipments from the factory, the light bulb *which we observe at time w* is proportionally more likely to be the one with the longer lifetime.

Formalizing this notion, we have

$$f_Y(s) = csf_X(s)$$

where c is a constant that will make f_Y integrate to 1. Setting this integral to 1, we find that $c = 1/\mu$, where $\mu = EX$.

Next, note that we will encounter this light bulb at *any* point in its lifetime with equal likelihood. In other words, given the value of Y, Z will be uniformly distributed on the interval $(0, Y)$:

$$f_{Z|Y}(t, s) = \frac{1}{s}, \quad 0 < t < s$$

Since we have f_Y and $f_{Z|Y}$, we can now find $f_{Y,Z}$ and then integrate to find f_Z [recall from Equation (4.1.9d) that we can obtain marginal densities by integrating joint densities].

$$f_{Y,Z}(s,t) = \frac{f_X(s)}{\mu} \qquad 0 < t < s$$

$$f_Z(t) = \int_{-\infty}^{\infty} f_{Y,Z}(s,t)ds = \int_t^{\infty} \frac{f_X(s)}{\mu} ds$$

Thus

$$f_Z(t) = \frac{P(X > t)}{\mu} = \frac{1 - F_X(t)}{\mu}$$

This result is also derivable for the case in which X is a discrete random variable, in which case the formula turns out to be

$$P(Z = i) = \frac{1 - F_X(i-1)}{\mu} = \frac{P(X \geq i)}{\mu}$$

This result has a nice intuitive interpretation: Suppose the time at which we walk into the room is, say 502. Then the event $\{Z = i\}$ occurs if and only if *both* of the following occur:

a. There is a bulb replacement at time $502 + i$.

b. The replaced bulb had lifetime at least i.

You should carefully check both the "if" and the "only if" parts of this assertion.

Now since replacements occur, on the average, once every μ time periods, then a proportion $1/\mu$ of all periods will be bulb replacement times. Thus, condition (a) has probability $1/\mu$. The probability of condition (b) is simply $P(X \geq i)$. Multiplying these last two quantities together yields

$$P(Z = i) = \frac{P(X \geq i)}{\mu}$$

It is also useful to define the concept of conditional expected value:

$$E(Y|X = s) = \sum_t t p_{Y|X}(t,s) \qquad \textbf{(4.1.14a)}$$

in the discrete X case, and

$$\int_{-\infty}^{\infty} t f_{Y|X}(t,s)dt \qquad \textbf{(4.1.14b)}$$

in the continuous X case.

The conditional and unconditional means are related as follows:

$$EY = \sum_s E(Y|X = s)p_X(s) \qquad \textbf{(4.1.15a)}$$

$$EY = \int_{-\infty}^{\infty} E(Y|X=s) f_X(s)\,ds \qquad \textbf{(4.1.15b)}$$

The proof of these relations is left for Exercise 4.5.

In the case that Y takes on only the values 1 and 0, the last two equations become relations concerning probabilities. To see this, let A denote some event, and let Y be either 1 or 0, depending on whether A occurs or not. Equations (4.1.15) then become

$$P(A) = \sum_s P(A|X=s) p_X(s) \qquad \textbf{(4.1.16a)}$$

$$P(A) = \int_{-\infty}^{\infty} P(A|X=s) f_X(s)\,ds \qquad \textbf{(4.1.16b)}$$

Example 4.1.11

Suppose the number of bits X in a message in a computer network has a geometric distribution with "success" probability 0.05. Each bit has probability 0.001 of being received incorrectly, independently of the others. Let us find the proportion of messages which are error-free.

Let A denote the event that a message is error-free. Then

$$P(A|X=s) = (0.999)^s$$

so

$$\begin{aligned} P(A) &= \sum_{s=1}^{\infty} P(A|X = s) p_X(s) \\ &= \sum_{s=1}^{\infty} P(A|X = s)(0.95)^{s-1}(0.05) \\ &= \frac{0.05}{0.95} \sum_{s=1}^{\infty} [(0.999)(0.95)]^s \\ &= \frac{(0.0526)(0.999)(0.95)}{1 - 0.94905} \\ &= 0.9804 \end{aligned}$$

Equations (4.1.15) also have analogs for variance. We present the continuous version only:

$$\mathrm{Var}(Y) = \int_{-\infty}^{\infty} \mathrm{Var}(Y|X=s) f_X(s)\,ds + \int_{-\infty}^{\infty} [E(Y|X=s) - EY]^2 f_X(s)\,ds \qquad \textbf{(4.1.17)}$$

Proof is left for Exercise 4.5.

Appendix

At the end of Example 4.1.6, we mentioned that because mathematical analysis of many complicated disk scheduling policies is extremely difficult, simulation provides an extremely valuable alternative method of analysis. In this Appendix, we present such a simulation analysis.

We are assuming the same parameters as in our previous examples regarding disk systems (Examples 4.1.6 and 2.3.9): 76 tracks (accessed with equal frequencies), 30 ms per rotation, seek speed of 3.2 ms per track, and a read/write time of 1.2 ms per sector.

We have modeled disk access requests as arriving according to a Poisson process; that is, times between arrivals are exponentially distributed, with mean value MeanArriv. We generate all the requests at the beginning of the program (in the "for" loop in the procedure Initialize), rather than generating them one at a time during the course of the execution of the program. This is rather wasteful of program memory since we must store all the requests at all times (even after a request has finished service), but we have designed the program in this way to make it more readable.

Another program design feature aimed at improving the readability of the program is the use of the variable DeleteEnd. When we get near the end of the simulation, we ordinarily have to watch out for things such as exceeding array bounds in expressions such as ReqstAry[NumArriv+1] in the procedure UpdateWaitingList. This would complicate the programming. So instead we decided to stay away from such "boundary problems," by the simple trick of avoiding the boundary. Specifically, even though we generate TotReqsts disk access requests in the procedure Initialize, we run the simulation only for the first TotReqsts – DeleteEnd of them (see "for" loop in the main program, and also the "writeln" statement at the very end). Again, this is a bit wasteful, but quite worthwhile in the sense of program development and debugging time, and (more importantly) the time it will take you to read and understand the program. Anyway, in a simulation one always has to choose the parameter determining run length (e.g., TotReps in Program 2.4.3); though we ideally would want an infinite run length, this is impossible. Thus, deleting the last few (DeleteEnd) requests is not a problem.

The program presented is a little longer than most of the programs we have presented so far, but should be fairly easy to understand if you read all the comments *first*, especially those in the main program. Results of running the program will follow it.

Program 4.1.1

```
program DiskSeek(input,output);
```

```
const TotTracks = 76;  {total number of tracks in disk}
      RotLat = 15.0;  {average rotational time, i.e. half
                          a rotation}
      SectorTime = 1.2;  {time to read or write sector}
      SeekSpeed = 3.2;  {time to do a seek of a distance
                          of one track}


 type ReqstRec =  {contains all the information for one
                          access request}
         record
            ArrivTime: real;  {arrival time of request}
            TrackNum: integer  {track number for the
                          request}
         end;


 var ReqstAry:  {array of records for all access requests}
        array[1..1000] of ReqstRec;
     {the next two items, WaitingList and NumWaiting, are
      used only for SSTF, not FCFS}
     WaitingList:  {array of numbers of waiting requests}
        array[1..1000] of integer;
     NumWaiting: integer;  {number of requests currently
                          waiting}
     NumArriv: integer;  {number of requests which have
                          arrived so far}
     StartTime: real;  {time at which the currently
                          processed request starts}
     TimeReqstDone: real;  {finish time for that request}
     MeanArriv: real;  {mean time between successive
                          arrivals}
     SchedType: char;  {schedule type, FCFS or SSTF}
     SumWait: real;  {sum of all waiting times, used at
                          the end to calculate mean wait}
     CurrReqst: integer;  {index of the currently
                          processed request}
     CurrTrack: integer;  {current track position of read/
                          write head}
     NumReqstsDone: integer;  {number of requests finished
                          so far}
     TotReqsts: integer;  {total number of requests to
                          simulate}
     DeleteEnd: integer;  {see introductory text}
     Seed: integer;


 function Expon(Mean: real; var Seed: integer): real;
 begin
    Expon := -Mean*ln(random(Seed))
 end;
```

```
procedure Initialize;
   var Reqst: integer;
begin
   writeln('enter MeanArriv,TotReqsts,DeleteEnd');
   readln(MeanArriv,TotReqsts,DeleteEnd);
   writeln('enter SchedType');  readln(SchedType);
   CurrTrack := 0;
   {generate all the requests:}
   for Reqst := 1 to TotReqsts do
      begin
      if Reqst = 1 then
         ReqstAry[Reqst].ArrivTime:=Expon(MeanArriv,Seed)
      else
         ReqstAry[Reqst].ArrivTime := ReqstAry[Reqst - 1].
            ArrivTime + Expon(MeanArriv,Seed);
      ReqstAry[Reqst].TrackNum := trunc(TotTracks*random
                          (Seed))
      end;
   NumWaiting := 0;
   SumWait := 0.0;
   {now simulate the start of the first request:}
   CurrReqst := 1;
   StartTime := ReqstAry[1].ArrivTime;
   NumArriv := 1
end;

procedure UpdateWaitingList;
begin
   if ReqstAry[NumArriv + 1].ArrivTime <= TimeReqstDone
                          then
      repeat
         NumArriv := NumArriv + 1;
         NumWaiting := NumWaiting + 1;
         WaitingList[NumWaiting] := NumArriv
      until (ReqstAry[NumArriv + 1].ArrivTime >
               TimeReqstDone) or (NumArriv+1 = 100)
end;

function SSTFSched: integer;
   var I,MinReqst,NewCurrReqst: integer;
       MinSeek,Temp: real;
begin
   if NumWaiting > 0 then
      begin
      {first search through waiting list, to find the
       waiting request which would result in shortest
       seek:}
      MinReqst := WaitingList[1];
```

```
      MinSeek := abs(CurrTrack - ReqstAry[WaitingList[1]].
                        TrackNum);
      for I := 2 to NumWaiting do
         begin
         Temp := abs(CurrTrack - ReqstAry[WaitingList[I]].
                        TrackNum);
         if Temp < MinSeek then
            begin
            MinReqst := WaitingList[I];
            MinSeek := Temp
            end
         end;
      NewCurrReqst := MinReqst;
      {now delete that request from the waiting list:}
      NumWaiting := NumWaiting - 1;
      for I := MinReqst to NumWaiting do
         WaitingList[I] := WaitingList[I + 1]
      end
   else
      begin
      {if no request is waiting, then the next request to
       be processed will be the next one to arrive:}
      NumArriv := NumArriv + 1;
      NewCurrReqst := NumArriv
      end;
   SSTFSched := NewCurrReqst
 end;

begin
  Initialize;
  for NumReqstsDone := 1 to TotReqsts - DeleteEnd do
     begin
     {determine when the currently processing request will
      finish:}
     TimeReqstDone := StartTime + SeekSpeed
        *abs(CurrTrack - ReqstAry[CurrReqst].TrackNum)
        + RotLat + SectorTime;
     {determine which new requests arrive during this
      processing time:}
     if SchedType = 's' then UpdateWaitingList;
     {simulate finish of current request, by doing the
      needed bookkeeping}
     CurrTrack := ReqstAry[CurrReqst].TrackNum;
     SumWait := SumWait + TimeReqstDone
                      - ReqstAry[CurrReqst].ArrivTime;
     {decide which request to process next, i.e. new value
      for CurrReqst:}
     case SchedType of
```

```
         'f': CurrReqst := CurrReqst + 1;
         's': CurrReqst := SSTFSched
        end;
      {start time for this new request will depend on
       whether this request arrives after or before the
       finish time of the last request to be processed:}

      if ReqstAry[CurrReqst].ArrivTime > TimeReqstDone then
         StartTime := ReqstAry[CurrReqst].ArrivTime
      else
         StartTime := TimeReqstDone
      end;

   writeln('mean wait time = ',SumWait/(TotReqsts-DeleteEnd)
                            :12:3)
end.
```

The program was run with MeanArriv = 150.0, for TotReqsts = 500 and DeleteEnd = 5. The results were that mean waiting time for a request under the FCFS policy was 198.4, while under SSTF it was 136.3. Thus SSTF yields a considerable improvement.

4.2 COVARIANCE, CORRELATION, AND INDEPENDENCE

This section is concerned with interactions between two or more random variables. The first concept is that of **covariance**. This is, as the name suggests, an attempt to adapt the notion of variance to the description of relations between two random variables. Just as variance acts as a summary measure of the variation of a single variable, covariance describes the degree to which two variables vary together.

Example 4.2.1

Consider the population of students at a large university. Let X be a student's score on the college admissions examination, and let Y be the student's first-year college grade point average. Think of the "experiment" as consisting of choosing a student at random, and recording his or her X and Y values. Intuitively, there should be a strong positive relation between X and Y, in the sense that most students having high values of X (relative to EX) also have high values of Y (relative to EY), and most students who have low values of X also have low values of Y. The covariance of X and Y, is designed to reflect this relation.

Definition 4.2.2

> The covariance between two random variables X and Y is denoted by $\mathrm{Cov}(X,Y)$, and is defined by
>
> $$\mathrm{Cov}(X,Y) = E[(X-b)(Y-c)] \qquad \textbf{(4.2.1)}$$
>
> where $b = EX$ and $c = EY$.

Equation (4.2.1) can be computed using the formulas for finding the expected value of a function of two random variables [Equation (4.1.6) or (4.1.7)], in this case the random variables X and Y, with $g(X,Y)=(X-b)(Y-c)$. Of course, b and c must be computed first. To compute b, just use Equation (4.1.6) or (4.1.7) with $g(X)=X$, and compute c similarly.

An alternative formula for Equation (4.2.1) is

$$E(XY)-(EX)\,(EY) \qquad \textbf{(4.2.2)}$$

(see Exercise 4.14); here $E(XY)$ would be computed using $g(X,Y)=XY$. This formula is the more convenient one for computation, but Equation (4.2.1) has more intuitive content.

Consider, for example, the four possible situations:

Situation	Sign of $(X-b)(Y-c)$
$X>b, Y>c$	$+$
$X<b, Y<c$	$+$
$X>b, Y<c$	$-$
$X<b, Y>c$	$-$

In Example 4.2.1, the first two situations should occur much more frequently than the last two, so $(X-b)(Y-c)$ is positive much more often than negative. Thus, from Equation (4.2.1) we might expect $\mathrm{Cov}(X, Y)$ to be positive in this example. On the other hand, if in an investigation of weather patterns, we let X denote temperature and Y denote rainfall, the third and fourth situations above might be more common, and a negative covariance will result.

This reasoning is, of course, meant to be only a rough intuitive introduction to the concept of covariance. Equation (4.2.1) is affected not only by the *sign* of $(X-b)(Y-c)$ but also by its *magnitude*, so it is possible for the covariance of a skewed bivariate distribution to have a surprising sign. For instance, in Example 4.2.1, even if the first two of the four "situations" tabulated occur more frequently than the other two, their total contribution to Equation (4.2.1) could be outweighed if $|X - b|$ and $|Y - c|$ are often extremely large in the third and fourth of those situations. The result would be a negative covariance, counterintuitive to our feeling about Example 4.2.1.

There are a number of important relations between expected values, variances, and covariances, many of them similar to those for the univariate case in Equations (3.2.15), (3.2.16), and (3.2.17), all of which should be reviewed before continuing. These new relations will be very useful in the rest of the book.

Miscellaneous Properties: Expected Value, Variance, Covariance

a. For any two random variables X and Y,

$$E(X+Y)=EX+EY \tag{4.2.3}$$

This extends to n variables:

$$E(X_1+\cdots+X_n)=EX_1+\cdots+EX_n \tag{4.2.4}$$

b. For any two random variables X and Y,

$$\text{Var}(X+Y)=\text{Var}(X)+\text{Var}(Y)+2\text{Cov}(X,Y) \tag{4.2.5}$$

In general,

$$\text{Var}(X_1+\cdots+X_n)=\sum_i \text{Var}(X_i)+2\sum_{i<j}\text{Cov}(X_i,X_j) \tag{4.2.6}$$

c. For any constants a, b, c, and d,

$$\begin{aligned}\text{Cov}(aX+bY,cU+dV) &= ac\text{Cov}(X,U)+ad\text{Cov}(X,V)\\ &\quad + bc\text{Cov}(Y,U)+bd\text{Cov}(Y,V)\end{aligned} \tag{4.2.7}$$

For (a), we will prove the continuous case only. The discrete case is parallel.

$$\begin{aligned}E(X+Y) &= \int_{-\infty}^{\infty}\int_{-\infty}^{\infty}(s+t)f_{X,Y}(s,t)\,ds\,dt\\ &= \int_{-\infty}^{\infty}\int_{-\infty}^{\infty}sf_{X,Y}(s,t)\,ds\,dt+\int_{-\infty}^{\infty}\int_{-\infty}^{\infty}tf_{X,Y}(s,t)\,ds\,dt\\ &= EX+EY\end{aligned} \tag{4.2.8}$$

where the last equality is obtained by thinking of EX and EY as instances of $E[g(X,Y)]$, with $g(X,Y)=X$ in the case of EX, for example.

To prove (b), we use Equation (3.2.16): $\mathrm{Var}(Z) = E(Z^2) - (EZ)^2$. Use this formula three times, with $X+Y$, X and Y playing the roles of Z. Also, use formula (4.2.2) for $\mathrm{Cov}(X,Y)$, and apply property (a). After like terms have been collected, the two sides of the equation will match.

The proof of (c) is similar to that of (b), using Equations (3.2.15) and (3.2.17) as well.

A positive sign in the covariance is an indication of correlation between two variables, but it does not tell us *how strong* the correlation is. Thus, we are interested in the magnitude of the covariance as well as the sign. However, it is not clear how large a covariance should be in order to indicate a strong correlation. In fact, the magnitude of a covariance is highly influenced by our choice of units of measurement.

Example 4.2.2

Suppose we wish to study the relationship between air temperature X and elevation Y. Let C_1 denote the covariance between X and Y if X is measured in Fahrenheit degrees and Y in feet, and let C_2 be the value if we use Celsius and meters. By using Equation (4.2.7), we see that $C_1 = kC_2$, where k is the product of the conversion factors for Centigrade to Fahrenheit and meters to feet (k will be approximately 6).

Thus, in order to give meaning to the magnitude of the covariance, a standardized form is used. This quantity will be called the **correlation between X and Y** (up to this point we have been using the term "correlation" informally). It is denoted by $\rho(X,Y)$, and is defined as follows:

$$\rho(X, Y) = \frac{\mathrm{Cov}(X, Y)}{\sigma_X \, \sigma_Y} \tag{4.2.9}$$

where σ_X and σ_Y denote the standard deviations of X and Y. It is clear from Equations (4.2.7) and (3.2.17) that ρ is dimensionless (i.e., it will have the same value no matter what units of measurement we use), so it does not present the problem encountered in Example 4.2.2.

To understand how to interpret the magnitude of ρ, consider the case in which X and Y are human height and weight. Consider a situation in which we know the height of some person but not the weight, and suppose we have decided to try to guess Y from X, with our guess taking the form $a+bX$ for some constants a and b. The error in our guess is then $Y - (a+bX)$, and the mean squared error is

$$E\left[\{Y - (a + bX)\}^2\right] \tag{4.2.10}$$

Of course, we should choose a and b to minimize expression (4.2.10), so we take partial derivatives with respect to a and b and set them to zero. It turns out that the optimal values of a and b are

$$a_1 = \frac{E(X^2)EY - EX\,E(XY)}{\mathrm{Var}(X)}$$

$$b_1 = \frac{E(XY) - EX\,EY}{\mathrm{Var}(X)} \tag{4.2.11}$$

On the other hand, suppose we did not know X. Then in trying to guess Y, we could not use X, which would force b to zero in Equation (4.2.10). Then we would take the derivative of Equation (4.2.10) under this constraint. The result would be

$$a_2 = EY$$

$$b_2 = 0 \tag{4.2.12}$$

The derivations of Equations (4.2.11) and (4.2.12) are left as Exercise 4.19. After a fair amount of algebraic manipulation, it can be shown that

$$\rho^2 = \frac{E[\{Y - (a_2 + b_2X)\}^2] - E[\{Y - (a_1 + b_1X)\}^2]}{E[\{Y - (a_2 + b_2X)\}^2]} \tag{4.2.13}$$

Thus, ρ^2 has a ready interpretation: It is the proportional improvement in mean squared error of a linear predictor based on X, compared to prediction *without* X (i.e., prediction by a constant). Since this proportion must be between 0 and 1, we now see that

a. $-1 \le \rho \le 1$.

b. The closer ρ is to 1 or -1, the stronger the degree of linear relation between X and Y.

The *sign* of ρ is loosely interpretable, with a positive ρ indicating that the events $\{X > EX, Y > EY\}$ and $\{X < EX, Y < EY\}$ happen more frequently than $\{X > EX, Y < EY\}$ and $\{X < EX, Y > EY\}$. However, the discussion of the effect of *magnitude* above holds here too, so that we must remind ourselves that, for example, a negative ρ could conceivably occur in a setting in which we might roughly describe X and Y as being "positively related." At the very least, one precise statement about the sign of ρ that we *can* make is that $\rho > 0$ means $b_1 > 0$ and $\rho < 0$ means that $b_1 < 0$.

It should also be emphasized that ρ measures only the degree of *linear* association between X and Y. It may be possible for X and Y to have a fairly strong dependence of some other kind, but with ρ near 0 or even equal to 0.

Example 4.2.3

Suppose the pair (X, Y) has a uniform distribution on the unit disk

$$\{(u, v) : u^2 + v^2 = 1\}$$

$E(XY)$, EX, and EY are all zero (this is clear without even computing these quantities, due to symmetry). Thus, from the definition of covariance, Equation (4.2.2), we see that $\text{Cov}(X,Y)=0$, and that $\rho = 0$ too. On the other hand, X and Y are certainly *not* unrelated. For example, if we know that $X > 0.5$, then we know that $|Y|$ definitely is less than $(0.75)^{1/2}$.

Since the last example demonstrated that $\rho = 0$ is not a strong enough condition for X and Y to be completely unrelated, we need some other criterion for expressing this idea.

Definition 4.2.3

Random variables S and T are said to be independent if for any two intervals (a, b) and (c, d), the events $\{a < S < b\}$ and $\{c < T < d\}$ are independent in the sense of Definition 2.3.2; that is,

$$P(a < S < b \text{ and } c < T < d) = P(a < S < b)P(c < T < d) \tag{4.2.14}$$

Independence of a set of k random variables is defined analogously, with k intervals.

Note that Equation (4.2.14) can also be written as

$$P(c < T < d | a < S < b) = P(c < T < d)$$

provided $P(a < S < b) > 0$. This says that even if we know that S happens to fall into a certain interval, this knowledge will not change our assessment of the probability that T falls into any given interval (this condition did *not* hold in Example 4.2.3). Note that the definition uses the concept of independent events in Section 2.3. There we used the example of tossing a fair coin twice, emphasizing the fact that knowledge of the outcome of the first toss does not change our assessment of the probability of heads on the second toss. In this definition, we are simply defining this notation in a random variable context.

Independent random variables have some very important properties, which we will use often in our study of sums of such variables. These properties should be memorized.

Properties of Independent Random Variables

a. If two discrete random variables X and Y are independent, then their bivariate probability mass function is simply the product of their individual (or **marginal**) probability mass functions.

$$p_{X,Y} = p_X \, p_Y \tag{4.2.15}$$

Similarly, in the continuous case,

$$f_{X,Y} = f_X \, f_Y \tag{4.2.16}$$

b. If X and Y are independent, then

$$E(XY) = EX \, EY \tag{4.2.17}$$

The expected value of a product of n independent random variables factors similarly.

c. If X and Y are independent, then $\mathrm{Cov}(X,Y)$ and $\rho\,(X,Y)$ are 0.

d. If $X_1, \ldots, X_n$ are independent random variables, then

$$\mathrm{Var}(X_1 + \cdots + X_n) = \mathrm{Var}(X_1) + \cdots + \mathrm{Var}(X_n) \tag{4.2.18}$$

e. If X and Y are independent and we form new random variables U and V by transformation; that is, $U = g(X)$ and $V = h(Y)$ for some functions g and h, then U and V form a new pair of independent random variables.

In (c) it is important to remember that the converse is not true, as shown in Example 4.2.3.

The proof of (a) is left to Exercise 4.15. We use (a) to prove (b). In the continuous case, for example,

$$\begin{aligned}
E(XY) &= \int_{-\infty}^{\infty}\int_{-\infty}^{\infty} st f_{X,Y}(s,t)\, ds\, dt \\
&= \int_{-\infty}^{\infty}\int_{-\infty}^{\infty} st f_X(s) f_Y(t)\, ds\, dt \\
&= \int_{-\infty}^{\infty} t f_Y(t) \left[\int_{-\infty}^{\infty} s f_X(s)\, ds\right] dt \\
&= \int_{-\infty}^{\infty} t\, EX f_Y(t)\, dt \\
&= EX\, EY
\end{aligned}$$

The proof of (c) follows immediately from (b) and Equation (4.2.2), and (d) follows from Equation (4.2.6). The result in (e) is very intuitive, so we omit a formal proof.

Example 4.2.4 illustrates the use and importance of these properties. Because this example makes quite a few intuitive statements about expected value and variance, you may wish to review these two concepts before continuing.

Example 4.2.4

Suppose we have a machine for making measurements, with the error in a measurement being a random variable V having mean 0 and variance 0.01. The fact that $E(V)=0$ is somewhat comforting, since it means (roughly) that while our measurement is liable to have some error, at least the machine has no tendency to consistently underreport or overreport the value of the quantity being measured. However, suppose the variance is too high for the degree of accuracy we need. We might try to obtain greater accuracy by making *two* measurements, and averaging the results. In effect, we are hoping that the two measurements have errors of opposite signs, thus partially canceling each other. Is this really a good idea? After all, the two errors could be of the same signs too, making the result even worse than that of the one-measurement case.

Let us investigate this averaging scheme. Let the two measurements be V_1 and V_2, and the resulting average be $U=(V_1+V_2)/2$. First consider $E(U)$. In the one-measurement case, at least we could say the mean error was 0. Is this still true for U? The answer is yes:

$$
\begin{aligned}
EU &= E[(V_1+V_2)/2] \\
&= \frac{1}{2}E(V_1+V_2) \qquad \text{[from (3.2.15)]} \\
&= \frac{1}{2}(EV_1+EV_2) \qquad \text{[from (4.2.3)]} \\
&= \frac{1}{2}(0+0) \\
&= 0
\end{aligned}
$$

Now how about Var(U)? If this is smaller than 0.01, then (again, roughly speaking) U will be close to its mean more often than the single measurement V_1 is to its mean. Since both U and V_1 have mean 0, U will close to 0 more often than does V_1, so our averaging scheme will be an improvement over the single measurement V_1. We can use our knowledge of properties of variance to determine whether this is indeed the case. Assume the two measurements to be independent.

$$
\begin{aligned}
\text{Var}(U) &= \text{Var}[(V_1+V_2)/2] \\
&= \frac{1}{4}\text{Var}(V_1+V_2) \qquad \text{[from (3.2.17)]} \\
&= \frac{1}{4}[\text{Var}(V_1)+\text{Var}(V_2)] \qquad \text{[from (4.2.18)]}
\end{aligned}
$$

$$= \frac{1}{4}(0.01 + 0.01)$$

$$= 0.005$$

Thus, our averaging scheme reduced variance by a factor of 2. However, keep in mind that U is an improvement over V_1 only "on the average"; it is still possible for V_1 to beat U in individual instances.

Example 4.2.5

As another illustration of the properties of mean, variance, and covariance developed in this section, consider the problem of deriving the mean and variance of a binomial distribution with parameters n and p. This can be done directly from the definitions of mean and variance, but only with some clever algebraic manipulations. Here is a much easier approach.

Let X have a binomial distribution with parameters n and p. Recall that X arises as the number of successes out of n independent success/failure trials, with probability p of success on each trial. Thus, we can write X as

$$X = X_1 + \cdots + X_n \tag{4.2.19}$$

where X_i is either 1 or 0, depending on whether the ith trial results in success or failure. We leave it to you to verify that $EX_i = p$ and $\text{Var}(X_i) = p - p^2$. Note too that since the trials are independent, the X_i are independent random variables.

Then the mean of X is easy to derive, using the fact that the expected value of a sum is the sum of the expected values of the individual terms in the sum [Equation (4.2.4)]:

$$EX = E(X_1 + \cdots + X_n) = EX_1 + \cdots + EX_n = np \tag{4.2.20}$$

verifying that $EX = np$, as claimed in Section 3.3. Note that the fact that we could represent X as a sum in Equation (4.2.19) was crucial to this derivation.

To derive the variance of X, we again exploit the fact that X can be written as a sum. In this case, we use the fact that the variance of a sum of *independent* random variables is equal to the sum of the variances of the terms [Equation (4.2.18)]. Since the terms in Equation (4.2.19) are independent, we have

$$\begin{aligned} \text{Var}(X) &= \text{Var}(X_1 + \cdots + X_n) \\ &= \text{Var}(X_1) + \cdots + \text{Var}(X_n) \end{aligned}$$

$$= n(p - p^2)$$
$$= np(1 - p) \tag{4.2.21}$$

again verifying the claim made in Section 3.3. Note that the independence of the X_i was important here, but was not used in Equation (4.2.20).

Here is an example of the use of Equation (4.2.16); it will also provide additional practice in finding probabilities for multivariate distributions.

Example 4.2.6

The following is a popular design for computer networks within a fairly small geographical area, say within one building. A cable is routed through various parts of the building, and the "stations" (computers, terminals, printers, etc.) attached to it. If two stations try to send a message on the cable at the same time, a **collision** will occur, and both messages will be garbled and will need to be retransmitted. For technical reasons, this will be true even if there is only a partial overlap of the two messages, even of only one bit.

For this reason, the system is designed to follow a "listen-before-send" rule: When a station has a message to send, it first senses the cable to check whether some other station currently has a transmission in progress; if so, it will not send yet, so as to avoid a certain collision.

However, collisions may still occur. For example, suppose Stations A and B have messages to send, but a transmission by Station C is currently in progress. A and B will hold off until C finishes, but if they start their own transmissions immediately after C finishes, they will collide with each other. The problem is that each A and B is unaware of the other's transmission (until after the collision occurs).

Thus, some systems adopt the following policy: Suppose a station, say S, wants to send but finds that another station is currently transmitting; let t_0 denote the time at which the current transmission finishes. S will *not* transmit at time t_0, but instead wait ("back off") until W time units after this time (i.e., it will start its own transmission at time $t_0 + W$). S goes through this backoff delay to try to avoid colliding with some other station that *might* wish to transmit at the same time. (Keep in mind that S has no way of knowing whether other stations have been waiting to transmit. The backoff time W is adopted to protect against the *possibility* of this occurring.)

Of course, W must be a random quantity—if it were a fixed quantity, then in the example with A and B above, A and B would perpetually collide!

Typically, the distribution of W is chosen by the designers to be exponential, so there is hardware or software in each station that will act like the function Expon presented in Section 3.4. Let us suppose the mean backoff time EW is chosen to be 15 ms.

Suppose transmissions last 10 ms each, and consider the example with Stations A and B above. Let X and Y be the random waiting times generated by A and B, respectively. What is the probability that A and B will collide?

Recall that if the two messages overlap by even one bit, a collision occurs. Thus, the desired probability is $P(|X - Y| < 10)$. Recall that this will be a double integral of the bivariate density $f_{X,Y}$ of X and Y, over the appropriate region. This region [the set E in Equation (4.1.2)] is

$$\{(u, v):0 < u, v < \infty \text{ and } |u-v| < 10\}$$

which is graphed in Figure 4.4.

This set is somewhat complicated, so instead of finding $P(|X - Y| < 10)$, let us find $P(|X - Y| > 10)$, and then subtract from 1.0. Now

$$P(|X-Y| > 10) = P(Y > X+10) + P(X > Y+10) = 2\ P(Y > X+10)$$

with the last equality due to symmetry.

Now we can carry out the computation. Recall that the density of an exponentially distributed random variable with mean m has the form $e^{-(t/m)}/m$. Thus, both X and Y have this density, with $m=15$. Moreover, because X and Y are independent, Equation (4.2.16) tells us that their bivariate density is the product of their individual (i.e., marginal) densities:

$$f_{X,Y}(u, v) = \frac{1}{15}e^{-(u/15)}\frac{1}{15}e^{-(v/15)} \qquad (u, v > 0)$$

Thus

$$P(Y > X + 10) = \int_0^\infty \int_{u+10}^\infty \frac{1}{15}e^{-(u/15)}\frac{1}{15}e^{-(v/15)}\, dv\, du = 0.26$$

Then the probability of a collision is

$$1-2(0.26) = 0.48$$

You may feel that this probability is too large. We could make it smaller by taking the mean backoff time to be larger, say 30 ms. However, recall that when A is waiting for C to finish, A does not know whether there are other stations waiting too; there may or may not be such stations. If we set A to have a long backoff time, think of what will happen in those cases in which no other stations are waiting with A. A will still go through its backoff delay, since it is not aware of the fact that no other station wishes to send; this is wasteful in any case, and the larger the backoff delay, the

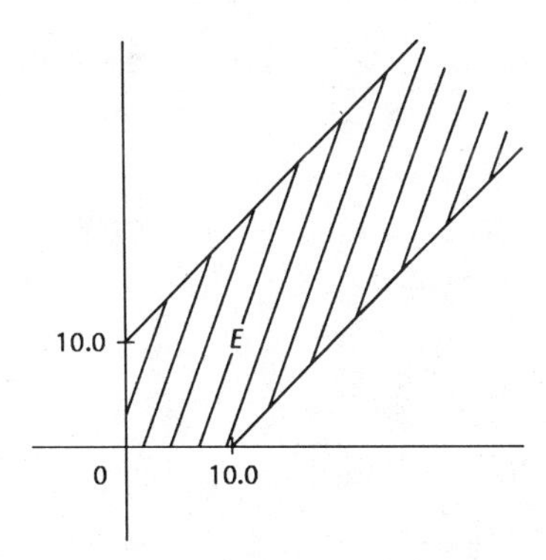

Figure 4.4 Region of interest in Example 4.2.6.

greater the waste. Thus, a large mean backoff time may not be desirable. An optimal choice for EW thus requires more detailed analysis and is beyond the scope of this text. However, the example should give you an idea as to how probability is used in the design of computer networks.

In many cases, integrals for multivariate probabilities, expected values, and so on are quite complicated, often far too complex to evaluate by hand. The next example gives us an opportunity to illustrate this point.

Example 4.2.7

In the last example, we found the probability of a collision between two waiting stations. What if there are three such stations? Four stations? n stations?

In principle, we can compute the probability of a collision in the same way as above, by setting up an integral. However, for n stations, we need an n-fold integral, which gets rather complicated, even for $n=3$. It is much more convenient to determine the desired probability via simulation, as shown in Program 4.2.1.

Program 4.2.1

```
program Collision(input,output);
   var NStns,Seed,Rep,NReps,I,Count: integer;
      X: array[1..10] of real;
      Collide: boolean;
```

```
   function Expon(B:real; var Seed: integer): real;
      var U: real;
   begin
      U := random(Seed);
      if U < 0.0001 then U := 0.0001;
         Expon := -ln(U)/B
   end;
procedure CheckCollide;
      label 99;
      var J,K: integer;
   begin
      Collide := false;
      for J := 1 to NStns - 1 do
         for K := J + 1 to NStns do
            if abs(X[J] - X[K]) < 10.0 then
               begin
               Collide := true;
               goto 99
               end;
      99:
   end;

begin
   writeln('enter NReps,NStns');
   readln(NReps,NStns);
   Seed := 9999;
   Count := 0;
   for Rep := 1 to NReps do
      begin
      for I := 1 to NStns do
         X[I] := Expon(1/15.0,Seed);
      CheckCollide;
      if Collide then Count := Count + 1
      end;
   writeln('probability of collision = ',Count/NReps:8:2)
end.
```

The author ran this program for NStns = 2, 3, and 4, with NReps = 1000. The run with NStns = 2 was just for verifying the program, since we already determined the exact answer of 0.48 mathematically. The simulation yielded the value 0.51, validating the program (since we had NReps only at 1000, we could not expect an exact match). For 3 stations, the probability of a collision was 0.88, and for 4 a collision was almost certain, with probability 0.98. Clearly a mean wait of only 15 milliseconds is insufficient here if we think there is a good chance that more than two stations will be waiting at a time. As before, much more detailed analysis is needed, which we will not present here.

It should be stressed again that a simulation solution here was much more convenient than a mathematical solution. We could vary the value of NStns at will, just by running the program again, in stark contrast to the mathematical approach, which would require carefully setting up and evaluating a complicated integral for each new value of NStns.

On the other hand, the mathematical approach is useful too. For example, we used it here to help validate the simulation program; in fact, the author's original version of the program did contain an error, which was exposed by the mathematical analysis. You are urged to reread the "philosophical" discussion on mathematics versus simulation, at the end of Section 2.4.

The next example will complete a topic started in Section 3.4; it will serve nicely as review, too, since it will use several concepts developed in the current chapter.

Example 4.2.8

Recall Program 3.4.10, which we reproduce here for convenience.

```
function X(A,B,C:real; var Seed: integer): real;
 var U1,U2: real;
begin
 repeat
    U1 := A + (B - A)*random(Seed);
    U2 := C*random(Seed)
 until U2 < h(U1);
 X := U1
end;
```

It was claimed that this program will generate a random variable X having density h. Let us verify this claim.

Consider any t in (a, b). Then

$$P(X < t) = P[U_1 < t | U_2 < h(U_1)] = \frac{P\,[U_1 < t \text{ and } U_2 < h(U_1)]}{P\,[U_2 < h(U_1)]} \tag{4.2.22}$$

Look at the numerator of this fraction. From Equation (4.1.16b), we have

$$P\,[U_1 < t \text{ and } U_2 < h(U_1)] = \int_a^t P\,[U_2 < h(U_1) \mid U_1 = s]\, f_{U_1}(s)ds$$

$$= \int_a^t P\,[U_2 < h(U_1) \mid U_1 = s]\, \frac{1}{b-a} ds$$

Since U_2 has a $U(0, c)$ distribution, we have

$$P\left[U_2 < h(U_1) \mid U_1 = s\right] = \frac{h(s)}{c}$$

Now look at the denominator of the fraction in Equation (4.2.22),

$$P[U_2 < h(U_1)] \tag{4.2.23}$$

This expression looks imposing, but it really isn't. It is a probability involving U_1 and U_2 so as usual it can be computed by integrating the bivariate density of U_1 and U_2 (i.e., f_{U_1,U_2}) over the appropriate region, as in Equation (4.1.2).

Since the density of U_1 is equal to $1/(a - b)$ on (a,b) and 0 elsewhere, and the density of U_2 is equal to $1/c$ on $(0,c)$ and 0 elsewhere, and since U_1 and U_2 are independent, their bivariate density is the product of their marginal densities, as in Equation (4.2.16).

Thus, the bivariate density of U_1 and U_2 is equal to $1/[c(b - a)]$ on the square

$$\{(x, y){:}a < x < b,\ 0 < y < c\}$$

and 0 elsewhere. In other words, the random point (U_1, U_2) is uniformly distributed over the square bounded by the points $(a,0)$,$(b,0)$,(a,c) and (b,c).

Thus, the probability in Equation (4.2.23) is equal to the area under the curve h—which must be equal to 1.0, since h is a density—divided by the area of the box, which is $c(b - a)$, so

$$P[U_2 < h(U_1)] = \frac{1}{c(b - a)}$$

Returning to Equation (4.2.22) with these discoveries, we have

$$P(X < t) = \int_a^t h(s)ds$$

which says that X does indeed have the density h.

4.3 SUMS OF INDEPENDENT RANDOM VARIABLES: EXACT METHODS

Very often we are interested in the distribution of a random variable obtained as a function of a random vector. For example, sums of random variables appear quite frequently in applications.

Example 4.3.1

A flashlight uses one battery. We have two new batteries. We will insert one battery in the flashlight now, and insert the second one when the first one dies. The overall usable time for the flashlight, for the set of two batteries, is then the sum of the two lifetimes.

Example 4.3.2

A manufacturer has just developed a new automobile tire. It will test the tire on a sample of 100 cars and use the resulting data to estimate the mean tread life of the tire. A natural estimator of this population mean is the mean life among our sample of cars,

$$\frac{X_1 + \cdots + X_{100}}{100} \tag{4.3.1}$$

where X_i is the tread life observed for the ith car in the sample. Assuming the cars in our sample are selected at random, the quantities X_i are random variables (there is randomness in the manufacturing process, producing variation from one tire to another). Thus, we are again interested in a sum of random variables.

To see how distributions of sums may be obtained, suppose that in Example 4.3.1 flashlight battery lifetimes have an exponential distribution with mean 50 hours. It is reasonable to assume that the lifetimes X and Y are independent. Let $W = X + Y$. We wish to find f_W, the density of W. We will do so by finding F_W first, and then taking the derivative to obtain f_W. Proceeding in this way, we have

$$F_W(t) = P(W \leq t) = P(X + Y \leq t) = P[(X, Y) \text{ is in } A] \tag{4.3.2}$$

where $A = \{(r, s): r, s \geq 0 \text{ and } r + s \leq t\}$. Recall again from Equation (4.1.2) that $P[(X, Y)$ is in A] is the double integral of $f_{X,Y}(r, s)$ over the region A.

Thus, we need to determine $f_{X,Y}$. For independent random variables, we know that the bivariate density is the product of the marginal densities [Equation (4.2.16)], so $f_{X,Y} = f_X f_Y$. Then continuing Equation (4.3.2),

$$\begin{aligned} F_W(t) &= \int_A\int f_X(r) f_Y(s)\, dr\, ds \\ &= \int_0^t \int_0^{t-s} 0.02e^{-0.02r} 0.02e^{-0.02s}\, dr\, ds \\ &= 1 - e^{-.02t} - t(0.02e^{-0.02t}) \end{aligned}$$

Thus

$$f_W(t) = \frac{d}{dt} F_W(t) = (0.02)^2 t\, e^{-0.02t} \tag{4.3.3}$$

You may recall that the sum of independent exponential random variables has an Erlang distribution (Section 3.3). Equation (4.3.3) verifies this for the special case considered here; a general proof of this property would follow the same pattern, using mathematical induction (see also Example 4.3.8).

In fact, we can generalize further: For *any* pair of continuous random variables X and Y, the density of $W = X + Y$ is given by

$$f_W(t) = \frac{d}{dt} \int_A\!\!\int f_{X,Y}(r,s)\, dr\, ds \tag{4.3.4}$$

where A again is $\{(r,s) : r + s \le t\}$. If X and Y are independent, *nonnegative* random variables, then Equation (4.3.4) reduces to

$$f_W(t) = \int_0^t f_X(r)\, f_Y(t-r)\, dr \tag{4.3.5}$$

(see Exercise 4.16). We say that f_W is the **convolution** of f_X and f_Y.

We can find the distribution of a sum of discrete random variables in a similar way.

Example 4.3.3

Suppose X and Y are independent Poisson-distributed random variables, with the parameter b in the formula for the Poisson probability mass function [Equation (3.3.7)] having the values c and d, respectively. Let $W = X + Y$. We can find the probability mass function of W as follows:

$$\begin{aligned} p_W(n) &= P(W = n) \\ &= P(X + Y = n) \\ &= \sum_{k=0}^{n} P(X = k \text{ and } Y = n - k) \\ &= \sum_{k=0}^{n} \left[e^{-c} \frac{c^k}{k!} \right] \left[e^{-d} \frac{d^{n-k}}{(n-k)!} \right] \end{aligned} \tag{4.3.6}$$

Reduction of the last expression to a simpler form requires some creative algebraic manipulation, which we omit here. The result turns out to be

$$p_W(n) = e^{-(c+d)} \frac{(c+d)^n}{n!} \tag{4.3.7}$$

Note that Equation (4.3.7) is again in the Poisson form (3.3.7), with $b = c + d$. Thus, we have found that the sum of two independent Poisson variables is itself Poisson-distributed. We also found that the b value for the sum is the sum of the individual b values. This latter finding should not be too surprising if we recall that for a Poisson variable, b is the expected value of that variable; Equation (4.2.3) thus *forces* the b values to add together, once we have found that $X + Y$ has a Poisson distribution. But it is rather surprising the same *distribution* type pops up here (i.e., the sum of independent Poisson random variables is again Poisson).

By induction, the analogous result for the sum of m independent Poisson random variables also holds.

In theory, the methods used here can be used to find the distribution of any sum. However, the derivations can get extremely complicated; imagine, for example, what the integrals in Equations (4.3.4) and (4.3.5) would look like if 20 random variables were involved instead of 2. Thus, it is important to have alternative methods available. We will discuss two such methods: moment-generating functions (in this section), and the Central Limit Theorem (in Section 4.4). Both are of great practical value.

Definition 4.3.1

For any random variable X, the function of t defined by

$$m_X(t) = E[e^{tX}] \tag{4.3.8}$$

is called the moment-generating function of X.

Example 4.3.4

We will find the moment-generating function for any normally distributed random variable V. Let μ and σ^2 be the mean and variance of V, and recall from Section 3.3 that the random variable $Z = (V - \mu)/\sigma$ has a normal distribution with mean 0 and variance 1. Then from Equation (3.2.11) for finding the expected value of a function of a random variable (in this case a function of Z), we have

$$m_Z(t) = \int_{-\infty}^{\infty} e^{tw} \frac{1}{\sqrt{2\pi}} e^{-0.5w^2}\, dw$$

By collecting powers of e, completing the square, and recalling that

$$\int_{-\infty}^{\infty} \frac{1}{\sqrt{2\pi}} e^{-0.5u^2} du = 1$$

(since the total area under a normal density, or any other density, must be 1), we find that

$$m_Z(t) = e^{0.5t^2} \tag{4.3.9}$$

Then since $V = \sigma Z + \mu$, we have

$$\begin{aligned} m_V(t) &= E\,[e^{tV}] \\ &= E\,[e^{t(\sigma Z + \mu)}] \\ &= e^{\mu t} E\,[e^{t\sigma Z}] \\ &= e^{\mu t} m_Z(t\sigma) \\ &= e^{\mu t} e^{0.5(t\sigma)^2} \end{aligned}$$

the last step coming from Equation (4.3.9). Thus

$$m_V(t) = e^{t\mu + t^2\sigma^2/2} \tag{4.3.10}$$

We will use moment-generating functions to find the distributions of sums of independent random variables. However, we should first discuss the source of the name moment-generating function, as it stems from another useful property.

Definition 4.3.2

> For any random variable X, the kth moment of X is defined to be $E(X^k)$.

The name used for m_X implies that we can obtain the moments of X from m_X. This is in fact the case. For example, we can obtain the first moment of X (i.e., the mean of X), by the following reasoning:

$$\frac{d}{dt} m_X(t) = \frac{d}{dt} E\left[e^{tX}\right]$$

$$= E\left[\frac{d}{dt}e^{tX}\right]$$

$$= E\left[Xe^{tX}\right] \qquad \textbf{(4.3.11)}$$

Evaluating this derivative at $t=0$, we have

$$m_X'(0)=EX$$

that is, the first moment of X can be obtained by evaluating the first derivative of the moment-generating function of X at $t=0$. Similarly, $m_X''(0) = E(X^2)$, and thus we can obtain the variance of X from the moment-generating function:

$$\text{Var}(X)=m_X''(0)-\left[m_X'(0)\right]^2$$

The moment-generating function can often save us some work when we need to find some moment of a random variable.

Example 4.3.5

Suppose we wish to find $E(V^4)$, where V has a normal distribution with mean μ and variance σ^2. The straightforward but tedious way to do this is to use the formula for the expected value of a function of a random variable [Equation (3.2.11)]: The function $g(t)$ would be t^4; the random variable X would be V; and the density f_X would be that of V, which is the normal density function [Equation (3.3.10)].

However, the resulting integral would be rather difficult to compute. It is much easier (although still a bit tedious) to differentiate Equation (4.3.10) four times and then evaluate at $t=0$.

However, our main use for moment-generating functions will be in saving work in another context, that of finding distributions of sums of independent random variables. To see how this works, suppose again that X and Y are independent random variables and $W=X+Y$. It turns out that m_W is simply related to m_X and m_Y:

$$\begin{aligned} m_W(t) &= E\left(e^{tW}\right) \\ &= E\left[e^{t(X+Y)}\right] \\ &= E\left[e^{tX}e^{tY}\right] \\ &= E\left[e^{tX}\right]E\left[e^{tY}\right] \end{aligned} \qquad \textbf{(4.3.12)}$$

where the last step follows from Equation (4.2.17) and the independence of X and Y. Thus, Equation (4.3.12) implies that

$$m_W = m_X \, m_Y \tag{4.3.13}$$

To see how Equation (4.3.13) is used for finding the distribution of $W = X + Y$, consider the following example.

Example 4.3.6

Suppose X and Y are independent and normally distributed, with means a and c and standard deviations b and d, respectively. From Equations (4.3.10) and (4.3.13), we know that

$$\begin{aligned} m_W(t) &= m_X(t) \, m_Y(t) \\ &= e^{ta + t^2b^2/2} e^{tc + t^2d^2/2} \\ &= e^{t(a+c) + t^2(b^2+d^2)/2} \end{aligned} \tag{4.3.14}$$

One more glance at Equation (4.3.10) reveals that the moment-generating function of W is that of the $N(a + c, b^2 + d^2)$ distribution. Since mathematicians have proved that there is a one-to-one correspondence between distributions and moment-generating functions, we thus see that W actually does have the normal distribution $N(a + c, b^2 + d^2)$. So just as in Example 4.3.3, in which we found that the sum of two independent Poisson-distributed random variables is itself Poisson-distributed, we have found a similar "reproductive" property for normal distributions; the sum of two independent normal variables is again normal. (By the way, Example 4.3.3 could also have been done with moment-generating functions in this manner.) Again as in the Poisson case, an analogous result holds for the sum of m independent normal random variables.

Here is an example of the consequences of this property of the normal distribution. It will also make concrete some of the intuitive statements in Example 4.2.4 by computing specific probabilities.

Example 4.3.7

Recall the errors-in-measurement situation in Example 4.2.4. There we assumed that the measurement errors had mean 0 and variance 0.01. Suppose we also know that the errors are normally distributed. We can compare the one-measurement and the averaging methods discussed there as follows.

First, from Example 4.3.6 we know that $V_1 + V_2$ is normally distributed with mean $0 + 0 = 0$, and variance $0.01 + 0.01 = 0.02$. Then from the Linearity Property of the Normal Family, we know that U, which is equal to $0.5(V_1 + V_2)$ has a normal distribution too, with mean $(0.5)0 = 0$ and variance $(0.5)^2 0.02 = 0.005$. Again, only the normality of U is new here; the mean and variance of U were determined in Example 4.2.4, just using properties of means and variances of sums. However, the normality is very important; if you know the *distribution* of a random variable, rather than merely its mean and variance, you can compute probabilities. For example, let us find the probabilities that our reported value will be within 0.12 of the true value, for V_1 versus U. Recalling the technique of evaluating normal probabilities by using the $N(0,1)$ table, we proceed:

$$P(|V_1| < 0.12) = P\left(|Z| < \frac{0.12 - 0}{\sqrt{0.01}}\right) = 0.7698$$

$$P(|U| < 0.12) = P\left(|Z| < \frac{0.12 - 0}{\sqrt{0.005}}\right) = 0.9108$$

Thus, the averaging method does have a substantially better chance of being within 0.12 of the true value than does a single measurement.

This last example used the fact that normal distributions "reproduced themselves"; that is, the sum of two (or more) normally distributed random variables is again normally distributed. We also used the Linearity Property of the Normal Family, from Chapter 3. The overall result is worth formalizing.

Sums of Independent, Normally Distributed Random Variables

> Let $X_1, \ldots, X_n$ be independent, normally distributed random variables, with X_i having mean μ_i and variance σ_i^2. Also, let $c_1, \ldots, c_n$ be any constants. Then the random variable
>
> $$T = c_1 X_1 + \cdots + c_n X_n$$
>
> is also normally distributed, with
>
> $$E(T) = c_1 \mu_1 + \cdots + c_n \mu_n \qquad \textbf{(4.3.15)}$$
>
> and variance
>
> $$\text{Var}(T) = c_1^2 \sigma_1^2 + \cdots + c_n^2 \sigma_n^2 \qquad \textbf{(4.3.16)}$$

Example 4.3.8

Suppose $X_1, \ldots, X_r$ are independent random variables, each exponentially distributed with the same value of the parameter b in the formula for the exponential density family, Equation (3.3.14). Let

$$X = X_1 + \cdots + X_r$$

It was mentioned without proof in Section 3.3 that X will then have an Erlang distribution with parameters r and b. We will now derive this fact.

The moment-generating function for X_i is

$$Ee^{uX_i} = \int_0^\infty e^{ut} b e^{-bt}\, dt = \frac{b}{b-u}$$

Thus

$$m_X(u) = \prod_{i=1}^{r} m_{X_i}(u) = \left(\frac{b}{b-u}\right)^r$$

But the moment-generating function for an Erlang distribution with parameters r and b is

$$\int_0^\infty e^{ut} c b^r t^{r-1} e^{-bt}\, dt = \left(\frac{b}{b-u}\right)^r \int_0^\infty e^{ut} c(b-u)^r t^{r-1} e^{-(b-u)t}\, dt$$
$$= \left(\frac{b}{b-u}\right)^r$$

matching the expression found for $m_X(u)$ (the integral evaluated to 1.0, since it is the total area under another Erlang density, with parameter $b - u$ instead of b). Thus, X has an Erlang distribution, as claimed.

We have found that the Poisson, normal, and Erlang distributions have reproductive properties. There are some other distributions that also have such properties. The binomial and negative binomial families also operate this way, as long as the success probability p is fixed. For example, if X and Y are independent with distributions $B(n,p)$ and $B(m,p)$, then $X+Y$ has distribution $B(n+m,p)$.

However, most distributions do *not* work like this. For example, suppose X and Y are independent $U(0,1)$ random variables. Then although the distribution of $X+Y$ is concentrated on the interval (0,2), the distribution is not uniform on that interval; that is, $X+Y$ does not have a $U(0,2)$ distribution. In fact, the distribution of $X+Y$ is not uniform at all, having instead the triangular shape in Figure 4.5 (Exercise 4.24).

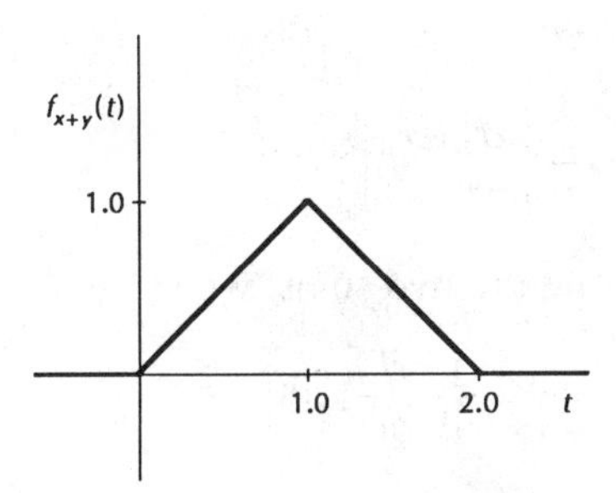

Figure 4.5 Distribution of the sums of two independent $U(0,1)$ random variables.

Before leaving this material on moment-generating functions, it should be mentioned that there are related functions, differing only by a change of variable. For example, the **Laplace transform**

$$l_X(t) = E(e^{-tX})$$

is often used in queueing theory; note that $l_X(t) = m_X(-t)$. Similarly, the **characteristic function**

$$c_X(t) = E(e^{itX})$$

is used in advanced probability theory. Its main advantage is that the expected value exists for any t (since it is the expected value of a bounded quantity); the disadvantage is that it uses complex numbers.

Another useful type of transform, for nonnegative integer-valued random variables, is the **generating function**:

$$g_X(t) = E(t^X) \tag{4.3.17}$$

It is related to the moment-generating function in that $g_x(t) = m_x(\ln t)$, and it exists for $-1 < t < 1$. Like the other transforms, we can use it to identify distributions, as we did with the moment-generating function in Examples 4.3.6 and 4.3.8, and generate moments of X. We can also find probabilities $P(X=n)$.

To prove this latter claim, match the computational expression for Equation (4.3.17)

$$\sum_{n=0}^{\infty} t^n P(X=n)$$

with the Taylor series expansion of $g_X(t)$ (considered as a function of t), which is

$$\sum_{n=0}^{\infty} a_n t^n$$

where

$$a_n = \frac{1}{n!}\left[\frac{d^n}{dt^n} g_X(t)\right]_{t=0}$$

After comparing the two sums, we see that

$$P(X=n) = \frac{1}{n!}\left[\frac{d^n}{dt^n} g_X(t)\right]_{t=0} \qquad \textbf{(4.3.18)}$$

Example 4.3.9

Suppose the number of orders N arriving in a given period of time at a factory has a Poisson distribution with mean 10. The number I of items in an order varies according to some distribution p_I. The total number of items ordered is then a sum of random variables, where even the number of terms is random:

$$T = I_1 + \cdots + I_N \qquad \textbf{(4.3.19)}$$

where I_j is the number of items requested in the jth order.

Since all information concerning the distribution of T can be obtained from its generating function g_T, let us compute this function. Equation (4.1.15a) will be useful here, with N playing the role of X there, and s^T playing the role of Y.

$$\begin{aligned} g_T(s) &= E(s^T) \\ &= \sum_{k=0}^{\infty} E(s^T|N = k)p_N(k) \\ &= \sum_{k=0}^{\infty} E(s^T|N = k)\frac{e^{-10}10^k}{k!} \end{aligned}$$

Now by the reasoning that produced Equation (4.3.13), we have from Equation (4.3.19) that

$$E(s^T|N=k) = [h(s)]^k$$

where h is the generating function of I. Thus

$$\begin{aligned} g_T(s) &= \sum_{k=0}^{\infty} \frac{e^{-10}10^k}{k!}[h(s)]^k \\ &= e^{-10}\sum_{k=0}^{\infty} \frac{[10h(s)]^k}{k!} \end{aligned}$$

$$=e^{-10+10h(s)}$$

using the fact that

$$e^z = \sum_{k=0}^{\infty} \frac{z^k}{k!}$$

Thus

$$g_T(s) = e^{-10+10h(s)} \tag{4.3.20}$$

Thus, if $h(s)$ is known, we can differentiate Equation (4.3.20) to find $P(T=k)$, $k=0, 1, 2, \ldots$.

4.4 SUMS OF INDEPENDENT RANDOM VARIABLES: THE CENTRAL LIMIT THEOREM

In the last section, we saw how to find the exact distributions of sums of independent random variables. These methods work in theory, but they may be impractical in some cases. There are two reasons for this:

a. The mathematics involved may be intractable. For example, the density for the sum of 10 independent variables is a complicated ninefold integral, which in most cases will not be capable of simplification. Or, using the moment-generating function approach, the product of the 10 moment-generating functions will in most instances not be identifiable in terms of the distribution to which it corresponds (i.e., we will not generally be as lucky as we were in Examples 4.3.6 and 4.3.8, in which m_W was seen to correspond to a known distribution).

b. We may not even know the densities of the random variables involved in the sum. This is in fact an even more commonly encountered problem than that in (a).

Because of the difficulties mentioned in (a), a simple approximating procedure would be very handy. Fortunately, such a procedure does in fact exist; it is based on the *Central Limit Theorem* presented below. Moreover, case (b), which might seem hopeless, can in certain senses also be handled in this way. The Central Limit Theorem is quite a remarkable result, and has very important implications.

The Central Limit Theorem

Suppose $X_1, X_2, X_3, \ldots$ are independent and identically distributed random variables, with $E(X_i) = \mu$ and $\mathrm{Var}(X_i) = \sigma^2$. Let

$$W_n = \frac{T_n - n\mu}{\sqrt{n}\sigma} \tag{4.4.1}$$

where $T_n = X_1 + \cdots + X_n$. Then for any real number u,

$$\lim_{n\to\infty} P(W_n \leq u) = P(Z \leq u) \tag{4.4.2}$$

where Z has a $N(0,1)$ distribution.

What does this theorem say, in practical terms? It says that for "large" n, the distribution of W_n is approximately $N(0,1)$, and thus (from the Linearity Property of the Normal Family presented in Section 3.3) the distribution of T_n is approximately $N(n\mu, n\sigma^2)$. Thus we can find approximate probabilities involving T_n by using the normal distribution table in Appendix III.

This is quite a remarkable phenonemon: No matter what the base distribution of the X_i is (i.e., no matter how "un-bell-shaped" f_{X_i} is), the distribution of T_n is approximately normal (i.e., f_{T_n} becomes approximately bell-shaped) and becomes closer and closer to the normal distribution as n grows. The theorem refers to cumulative distribution functions, not densities, but there are similar theorems for densities.

Let us see an example of how this result is applied.

Example 4.4.1

Suppose flashlight battery lifetimes are exponentially distributed with mean 50 hours; then a lifetime has the density function (3.3.14), with $b = 1/50 = 0.02$:

$$f_L(t) = 0.02e^{-0.02t}$$

Suppose we have 20 batteries and use them one at a time in a one-cell flashlight. What is the probability that we are able to use the flashlight more than 1200 hours?

Letting L_i denote the lifetime of the ith battery ($i = 1, \ldots, 20$), we see that the desired probability is $P(S > 1200)$, where $S = L_1 + \cdots + L_{20}$. We know from Section 3.3 that the exact distribution of S is Erlang with $r=20$ and $b=0.02$. Thus

$$P(S > 1200) = \int_{1200}^{\infty} c0.02^{20} t^{19} e^{-0.02t} dt$$

(where c would have to be found separately; see Section 3.3). However, to evaluate this integral, we would need to integrate by parts 19 times—not a very pleasant task!

The Central Limit Theorem makes our job easy. The X_i in the statement of the theorem are the L_i here, and T_n is S (with $n=20$), so

$$\mu = E(L_i) = 50$$

and

$$\sigma^2 = 50^2 = 2500$$

(recall that the mean and variance of an exponential distribution with parameter b are $1/b$ and $1/b^2$).

Thus, the distribution of T is approximately normal with mean $(20)(50)=1000$, and variance $(20)(2500)=50{,}000$; the standard deviation is $\sqrt{50000}=223.61$. Then

$$P(S > 1200) \approx P\left(Z > \frac{1200 - 1000}{223.61}\right) = P(Z > 0.89) = 0.1867$$

The Central Limit Theorem saved us quite a bit of tedious work in the last example. Note that it would have been even more useful if the base distribution (i.e., the distribution of the individual terms L_i) had been something other than the exponential. In the last example, at least we knew the exact distribution of S, but this would not have been true if the base distribution had been, say, $U(25, 75)$. As outlined in Exercise 4.24, even the case $n=2$ is difficult for finding densities of sums of uniform random variables.

The Central Limit Theorem is remarkable in that it is often useful even if we do not know the base distribution of the summands, as shown in the following example.

Example 4.4.2

Consider Example 4.3.2. Suppose we know nothing of the distribution of the X_i; after all, it is a *new* tire that is being studied, so no previous data are available. We don't even know the values of $EX_i = \mu$ and $\text{Var}(X_i) = \sigma^2$; in fact, estimating μ is the goal of the study. However, the developers of the tire may feel sure that σ is no larger than 5000 miles. This knowledge can then be used as follows.

Let $\overline{X}$ denote the sample mean (4.3.1). $\overline{X}$ will be our estimate of μ, so we hope $|\overline{X}-\mu|$ is small. Thus it is of interest to find probabilities such as

$$P(|\overline{X}-\mu|) < 1000) \quad \textbf{(4.4.3)}$$

Assuming for the moment that $\sigma \leq 5000$, we can use the Central Limit Theorem to find the approximate value of (4.4.3): The theorem says that $X_1 + \cdots + X_{100}$ is approximately normally distributed with mean 100μ and variance $100\sigma^2$. Then since $\overline{X} = \frac{1}{100}(X_1 + \cdots + X_{100})$, the Linearity Property of the Normal Family tells us that $\overline{X}$ is also approximately normally distributed, with mean

$$\frac{1}{100}(100\mu) = \mu$$

and variance

$$\left(\frac{1}{100}\right)^2 (100\sigma^2) = \frac{\sigma^2}{100}$$

Then we have in the worst case ($\sigma = 5000$)

$$\begin{aligned} P(|\overline{X} - \mu| < 1000) &= P\left(\frac{\overline{X} - \mu}{\sqrt{250000}} < \frac{1000}{\sqrt{250000}}\right) \\ &= P(|Z| < 2.0) \\ &= 0.9544 \end{aligned}$$

where Z has a $N(0,1)$ distribution.

Now if σ is actually less than 5000, the preceding probability computation will result in $P(|Z| < c)$ for some c larger than 2.0, so the probability will be even larger than 0.9544. Thus, if the developers of the tire are correct in their assumption that σ is at most 5000, then this sample of size 100 is almost certain to give us the desired accuracy, which is

$$|\overline{X}-\mu| < 1000$$

As mentioned in Chapter 3, the Central Limit Theorem is very useful in the approximation of binomial probabilities. The technique is explained in the following example.

Example 4.4.3

Suppose a certain electronic part is very inexpensive to produce, but is often made defectively. In fact, only 1/3 of the parts produced are good. Suppose it is extremely expensive to inspect for defects. The manufacturer is thus considering selling these parts in packages of 50, intended for an application requiring 12 good parts. Under this plan, what proportion of the packages will not contain at least 12 good parts?

Let T be the number of good parts in a package of 50. Assuming that the parts act independently, T then has a binomial distribution with $n=50$ and $p=1/3$, so the probability of interest $P(T \leq 11)$ is equal to

$$\sum_{i=0}^{11} \binom{50}{i} \left(\frac{1}{3}\right)^{i} \left(\frac{2}{3}\right)^{50-i} \tag{4.4.4}$$

The arithmetic in Equation (4.4.4) is certainly not something we would want to do by hand, or even with a calculator. We *could* use a computer to do the work; however, an even simpler solution is to use the Central Limit Theorem. To do this, write

$$T = X_1 + \cdots + X_{50} \tag{4.4.5}$$

where X_i is equal to either 1 or 0, depending on whether or not the ith part in the package is good (recall that the same kind of representation was very useful in Example 4.2.5, in which the mean and variance of the binomial distribution were derived). This representation is perfect for the Central Limit Theorem, since it is in the form of a sum of identically distributed terms.

From the theorem, we have that T is approximately normally distributed with mean $np=50/3=16.67$ and variance $np(1-p)=100/9=11.11$ (again, the mean and variance were found in Example 4.2.5). Thus $P(T \leq 11)$ is approximately equal to

$$P\left(Z \leq \frac{11 - 16.67}{\sqrt{11.11}}\right) = P(Z \leq -1.70) = 0.0446 \tag{4.4.6}$$

Thus, only about 4 or 5% of the packages will lack the sufficient number of good parts.

Note that this reasoning generalizes to all binomial distributions.

Approximate Distribution of Binomial Random Variables

> If the random variable X is binomially distributed (i.e., it has the probability mass function (3.3.2)), then X is approximately normally distributed with mean np and variance $np(1 - p)$.

We have emphasized that the Central Limit Theorem is an approximation, one which gets better with increasing n. It is then important to ask how large n must be for reasonably good accuracy. There is no simple answer to this question, although various investigations have shown that in most applications $n = 20$ or so will be large enough for good results (more is known for the binomial case; see Exercise 4.29). As an example, let us see how accurate the approximation in Example 4.4.1 was. There we found $P(S > 1200)$ to be approximately 0.1867; we will now find the exact answer and compare. Rather than do the 19 integrations-by-parts mentioned in that example, we can find the answer by simulation. Of course, this too is an approximation, but with NReps equal to 20,000, the simulation output can be considered correct to two decimal places (see Section 6.5).

Program 4.4.1

```
program S1200(input,output);
   const NReps = 20000;
   var I,Rep,Seed: integer;
       S,ProbOfS: real;
     {'random' and Expon are then declared here, from Ch.3}

begin
   Count := 0;
   for Rep := 1 to NReps do
      begin
      S := 0;
      for I := 1 to 20 do
         S := S + Expon(0.02,Seed);
      if S > 1200 then Count := Count + 1
      end;
   ProbOfS := Count/NReps;
   writeln('P(S > 1200) = ',ProbOfS:5:2)
end.
```

The output from this program gave $P(S > 1200) = 0.18$. Thus, the use of the Central Limit Theorem in Example 4.4.1, which resulted in an answer of 0.1867, gave excellent accuracy.

To investigate the accuracy of the Central Limit Theorem a little further, consider Example 4.4.3. We can find the exact answer by evaluating Equation (4.4.4). This value is 0.0571. Thus the approximate value 0.0446 is fairly good in terms of absolute error, which is only 0.0125, although not quite so good in terms of the relative error 0.0125/0.0571, which is about 22% (of course, the relative error will be much smaller for the larger probabilities, e.g. $P(T \leq 20)$). The approximation can be improved by using the **correction for continuity** in Exercise 4.29; the approximate value then becomes 0.0606, which is a relative error of only about 6%.

We have been discussing the Central Limit Theorem in its simplest form, that of sums of independent and identically distributed terms. There are versions of the theorem for some more general situations. The statements of these theorems are very complicated, so will not be discussed further here, aside from the informal statement that most sums occurring in practice are approximately normal.

4.5 PARAMETRIC FAMILIES OF MULTIVARIATE DISTRIBUTIONS

There are relatively few parametric families of multivariate distributions used commonly in practice. However, there are two such families which are quite useful: the **multinomial** distributions and the **multivariate normal** distributions. We will present these in this section.

The family of multinomial distributions is a generalization of the binomial family. Recall conditions (a)–(e) following Equation (3.3.2), which gave rise to a binomial distribution. These conditions state that we have n independent trials, each of which has two possible outcomes. Let us refer to these two outcomes, which were called "success" and "failure" at that time, as Outcome 1 and Outcome 2. In Section 3.3 we were interested in the distribution of the total numbers of occurrences of Outcomes 1 and 2 out of the n trials (although we concentrated only on Outcome 1, since the number of occurrences of Outcome 2 could be obtained easily by subtraction from n).

In the case of a multinomial distribution, we have r possible outcomes, rather than just two. Let T_i denote the total number of occurrences of Outcome i in the n trials, and let p_i be the probability of that outcome on any particular trial. We then say that the following vector has a multinomial distribution.

$$T = \begin{bmatrix} T_1 \\ \cdot \\ \cdot \\ \cdot \\ T_r \end{bmatrix}$$

For example, let X, n and p be as in conditions (a)–(e) following Equation (3.3.2). Then we have a multinomial distribution with $r = 2$,

$$T = \begin{bmatrix} X \\ n - X \end{bmatrix}$$

and

$$\begin{bmatrix} p_1 \\ p_2 \end{bmatrix} = \begin{bmatrix} p \\ 1 - p \end{bmatrix}$$

We can derive the r-variate probability mass function of T in a way similar to that which led to Equation (3.3.1). (You may wish to review Example 2.3.4 before continuing.) It will be much easier if we use a specific example to guide our thinking:

Example 4.5.1

Suppose we roll a fair die five times. Let us find the probability that we obtain one 2, two 3s and two 6s—the probability that

$$T = \begin{bmatrix} 0 \\ 1 \\ 2 \\ 0 \\ 0 \\ 2 \end{bmatrix}$$

First, note that the probability of any individual *sequence* of outcomes (e.g., 1, 6, 4, 4, 2) is $(1/6)^5$. Thus, we need only determine the total number of orderings possible in which one 2, two 3s and two 6s occur. Proceeding as in Example 2.3.4, we have five slots to fill with these objects. There are $\binom{5}{1} = 5$ ways to choose the position for our one 2. Once that position is chosen, there will be $\binom{4}{2} = 6$ ways to choose the positions for our two 3s. Finally, there will be $\binom{2}{2}$ ways to place our two 6s.

Thus, the total number of orderings possible is

$$\binom{5}{1}\binom{4}{2}\binom{2}{2}$$

which after some algebra turns out to be

$$\frac{5!}{1!2!2!}$$

Thus, the probability we are seeking is

$$\frac{5!}{1!2!2!}\left(\frac{1}{6}\right)^5 = 0.004$$

Generalizing the results of the last example, we have

$$P(T_1 = a_1, \ldots, T_r = a_r) = \frac{n!}{a_1! \cdots a_r!} p_1^{a_1} \cdots p_r^{a_r} \tag{4.5.1}$$

Note that the components T_i are not independent. For example, in the context of Example 4.5.1, if we know that $T_5 = 4$, then we know that T_2 is at most 1. The covariances $\mathrm{Cov}(T_i, T_j)$ are derived in Exercise 4.21.

For the next parametric family of multivariate distributions, we need to review some matrix notation. First, recall that if we have square matrices A and B such that $AB = BA = I$ (where I is the identity matrix, with its diagonal elements all equal to one, and its off-diagonal elements all zero), then we write $B = A^{-1}$. Also, recall that for any $m \times n$ matrix C, the transpose of C, denoted C^t, is the $n \times m$ matrix obtained by interchanging rows and columns of C (i.e., the ith row of C becomes the ith column of C, $i = 1, \ldots, m$).

The family of multivariate normal distributions has great practical value. Its joint density function has the form

$$f_{X_1, \ldots, X_p}(s_1, \ldots, s_p) = \frac{1}{(2\pi)^{p/2}\sqrt{\det(\Sigma)}} \exp\left[-\frac{1}{2}(s - \mu)^t \Sigma^{-1} (s - \mu)\right] \tag{4.5.2}$$

where

$$s = \begin{pmatrix} s_1 \\ \cdot \\ \cdot \\ \cdot \\ s_p \end{pmatrix} \text{ and } \mu = \begin{pmatrix} \mu_1 \\ \cdot \\ \cdot \\ \cdot \\ \mu_p \end{pmatrix}$$

and t denotes matrix transpose. The $p \times 1$ vector μ and the $p \times p$ positive definite matrix Σ are parameters for the family.

The density in Equation (4.5.2) looks quite formidable, but is actually a natural generalization of the ordinary univariate normal density that we have used before. The vector μ consists simply of the means of the individual components; that is, $\mu_i = EX_i$. The matrix Σ is called the **covariance matrix**, because its (i, j) element is $\text{Cov}(X_i, X_j)$. Note, by the way, that for any random variable W, $\text{Cov}(W,W)=\text{Var}(W)$ (just check the definitions to verify), so we see that the diagonal elements of Σ are the variances of the components X_i.

Linearity Property of Multivariate Normal Distributions

If $(X_1,\ldots,X_p)$ has the density (4.5.2), then any linear combination $L = a_1X_1 + \cdots + a_pX_p$ has a univariate normal distribution. Furthermore,

$$EL = a_1\mu_1 + \cdots + a_p\mu_p \qquad \textbf{(4.5.3a)}$$

and

$$\text{Var}(L) = a^t\Sigma a \qquad \textbf{(4.5.3b)}$$

where $a = (a_1,\ldots,a_p)^t$

Example 4.5.2

Suppose a team of environmental engineers and public health specialists have found that air conditions are unhealthful if $1.56X_1 + 2.12X_2 > 10.62$, where X_1 and X_2 are the concentrations of ozone and sulfur. Suppose also that in a certain region, the vector (X_1, X_2) has a bivariate normal distribution with mean (3.10,1.14) and covariance matrix

$$\begin{pmatrix} 1.61 & 0.63 \\ 0.63 & 0.82 \end{pmatrix}$$

We can use this information to find the proportion of days in which air conditions are unhealthful. Let $L = 1.56X_1 + 2.12X_2$. Then from Equations (4.5.3), $EL = (1.56)(3.10) + (2.12)(1.14) = 7.25$, and

$$\text{Var}(L) = (1.56 \;\; 2.12)\begin{pmatrix} 1.61 & 0.63 \\ 0.63 & 0.82 \end{pmatrix}\begin{pmatrix} 1.56 \\ 2.12 \end{pmatrix} = 11.77$$

Thus

$$P(L > 10.63) = P\left(\frac{L - 7.25}{\sqrt{11.77}} > \frac{10.63 - 7.25}{\sqrt{11.77}}\right) = 0.1635$$

from Appendix B. Thus, about 16% of the days will have severe pollution problems.

Just as the Central Limit Theorem makes the univariate normal distribution so useful, there also is a Multivariate Central Limit Theorem which similarly makes Equation (4.5.2) useful. Here is an informal statement of that theorem.

Multivariate Central Limit Theorem

> Let $X_1, \ldots, X_n$ be independent, identically distributed random *vectors*, each having mean μ and covariance matrix Σ, and let $T = X_1 + \cdots + X_n$. Then T has an approximate multivariate normal distribution with mean $n\mu$ and covariance matrix $n\Sigma$.

Example 4.5.3

Let $T = (T_1, T_2)^t$ be the total daily sales of a computing equipment manufacturer, where T_1 is the hardware total and T_2 is the software total, and let $X_i = (X_{i1}, X_{i2})^t$ represent the hardware and software sales for the ith of the n customers. Then since

$$T = \sum_{i=1}^{n} X_i$$

the multivariate Central Limit Theorem says that T has an approximate bivariate normal distribution.

Example 4.5.4

A multinomial random vector T can be expressed as a sum, just as was done for the binomial case in Equation (4.2.19): Consider n repetitions of an experiment with r possible outcomes. Let X_{ij} be either 1 or 0, according to whether the ith repetition results in the outcome of type j, $i = 1, \ldots, n$; $j = 1, \ldots, r$. Define the vector

$$X_i = \begin{bmatrix} X_{i1} \\ X_{i2} \\ \cdots \\ X_{ir} \end{bmatrix} \tag{4.5.4}$$

Then for T_j and T described at the beginning of this section, we have

$$T_j = \sum_{i=1}^{n} X_{ij} \tag{4.5.5}$$

and

$$T = \sum_{i=1}^{n} X_i \tag{4.5.6}$$

This representation of T as a sum allows us to conclude that T has an approximately multivariate normal distribution. Thus, T has an approximate r-variate normal distribution, with mean $n\mu$ and covariance matrix $n\Sigma$, where μ and Σ are determined as follows:

$$\mu = EX_i = \begin{bmatrix} EX_{i1} \\ EX_{i2} \\ \cdots \\ EX_{ir} \end{bmatrix} = \begin{bmatrix} p_1 \\ p_2 \\ \cdots \\ p_r \end{bmatrix}$$

since

$$EX_{ij} = 0P(X_{ij} = 0) + 1P(X_{ij} = 1) = p_j$$

For the covariance matrix Σ, the (j,j) element is equal to $p_j(1 - p_j)$, and for $j \neq k$, the (j,k) element is $-p_j p_k$ (Exercise 4.21).

Here is an example of the usefulness of this approximation.

Example 4.5.5

Traffic going past a certain intersection consists of 54% cars, 26% trucks, and 20% buses. Among the next 100 vehicles to pass, what is the approximate probability that buses outnumber trucks?

Let T_1, T_2, and T_3 be the number of cars, trucks, and buses, respectively. For the probability in question, we should set the vector a in Equation (4.5.3b) to

$$a = \begin{bmatrix} 0 \\ -1 \\ 1 \end{bmatrix}$$

and find $P(L > 0)$. Then from Equations (4.5.3)

$$EL = (0)(100)(0.54) + (-1)(100)(0.26) + (1)(100)(0.20) = -6$$

and

$$\text{Var}(L) = [0\ \ -1\ \ 1]\begin{bmatrix}(100)(0.54)(0.46) & (100)(-0.54)(0.26) & (100)(-0.54)(0.20)\\ (100)(-0.54)(0.26) & (100)(0.26)(0.74) & (100)(-0.26)(0.20)\\ (100)(-0.54)(0.20) & (100)(-0.26)(0.20) & (100)(0.20)(0.80)\end{bmatrix}\begin{bmatrix}0\\ -1\\ 1\end{bmatrix}$$

$$= 45.64$$

Thus

$$P(L > 0) \approx P\left[Z > \frac{0-(-6)}{\sqrt{45.64}}\right] = 0.19$$

where Z has a N(0,1) distribution.

4.6 SIMULATION OF RANDOM VECTORS

Except for the case of independent random variables, there is no general method for exact simulation of random vectors. For example, there is no analog of the inverse c.d.f. method used in the univariate case in Section 3.4. Thus, usually some *ad hoc* method must be improvised.

Example 4.6.1

Suppose we wish to simulate (X,Y), having the density in Example 4.1.7. We can use the information derived in Example 4.1.9 to write a simulator for (X,Y).

First, we use the inverse c.d.f. method of Section 3.4 to generate X. To do so, we first recall from Example 4.1.7 that $f_X(s) = 2-2s$ for $0 < s < 1$, and then find the cumulative distribution function H from this. This must be done carefully (see Exercise 4.10); we find that $H^{-1}(w) = w - \sqrt{1-w}$. We then set $X = H^{-1}(U)$, where U has a $U(0,1)$ distribution.

Then we can generate Y as a uniform random variable on the interval $(0,1-X)$, since we found in Example 4.1.9 that the conditional distribution of Y given X was uniform on this interval. The result is the following program.

Program 4.6.1

```
procedure Triangle(var X,Y: real; var Seed: integer);
```

```
   var Temp: real;
begin
   Temp := random(Seed);
   X := Temp - sqrt(1 - Temp);
   Y := (1 - X)*random(Seed)
end;
```

Another example of *ad hoc* methods for simulating random vectors is that of the multivariate normal distribution. This technique requires some knowledge of linear algebra. Suppose we wish to generate random vectors that are multivariate normal with mean vector μ and covariance matrix Σ. From the theory of linear algebra, we know that there are $p \times p$ matrices P and D such that $P^tDP = \Sigma$, where $P^{-1} = P^t$ and D is a diagonal matrix (one whose off-diagonal entries are all 0).

Let $Y_1, \ldots, Y_p$ be independent univariate normal random variables, with Y_i having mean 0 and variance equal to d_{ii}, the (i, i) element of D. Let $Y = (Y_1, \ldots, Y_p)^t$. Then Y has a multivariate distribution, with mean equal to the $p \times 1$ zero vector, and covariance matrix D. Define the $p \times 1$ vector X to be $P^tY + \mu$. Then from the linearity property (actually, an extension of the version given in Section 4.5), it can be shown that X too has a multivariate normal distribution, with mean vector equal to μ and covariance matrix Σ.

In this way, we can generate the desired X. We simply generate the Y_i in the usual way (Chapter 3) and then generate X from Y, by $X = P^tY + \mu$.

FURTHER READING

Richard Johnson and Dean Wichern, *Applied Multivariate Statistical Analysis*, Prentice-Hall, 1982. One of the few books on multivariate analysis that does not limit itself to the multivariate normal distribution. Requires material from Chapters 5, 6, and 7 of this text.

Sheldon Ross, *A First Course in Probability*, Macmillan, 1976. A deeper, more theoretical treatment of probability modeling.

Kishor Trivedi, *Probability and Statistics, with Reliability, Queueing and Computer Science Applications*, Prentice-Hall, 1982. Many interesting applications. Excellent treatment of conditional expectation (most books do not cover this area much), and of generating function/transform methods.

EXERCISES

4.1 (M, S) Review the setting of Example 4.1.4, involving lateness of buses.

a. Find the mean waiting time EW of the person arriving on Line B. *Hints*: Note that W is a function of X and Y. Keep in mind that the waiting time is 0 if this person arrives before the one coming on Line A. Then use Equation (4.1.7).

b. Find the correlation between X and Y, the arrival times.

4.2 Example 4.1.6 concerns seek time for a computer disk drive. Let us continue to represent the position of the read/write head as a variable taking on values in the range [0,1].

a. Fill in the blank, and justify your answer: In that example, we assumed the joint density of X and Y to have the constant value 1 over the unit square. This would imply that X and Y are

__________.

b. (S) Find the proportion of seeks that involve a distance of less than 0.4, that is $P(|X - Y| < 0.4)$.

c. (M) Review the unnumbered example leading up to Equation (4.3.3), and then find the density of $|X - Y|$. Verify your answer by using it to compute the probability in (b).

4.3 (M) In Example 4.2.5, we were able to find the mean and variance of a binomially distributed random variable X, by expressing X as a sum. We also used this representation of X as a sum to apply the Central Limit Theorem in Example 4.4.3. In this problem, we will do similar things for the negative binomial distribution.

a. Review the definition of the negative binomial family in Section 3.3, and then express such a random variable as a sum of geometrically distributed terms.

b. (M) Derive the mean and variance for this distribution, in analogy with Example 4.2.5.

c. (M) Use the Central Limit Theorem to get an approximate value for the probability in Exercise 3.31. Check the accuracy of this approximation by finding the exact value.

4.4 (S) Modify Program 4.2.1, to find the mean *number* of collisions arising from NStns waiting stations.

4.5 (M) Prove Equation (4.1.15b), for the case in which Y is also continuous. (*Hint*: Just apply definitions to both sides of the equation.) Then use Equations (4.1.15) to prove Equation (4.1.17).

4.6 (M) This problem continues Example 4.4.3. The manufacturer wants to package the parts in boxes of $n > 12$ parts, in order to have a probability of 98% that any one box contains at least 12 good parts. Find the minimum n (*approximately*), using the Central Limit Theorem.

4.7 (M) In Example 4.3.1 and its continuation near Equation (4.3.2), suppose the backup battery is cheaper and comes from a population having mean lifetime of only 25 hours. Find f_W, EW and $Var(W)$.

4.8 (M, S) In Example 4.1.3, find $P(Y = b|X = 1)$.

4.9 Suppose radioactive particles are emitted according to a Poisson process with parameter 5.0. Then the number of particles emitted during the time interval $[0,u]$ has the probability function

$$p_X(s) = \frac{e^{-5.0u}(5.0u)^s}{s!} \quad (s = 0, 1, 2, \ldots)$$

However, suppose that 10% of all particles emitted are missed by our recording instrument. Let Y denote the number of particles actually registered.

a. (M) Mimic Example 4.3.9 to show that Y has a Poisson distribution with mean 4.5. *Hint*: Think of each arriving particle as an "arriving order," and think of the number of particles *recorded* when a particle arrives (0 or 1) as the number of "items" in the order.

b. (M, S) Find $P(4 \leq Y \leq 7)$.

4.10 (M) In Example 4.6.1, show that $H^{-1}(w) = w - \sqrt{1-w}$ for $0 < w < 1$. *Hint*: You will need to solve a quadratic equation, which will have two roots. One of them is extraneous.

4.11 (M) In Example 4.3.9, suppose the number of items in an order is either 1, 2, 3, or 4, with probability 0.25 each. Use Equation (4.3.18) on Equation (4.3.20) to find the first few values of $P(T=k)$. Verify by computing $P(T=k)$ directly (recall the idea to break big events into small events in Chapter 2).

4.12 (M, S) In Example 4.1.2 concerning the rolling of two dice, find the correlation between the sum S and the product Z.

4.13 (M, S) In Example 4.2.4 concerning the averaging of two measurements, find the correlation between the averaged measurement U and the first individual measurement, V_1.

4.14 (M) Derive Equation (4.2.2), $\mathrm{Cov}(X,Y) = E(XY) - (EX)(EY)$.

4.15 (M) Derive Equations (4.2.15) and (4.2.16): Show that for independent random variables X and Y, we have $p_{X,Y} = p_X p_Y$ in the discrete case, and $f_{X,Y} = f_X f_Y$ in the continuous case. *Hint* for Equation (4.2.15):

$$f_{X,Y}(s,t) = \frac{\partial^2}{\partial s \, \partial t} P(X \leq s \text{ and } Y \leq t)$$

in analogy with the one-dimensional case, in which

$$f_W(t) = \frac{d}{dt} P(W \leq t)$$

4.16 (M) Using Equation (4.3.4), derive Equation (4.3.5): Show that for independent, continuous, nonnegative random variables X and Y, we have

$$f_W(t) = \int_0^t f_X(r) f_Y(t-r)dr$$

4.17 (M) Show that for a nonnegative integer-valued random variable X,

$$EX = g_X'(1)$$

where g_X is the generating function of X. How can Var(X) be obtained from g_X?

4.18 (S) Suppose we are interested in finding the expected value of some random variable Y, and that we know the density functions f_X and $h_{Y|X}$ for some other variable X. Make a program template, similar to that of Program 3.2.1, which could be used to find EY.

4.19 (M) Derive Equations (4.2.11) and (4.2.12).

4.20 (M) Derive a formula similar to that of Equation (4.3.5) for the product XY of positive random variables X and Y.

4.21 (M) Show that the covariance matrix of a multinomially distributed random vector with $n=1$ has as its (j,j) element $p_j(1 - p_j)$, and as its (j,k) element $(j \neq k)$ $-p_j p_k$.

4.22 (M) Argue that an Erlang random variable Y (discussed in Section 3.3) acts like a constant if n is large. First do this on an intuitive level, based on the notion that a random variable whose standard deviation is small relative to its mean is in some sense close to being a constant. Then back up your argument by using the Central Limit Theorem to evaluate

$$\lim_{n\to\infty} P(|Y-EY| > 0.0001EY)$$

(of course, the 0.0001 can be replaced by even smaller numbers).

4.23 (S) Some science museums have a dramatic display of the Central Limit Theorem, arranged in the following way: They have m marbles (or billiard balls) in a chute, above an array of r rows of pegs (depicted for $r=4$). When the chute is opened, the marbles bounce down the array of pegs.

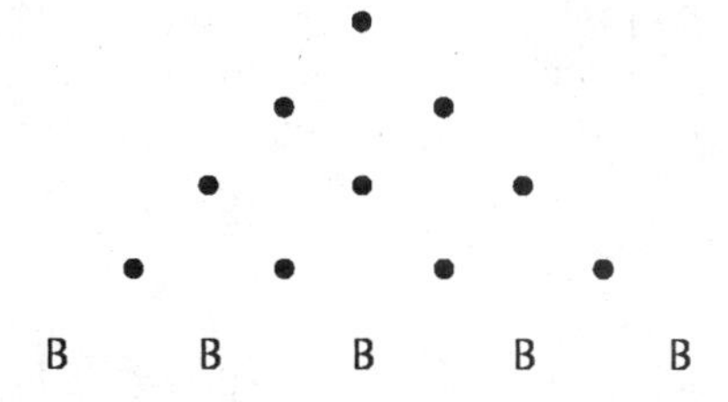

The opening of the chute is located directly above the top peg, so a marble will bounce off that peg to the left or right with probability

0.5 each. Then the marble will bounce to the left or right from a peg in the second row, with probability 0.5 each, and so on. The marbles are collected in $r+1$ bins beneath the bottom row (denoted by *B*s in the picture above). Each bin is the width of one marble, so the marbles in any bin stack up in a vertical column. After all the marbles have fallen, the pattern of the collected marbles stacked in the bins turns out to be bell-shaped!

Write a program to simulate this process, including printer or terminal screen output, with stars representing the marbles. Then carefully explain the connection with the Central Limit Theorem here.

4.24 (M) In a certain process for manufacturing metal rods, the finished product sometimes breaks under test. The metal rods are one foot long, and the breaks occur randomly along that one-foot length (i.e., the breakpoint position has a $U(0,1)$ distribution). Suppose two pieces are chosen at random from a collection of broken rods. Let the lengths of the two pieces chosen be X and Y. Suppose they are to be welded together to produce a rod of length $X+Y$. Find the density of $X+Y$. *Hint*: In deriving $f_{X+Y}(t)$, consider the cases $0 < t < 1$ and $1 < t < 2$ separately.

4.25 (S) Before the Box-Mueller method became widely used, it was common to generate a $N(0,1)$ random variate by first finding

$$U = U_1 + \cdots + U_{12}$$

where $U_1, \ldots, U_{12}$ are the results of 12 successive calls to a $U(0,1)$ generator, and then applying a certain transformation.

a. State what transformation is applied, and explain the motivation behind the entire process, including the choice of the number 12.

b. Investigate this method with respect to accuracy and computation time, and compare the results to those for the Box-Mueller method.

4.26 (M) Suppose we wish to compute the sum

$$\sqrt{a_1} + \cdots + \sqrt{a_{25}}$$

on a computer. The computation of each square root will have some roundoff error. Assume that the errors are independent with distribution $U(-0.5,0.5)$. Find a number c such that the probability is approximately 95% that the sum of square roots above is in error by no more than an amount d.

4.27 (M) Recall from Section 3.3 that a chi-square random variable has the same distribution as that of a sum of squares of independent $N(0,1)$ random variables. Thus a chi-square distribution should be approximately normal, due to the Central Limit Theorem. Find the mean and variance of this approximating normal distribution. (You will find Example 4.3.5 helpful at some point.) Then check the accuracy of the approximation by consulting the chi-square distribution values in Table 4 of Appendix B.

4.28 Consider Example 3.3.11, concerning the error in pin insertion.

a. (M) In that example, we found $P(W > 7.5)$ by exploiting the fact that $W^2/9$ had a chi-square distribution, for which we had a table. Suppose we had not had that knowledge: Set up the double integral needed for that problem, and comment on the feasibility of computing the integral.

b. (S) Suppose we wanted to find $P(X > 2$ and $Y > 2)$. Set up this integral, and comment on the feasibility of computing it. Then evaluate the probability using simulation instead.

c. (M, S) Suppose that X and Y are correlated, with bivariate normal density, and with $\mathrm{Cov}(X,Y)=2$, still with X and Y having means 0 and standard deviations 3. Also, assume that the only near edge of the surface (which is to the right and above the point P) forms the line $X+1.5Y=4$. Find the probability that the insertion misses the surface entirely. (Review Example 4.5.2 before starting.)

4.29 (M) Consider the use of the Central Limit Theorem as an approximation to a binomial random variable T, as in Example 4.4.3. The fact that we are approximating a discrete distribution by a continuous one suggests that an improvement to the approximation could be made as follows (for concreteness, assume the context of Example 4.4.3): Suppose we were interested in finding a probability of a single value of T, $P(T=4)$, for example. Temporarily acting as if T were continuous, we might convert this probability to the form $P(3.5 \leq T \leq 4.5)$, and then use the $N(0,1)$ table on this probability.

a. Compute $P(3.5 \leq T \leq 4.5)$ in this way, and check its accuracy against the exact value of $P(T = 4)$.

b. Similarly, compute $P(T \leq 11)$ as

$$\sum_{i=0}^{11} P(T=i) = \sum_{i=0}^{11} P(i-0.5 \leq T \leq i+0.5) = P(-0.5 \leq T \leq 11.5)$$

and check its accuracy, relative to both the exact value and the value obtained by applying the normal approximation to the expression $P(0 \leq T \leq 11)$.

4.30 (M, S) Consider the following game: We roll a fair die first. If the outcome is 1 or 2, we then toss a fair coin twice; otherwise we toss the coin three times. We win one dollar for each head tossed. Let X be the outcome of the toss, and Y be the number of dollars we win. Find the probability mass function p_Y of Y, and then find $\mathrm{Var}(Y)$ from it.

Next, find Var(Y) using simulation. Then find Var(Y) using the discrete analog of Equation (4.1.17). Verify that all three answers are the same.

4.31 (M, S) Recall Example 3.3.8. If we need 80 hours of use, it is clear that one battery won't be enough. How many batteries should we bring in order to have a 90% chance that we will get a total of at least 80 hours of use?

4.32 (M, S) Recall Example 3.2.5, in which Huffman coding was used to compactify data in computer databases. Suppose we have 500 letters stored in the database, with the same coding and frequencies as in the example. We are interested in the total number of bits of storage used—$B_1 + \cdots + B_{500}$ in the notation of the example. Use the Central Limit Theorem to find an approximate probability that more than 715 bits are needed. Find the actual probability using simulation.

4.33 Review Example 4.1.10, which concerns the lifetimes of a sequence of light bulbs. Let A denote the "accumulated" lifetime, the amount of time the current bulb has been burning, at time w.

a. (M) Derive f_A (it will turn out to be the same as f_Z from that example).

For the remaining parts of this problem, assume $f_X(t) = 2t$ for $0 < t < 1$ (and 0 elsewhere), and let $w = 104$. (The latter assumption is so that the process will have been running "a long time" at time w.)

b. (M, S) Use f_Z to find the probability that the next bulb installation will occur at least 0.3 time units from now (i.e., at or after time $w+0.3$), and check using simulation.

c. (M, S) Find the probability that the last installation was more than 0.1 units earlier (i.e., before time $w - 0.1$.)

d. (M, S) Find the probability that both conditions in (b) and (c) occur. *Hint* for the "M" part: A and Z are not independent. Adapt the derivation of f_Z for use here.

4.34 (S) Write a generalization of Program 3.4.5 for generating multinomial random vectors. Note that since we are generating a vector, rather than a single value, you will write a Pascal procedure rather than a Pascal function. The input parameters should consist of r, n, and an array containing $p_1, \ldots, p_r$. Test your program by using it to find the probability in Example 4.5.1. Show how, for greater efficiency, the array containing the values p_j might instead be $q_1, \ldots, q_r$, where

$$q_j = \sum_{k=1}^{j} p_k$$

CHAPTER 5

Concepts in Sampling

This chapter lays the groundwork for the material in Chapters 6 and 7, which serve as an introduction to the analysis of sample data. The concepts in this chapter form the foundation for a quite substantial portion of statistical applications. We have kept the chapter brief, to enable us to quickly get to the material in Chapters 6 and 7. You will get additional practice with sampling concepts in those chapters, so do not worry about the brevity here.

Frequently we need to make a decision about some population, based on sample data. Our task is to use the sample data to infer information concerning some population characteristic. A difficult question is involved here: How much can we extrapolate about the population from the sample data? Here are some examples:

- A company has an idea for developing a new service, and surveys 50 people as to their interest in the service. Suppose 32% of those surveyed say that they are interested in the service. However, 32% is only the *sample* proportion, whereas we are really interested in the *population* proportion, that is, the proportion of those interested among the general population of potential customers. Should we conclude that the population proportion is near 32% too?
- A tire manufacturer wants to measure the durability of one of the types of tires it produces. It gives free tires to a sample of 200 car owners, and measures tread wear after 5000 miles. How relevant is this result to the general population of car owners? In particular,

how close is the mean tread wear among these 200 cars to the overall mean tread wear that would occur if these tires were to be used in all cars?

- The computer center manager in a large company is trying to decide which of two Pascal compilers to use in the center. She tries each of the compilers on a sample of 100 user programs, and then records the run time of the resulting object codes. She finds that, *in this sample*, Compiler A produced code averaging 5.6 seconds faster than that of Compiler B. Is this value near the average difference among *all* user programs? In fact, can we even be confident that Compiler A actually is faster than B?
- In a study of air pollution, data are collected on the levels of 20 different pollutants and 35 environmental factors, such as weather variables, traffic volume, and factory activity. The correlation coefficient for each pollutant/environmental factor pair is computed on each of 15 days. Suppose it is found that the correlation between sulfur concentration and average vehicle speed is surprisingly high. Is this relationship "real"; that is, is it indicative of the long-term situation, or is it just the result of an accidental juxtaposition of data for the 15 days studied?

In each of the examples above, there is some kind of population involved. In the first example, we sample 50 people from the population of potential customers for the new service. The second and third examples are similar. However, the last example is harder to describe in terms of sampling from a population; here the population is more conceptual than real—we can think of sampling 15 days worth of data from the "population" of all daily weather conditions (note that this population is also what we referred to as the "long term" in the example). *From this point on, whenever we talk about a "population" being sampled, the reader should keep in mind that the population might only be conceptual in nature.*

Whether the population is real or conceptual, our analysis is always based on the model of sampling from some population. The analysis then is concerned with how well the sample represents the population. We are mainly interested in the accuracy of summary statistics, such as the 32% figure in the first example, the mean tread wear for the 200 cars in the second example, the value of 5.6 for the mean difference in code speed in the third example, and the 15-day correlation in the fourth example. In each case, we are interested in knowing how close these *sample* estimates are to the corresponding *population* values.

We will discuss accuracy in Chapter 6, but before doing so we need take a careful look at the sampling process itself; that is the purpose of this short chapter. First, we discuss general goals in Section 5.1. Then in Section 5.2, we will see how these goals can be described precisely—a requirement for good analysis—using the probability concepts developed in the earlier chapters. Finally, in Section 5.3, we will begin our preparations for analysis of the properties of sample estimates, by looking at the important special case of sample means.

5.1 PROBLEMS OF EXTRAPOLATING FROM SAMPLES TO POPULATIONS

Think about the tire company example above. Let $\overline{X}$ denote the mean tread wear in the sample, and let μ be the mean wear in the population. Is $\overline{X}$ an accurate estimator of μ? If not, what could go wrong, and how could we try to avoid it?

First, there is the problem of **sampling variability**. The value of $\overline{X}$ does of course vary from one sample to another, and "accidents" can happen. For example, our sample of 200 car owners might accidentally contain a large number of people whose driving habits tend to result in rapid tread wear. This would make the new brand of tires look worse than it actually is, that is, $\overline{X}-\mu < 0$. On the other hand, our sample might instead happen to contain a disproportionate number of car owners whose driving habits conserve tire life. The resulting value of $\overline{X}$ would then be overly optimistic, that is, $\overline{X}-\mu > 0$. Keep in mind that we do not know the value of μ; after all, that is why we are sampling in the first place. However, μ does exist, and we can ask ourselves "what if" questions; for example, what if our sample were accidentally to contain a large number of car owners who abuse tires?

The number of observations in the sample is called the **sample size**. For instance, in the tire example, the sample size is 200. It is intuitive that "accidents" are more likely to occur with small samples than with larger samples. In Section 5.3 this will be verified, and we will see how this is reflected in a decrease in the sampling variability of $\overline{X}$ as the sample size grows.

A second major problem is **sampling bias**. This occurs when the sampling procedure does not choose all members of the target population with equal probability. To see this problem, recall Example 3.1.3, in which we considered sampling from a large bag filled with pieces of wire of different lengths. Suppose we wish to estimate μ, the mean length of wires in the bag, and our estimator will be based on $\overline{X}$, the mean length of a sample of 50 wires drawn from the bag. Let us continue to assume, as in Example 3.1.3, that in the sampling process we are more likely to draw long wires than short ones. This of course will result in a tendency to overestimate μ, that is, a tendency for $\overline{X}-\mu$ to be greater than 0. However, this is qualitatively very different from the problem of sampling variability discussed above. Here the sampling process induces a built-in force in the direction of overestimation. This is in contrast to the tire company example, in which the sampling process does not inherently favor overestimation or underestimation, as long as all car owners are equally likely to be sampled. Another difference between these two concepts is that the problem of sampling variability can be solved by increasing the sample size, while this remedy would *not* work for a sampling bias problem.

The problem of sampling bias occurs quite frequently in applications. A famous example involves a pre-election presidential poll in the 1930s. The poll took the form of a telephone survey. At that time many households could not afford telephones, so the survey had a built-in bias toward the more affluent sections of the population. Some of the details of this incident are now being clarified by statistical historians, but the incident certainly makes it clear that sampling bias is something that must be very carefully avoided in applications.

Another source of error is **estimator bias**, similar in effect to sampling bias, but from a completely different source: the type of estimator used. At this point, it is not easy to explain this kind of bias, since the only estimator discussed so far, $\overline{X}$, is **unbiased**. However, even with equal-probability sampling, many types of estimators do include some bias, although usually small. All of these points will be treated in Chapter 6.

5.2 SAMPLING MODELS

We will first look at the process of sampling from a finite population. The example below will be frequently used. It is quite artificial, but will be very helpful and convenient in illustrating the concepts.

Example 5.2.1

Suppose we wish to estimate the mean height of people in a certain town. There are only five people in the town, to be referred to as A, B, C, D, and E. Their heights in inches are 62, 65, 70, 71 and 73; keep in mind that at the time we sample, we do not know any of these values. Suppose our sampling plan is to choose two people without replacement; we will then estimate the mean height of all five people, by using the mean of the heights of the two sampled people.

In accordance with our discussion in Section 5.1 regarding sampling bias, suppose that we sample in such a way that all members are equally likely to be chosen; each of the five residents has probability 1/5 of being chosen first, and given that, say, B was chosen first, each of A, C, D, and E has probability 1/4 of being chosen second.

The first important point is that the observations in our sample are random variables. This does not mean that the height of, say, Person B is a random quantity, changing through time. Instead, the randomness comes from the fact that our sampling is random. Let S be the height of the first person drawn in our sample, and let T be the height of the second person drawn. It may be that the first person turns out to be B, and the second E; then $S=65$ and $T=73$. However, if we repeat the experiment (our experiment is to draw two people at random from the population), then we may get Persons C and A, in which case $S = 70$ and $T = 62$. S and T change with each repetition of the experiment and thus are indeed random variables.

Since S and T are random variables, they have distributions. What are they? Well, S can take on the values 62, 65, 70, 71, and 73 with probability 1/5 each, so $p_S(62) = 1/5$, and so on. What about T? The answer is the same: for example, if we repeat the experiment over and over again, Person A will happen to be the second person chosen in 1/5 of the repetitions, so $P(T = 62) = 1/5$. Of course, there is nothing special about Person A; the same situation holds for Persons B, C, D, and E, so T takes on the values 62, 65, 70, 71, and 73 with probability 1/5 each. Thus $p_S = p_T$. [Don't confuse these *unconditional* probabilities with *conditional* probabilities. For example, while $P(T = 62)$ is equal to 1/5, the conditional probability $P(T = 62|S = 71)$ is 1/4. In other words, $p_T \neq p_{T|S}$.]

It is important to note that not only do S and T have the same distribution, but they also have the same distribution as the population. For example, $P(S = 71) = 1/5$, and the proportion of people in the population who are of height 71 is also 1/5. Then, since the distributions are the same, the means and variances are the same; for example, the mean of S is

$$\begin{aligned} ES &= \sum r p_S(r) \\ &= 62\left(\frac{1}{5}\right) + 65\left(\frac{1}{5}\right) + 70\left(\frac{1}{5}\right) + 71\left(\frac{1}{5}\right) + 73\left(\frac{1}{5}\right) \\ &= 68.2 \end{aligned}$$

while the mean of the population is

$$\mu = \frac{62 + 65 + 70 + 71 + 73}{5} = 68.2$$

Note that S and T are not independent; for example, if we know that $S = 71$, then we know definitely that $T \neq 71$. However, if we were to sample with replacement, then S and T would be independent.

We extend and summarize the above discoveries as follows:

Sampling from Finite Populations

Suppose we sample n items without replacement from a population of size N, and that at the k-th of the n sampling stages, each of the $N - (k - 1)$ units not yet chosen has probability $1/[N - (k - 1)]$ of being selected ($k = 1, \ldots, n$). The collection of n data points $X_1, \ldots, X_n$ is called a **simple random sample of size *n*.**

Let X_k be the value of the unit selected at the k-th stage. Then we can generalize what we found for the case $n = 2$, $N = 5$ above to the following properties:

a. $X_1, \ldots, X_n$ are random variables.

b. The unconditional distribution of X_k is the same as the distribution of the population ($k = 1, \ldots, n$).

c. $E(X_k) = \mu$ and $\text{Var}(X_k) = \sigma^2$ ($k = 1, \ldots, n$), where μ and σ^2 are the mean and the variance of the population, respectively.

d. $X_1, \ldots, X_n$ are not independent.

Furthermore, suppose that we sample *with* replacement, and that at each stage of the sampling, each of the N units in the population has equal probability ($1/N$) of being chosen. Then (a), (b), and (c) above hold, and instead of (d), we have that $X_1, \ldots, X_n$ are independent. The set of n data points in this situation is called a **random sample of size *n*.**

Most sampling from finite populations is without replacement. However, except for very advanced and specialized sampling methods, most statistical analysis tools still assume that the sampling is with replacement. There are several reasons for this. First, the resulting independence of $X_1, \ldots, X_n$ greatly simplifies our work. Second, in most cases n is small relative to N, so that the difference between sampling with and without replacement is negligible, since the probability of some unit being selected twice in a with-replacement sample is quite small. Third, the independence assumption results in the same formulas as used with the infinite population model described below.

Here is the structure in infinite population sampling:

Sampling in Infinite Populations

The sample data $X_1, \ldots, X_n$ have the following properties:

a. $X_1, \ldots, X_n$ are random variables.

b. The density or probability mass function of each X_k is the same as that of the population.

c. The mean and variance of each X_k are the same as those of the population.

d. $X_1, \ldots, X_n$ are independent.

Again, the set of n data points is called a random sample of size n.

So, the only real difference between the two sampling models lies in Property (d). Note, by the way, that (d) is really an assumption, which usually holds, but occasionally fails. For example, suppose we are making measurements on tree leaves. Our sample may have 400 data points in it, collected as follows: Twenty trees are selected at random from the forest under investigation, and twenty leaves are sampled from each tree. Then since leaves from the same tree have something in common, we have, for example, that X_3 and X_{19} are not independent, coming from the same tree, although X_3 and X_{21} are independent—X_3 comes from the first tree sampled, and X_{21} comes from the second tree sampled. In such a situaion, (d) does not hold, and more advanced methods must be used in the data analysis.

Sampling theory can be quite involved, especially for what is called **survey sampling**. For more information, see the text by Cochran listed in the references at the end of this chapter.

5.3 SMALL AND LARGE SAMPLE DISTRIBUTIONS OF SAMPLE MEANS

A large number of statistical applications involve some kind of mean. Thus, the distributional theory for sample means developed in this section is quite important, and will be used heavily in Chapters 6 and 7.

Using the random variable structure established in the last section, we can now more precisely describe the concept of sampling variability discussed in Section 5.1.

As before, let $\overline{X}$ denote the sample mean

$$\frac{X_1 + \cdots + X_n}{n} \tag{5.3.1}$$

Since each X_k is a random variable, $\overline{X}$ is also a random variable. This is a very important concept, but it is really simple: Since our "experiment" consists of collecting our random sample of size n, to say that $\overline{X}$ is a random variable is simply to say that if we repeatedly take random samples of size n, the value of $\overline{X}$ will vary from sample to sample.

Since $\overline{X}$ is a random variable, we can talk about its distribution. Again, in the context of repeated sampling, the distribution of $\overline{X}$ is nothing more than a description of *how* $\overline{X}$ varies from one sample to the next. For example, it can tell us how often $\overline{X}$ takes on values in some given range.

Example 5.3.1

Recall Example 5.2.1. There n was equal to 2, and, using the notation following that example, $\overline{X} = (S + T)/2$. There are (5)(4) = 20 possible samples (5 choices for S, and then, once S is chosen, 4 remaining choices for T). This means that there are 10 possible values for $\overline{X}$, since each value appears in 2 of the 20 samples; for example, the two samples $(S, T) = (65,71)$ and $(S, T) = (71,65)$ both give $\overline{X} = 68.0$. The 10 values are 63.5, 66.0, 66.5, 67.5, 67.5, 68.0, 69.0, 70.5, 71.5, and 72.0. The value 67.5 still appears twice, from $(65+70)/2$ and $(62+73)/2$. Thus the probability mass function of $\overline{X}$ is $p(t) = 1/10$ for t = 63.5, 66.0, 66.5, 68.0, 69.0, 70.5, 71.5, and 72.0, $p(67.5) = 2/10 = 1/5$, and $p(t) = 0$ for any other t.

Of course, the above example is highly artificial. However, it serves a purpose: It makes concrete the notion that $\overline{X}$ is a random variable, with its own distribution. The distribution will be a central concept in the next two chapters. For example, it will allow us to discuss such topics as the accuracy of $\overline{X}$ as an estimator of μ. This section will thus be devoted to presenting important technical material about the distribution of $\overline{X}$.

Note also that in this example, since we know all of the values in the population, we can compute μ; the value is 68.2. Since we know the value of μ here, there is no need for us to estimate it. Thus, the description of $\overline{X}$ as an estimator of μ may seem strange to you. However, it does make sense to someone who does *not* know the value of μ. What we are asking is the following: *Suppose someone without knowledge of μ were to sample two observations from this population, in order to estimate μ using $\overline{X}$. How well would this person do? For example, what would be the chance of $\overline{X}$ being in error by more than 1.0 unit?*

We ask this question because we ourselves are usually in the situation of the person above, as in the four examples at the beginning of this chapter. In each of those examples, the analyst does not know about any population quantities, but rather only knows about quantities computed from a sample drawn from the population.

As a quick introduction to the analysis below, let us answer the question posed above. In Example 5.2.1, what is the probability that $\overline{X}$ is in error by more than 1.0 unit? Since we evaluated the distribution of $\overline{X}$ previously (Example 5.3.1), answering this question is quite straightforward:

$$\begin{aligned} P(|\overline{X} - \mu| > 1.0) &= 1 - P(|\overline{X} - \mu| \leq 1.0) \\ &= 1 - P(\overline{X} = 67.5 \text{ or } \overline{X} = 68.0 \text{ or } \overline{X} = 69.0) \\ &= 1 - \left(\frac{2}{10} + \frac{1}{10} + \frac{1}{10}\right) \\ &= 0.6 \end{aligned}$$

Thus we will have a 60% chance of being in error by more than 1.0 inch. Viewed in another way, if we were to repeat this experiment again and again, each time computing our estimate $\overline{X}$ of μ, then 60% of the time our estimate would be in error by more than 1.0 inch.

Next, we look at two particular aspects of the distribution of $\overline{X}$, namely its expected value and variance. Here we will draw heavily on Equations (3.2.15), (3.2.16), and (3.2.17), as well as the fact that each sample observation X_k has the same mean μ and variance σ^2 as the population being sampled (the reader may wish to review these facts before continuing).

To find $E(\overline{X})$, write

$$\begin{aligned} E(\overline{X}) &= E\left[\frac{X_1 + \cdots + X_n}{n}\right] \\ &= \frac{1}{n}[EX_1 + \cdots + EX_n] \\ &= \frac{1}{n}(\mu + \cdots + \mu) \\ &= \mu \end{aligned} \qquad \textbf{(5.3.2)}$$

Thus the expected value of $\overline{X}$ is the population mean μ.

We will find the variance in the case of the sample data $X_1, \ldots, X_n$ being independent, which will be the main situation considered in this book [note that we did not need to assume independence in Equation (5.3.2)]. Recall from Equation (4.2.18) that variances add for independent random variables. Thus:

$$\begin{aligned} \text{Var}(\overline{X}) &= \text{Var}\left[\frac{X_1 + \cdots + X_n}{n}\right] \\ &= \frac{1}{n^2}\text{Var}(X_1 + \cdots + X_n) \\ &= \frac{1}{n^2}[\text{Var}(X_1) + \cdots + \text{Var}(X_n)] \end{aligned}$$

$$= \frac{1}{n^2}(n\sigma^2)$$

$$= \frac{\sigma^2}{n} \tag{5.3.3}$$

Thus the variance of $\overline{X}$ is the population variance divided by the sample size n.

From our original definition of variance (Section 3.2), we have

$$\text{Var}(\overline{X}) = E[(\overline{X} - E\overline{X})^2]$$

Now substituting Equations (5.3.2) and (5.3.3) into this equation, we see that

$$E[(\overline{X} - \mu)^2] = \frac{\sigma^2}{n}$$

This says that for large n, $\overline{X} - \mu$ is "usually" small, that is, $\overline{X}$ tends to stay closer to μ. This confirms our intuitive idea that $\overline{X}$ should be more accurate for larger n. However, keep in mind that this is true only in a probabilistic sense; it still is possible—though less likely—for $\overline{X}$ to have a substantial error even though n is large, because it is always possible for "accidents" to occur, no matter how large n is. The point is that with a larger n, these accidents have a much lower probability of occurring. This is intuitively clear, and Equation (5.3.3) used with Chebychev's inequality (Exercise 3.40) verifies it: for any potential error size ϵ,

$$P(|\overline{X} - \mu| > \epsilon) \leq \frac{\text{Var}(\overline{X})}{\epsilon^2} = \frac{\sigma^2}{n\epsilon^2}$$

so that

$$\lim_{n \to \infty} P(|\overline{X} - \mu| > \epsilon) = 0$$

In other words, the probability of an error occurring of size greater than any given tolerance level goes to 0 as the sample size goes to infinity.

Equation (5.3.3) and the discussion following it are of absolutely crucial importance, and clear up an extremely common misconception. Suppose we have a finite population, say the population of a city, state, or nation. Let N be the size of the population. As mentioned before, unless N is extremely small, the difference between sampling with and without replacement is quite negligible. Thus we may assume sampling with replacement, from which we get Equation (5.3.3), which says that the variance of $\overline{X}$—which is a measure of the accuracy of $\overline{X}$ as an estimator of μ—depends only on the sample size n, *not on the population size* N.

Thus, for example, a sample of only $n = 1000$ people in a national opinion poll in which N could be more than 100,000,000 may be quite satisfactory—even though only 0.001% of the population is being polled!

This tends to cause quite a bit of confusion (and skepticism) among the consumers of such information, who feel that the ratio n/N needs to be fairly substantial to provide any accuracy in $\overline{X}$. This confusion arises out of ignorance of the fact that n is what is important here, not n/N. (Of course, people's opinions change over the course of time, so the results of an opinion poll in August may or may not tell us what will happen in November. Thus there is a chance for error in this sense, but the point is that n/N is not the source of "error" if this happens.)

A similar situation occurs after the election itself, when television reports can usually "announce" the winner of the election a few minutes after the polls close, even though only a small fraction of the votes have been counted. Viewers of these announcements react with amazement, and even suspicion. Here again, though, the fraction n/N is irrelevant, with n being the only quantity which affects accuracy.

The idea that $\mathrm{Var}(\overline{X})$ represents the accuracy of $\overline{X}$ as an estimator of μ is made even clearer by the Central Limit Theorem, as we will see in the following developments.

In Example 5.3.1, we found the distribution of $\overline{X}$ "the hard way," by enumerating all the possibilities; this clearly is infeasible for finite populations with large N, and impossible for infinite populations with continuous X. Thus, we need a better way to find the distribution of $\overline{X}$.

Since $\overline{X}$ is basically a sum of random variables, we can draw upon the material we developed in Sections 4.3 and 4.4. An important special case is that in which the sampled population has a normal distribution.

Example 5.3.2

Suppose that the heights of residents of a certain large city are normally distributed, with mean 68 inches and standard deviation 3 inches. Of course this represents an idealization. For example, any normal distribution extends from $-\infty$ to ∞, yet nobody is going to have a negative height or a height of 5000 inches. However, empirical data on height frequencies have been found to be well approximated by such a curve.

Getting back to the question of the distribution of $\overline{X}$, suppose $n = 10$. Since the observations X_i have normal distributions, and since we know from Section 4.3 that sums of independent normal random variables are themselves normally distributed, we then know that $T = X_1 + \cdots + X_n$ has a normal distribution. Then from the linearity property of the normal family, we know that $\overline{X} = (1/n)T$ has a normal distribution too. From Equations (5.3.2) and (5.3.3), we know that the mean and variance of this distribution are 68 and $3^2/10 = 0.9$. In summary, $\overline{X}$ has a $N(68,0.9)$ distribution.

The densities of X and $\overline{X}$ are drawn in Figure 5.1. Note that both are centered around 68, but the density of $\overline{X}$ is much narrower (due to its smaller standard deviation), and much taller (since the total area must still be 1.0).

The density of $\overline{X}$ is quite useful. For example, the probability that $\overline{X}$ is in error (in estimating μ) by more than 1 inch is

$$\begin{aligned} P(|\overline{X} - \mu| > 1) &= 1 - P(|\overline{X} - \mu| \leq 1) \\ &= 1 - P(-1 \leq \overline{X} - \mu \leq 1) \\ &= 1 - P\left(\frac{-1}{0.95} \leq \frac{\overline{X} - \mu}{0.95} \leq \frac{1}{0.95}\right) \\ &= 1 - P(-1.05 \leq Z \leq 1.05) \\ &= 0.29 \end{aligned}$$

Thus we have almost a 30% chance of erring more than an inch in our estimate of μ. If this is unacceptable, we will need a larger sample.

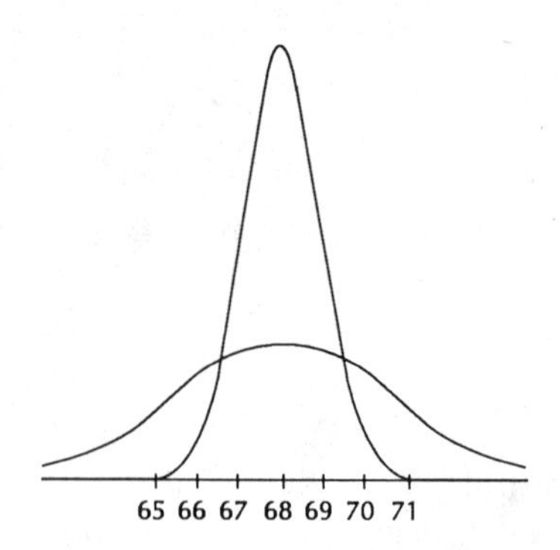

Figure 5.1 Distributions of X and $\overline{X}$ for Example 5.3.2.

In the preceding example, we treated a case in which the population distribution was known to be normal. In the cases in which this is not true, the exact distribution of $\overline{X}$ is difficult to obtain. However, from Section 4.4, we know that we can get a good approximation using the Central Limit Theorem. This approximation is used extremely frequently, and does in fact play a "Central" role in statistics.

Example 5.3.3

Suppose in Example 5.3.2 it was known that while the mean and standard deviation of the population were 68 and 3, the distribution was not normal. Even so, the central limit theorem *still* says that the distribution of $\overline{X}$ would be approximately normal, with the mean and variance found above.

In general, we have the following:

Large-Sample Distribution of the Sample Mean

Let $X_1, \ldots, X_n$ be a random sample from a population with mean μ and variance σ^2. Then the distribution of $\overline{X}$ is approximately normal, with mean μ and variance σ^2/n.

Also, the distribution of the quantity

$$\frac{\overline{X} - \mu}{\sigma/\sqrt{n}} \tag{5.3.4}$$

is approximately $N(0,1)$.

FURTHER READING

William Cochran, *Sampling Techniques*, John Wiley, 1977.

EXERCISES

5.1 (M, S) In Examples 5.2.1 and 5.3.1, compute the following:

a. The probability that $\overline{X}$ underestimates μ by more than 0.5 unit.

b. The mean absolute error $E[|\overline{X} - \mu|]$.

5.2 (M, S) In Example 5.3.2, find the probability requested in Problem 5.1a above. Also, find the increase in sample size needed (from n=10) so that this probability is only 0.05.

5.3 (M, S) Suppose we have a coin, whose probability of heads p is unknown to us. We estimate p by tossing the coin 100 times and taking as our estimator the quantity $X/100$, where X is the number of heads we obtain in the 100 tosses. Suppose that, unknown to us, the true value of p is 0.45. What is the approximate probability that we overestimate p by more than 0.08? (Hint: Citing material from Section 4.4, justify the use of the Central Limit Theorem in this problem.)

5.4 In Section 5.3 we derived the mean and variance of $\overline{X}$ under with-replacement sampling to be μ and σ^2/n. Consider sampling without replacement, from a population of size N.

a. (M) Show that $E[\bar{X}]$ is still μ. [Hint: The derivation in (5.3.2) is still valid in this case. Why?]

b. (M, S) Show that

$$\mathrm{Var}(\bar{X}) = \frac{\sigma^2}{n} \frac{N}{N-1} \left(1 - \frac{n}{N}\right)$$

Give an intuitive reason why the variance under the no-replacement sampling should be smaller. (Note: This problem requires some fairly intricate algebra to do mathematically. A valuable alternative would be to verify these values for the population in Example 5.2.1, both mathematically and by simulation.)

CHAPTER 6

Analysis of Sample Data: Estimation

6.1 METHODS OF ESTIMATION

In our look at sampling in Chapter 5, the usual example was estimation of a mean. In this case the method of estimation is very natural and obvious: To estimate the *population* mean μ using the sample data $X_1, \ldots, X_n$, simply take the estimator to be the *sample* mean $\overline{X}$. One might call this method of estimation the analog method, in which the estimator is taken to be the sample analog of the population quantity we wish to estimate (here the sample mean is the analog of the population mean). There are many examples of this method.

Example 6.1.1

Suppose we wish to estimate the population proportion π of individuals possessing a certain attribute (as in the survey example at the beginning of Chapter 5). Define notation for the sample data as follows: For the *i*th individual in the sample, let X_i be either 1 or 0, according to whether this individual possesses the attribute of interest ($i = 1, \ldots, n$). Then the analog estimator of the population prevalence π is the sample prevalence p, that is,

$$p = \frac{\text{number of } X_i \text{ which equal 1}}{n} \tag{6.1.1}$$

Note, however, that this is really just a special case of $\overline{X}$; that is, $p = \overline{X}$ here, since the numerator in Equation (6.1.1) is the sum $X_1 + \cdots + X_n$, as in Equation (4.4.5).

Example 6.1.2

Suppose we wish to estimate a population variance σ^2. Recall that $\sigma^2 = E[(X - \mu)^2]$; that is, the population variance is the averaged squared distance from X to the population mean, averaged over the X values of all individuals in the population. The sample analog is the averaged squared distance from X to the *sample* mean, averaged over the X values of all the individuals in the *sample*:

$$S^2 = \frac{1}{n}\sum_{i=1}^{n}(X_i - \overline{X})^2 \tag{6.1.2}$$

(We will change this definition slightly in Section 6.2.)

Finding an analog estimator in the next example is possibly a bit more subtle:

Example 6.1.3

We are attending the drawing for awarding 10 prizes in a local raffle. We are wondering how many tickets were sold; let θ denote this number. We wish to estimate θ, based on our knowledge of the announced numbers of the 10 winning tickets, $X_1, \ldots, X_{10}$. How might we use this information for guessing θ?

In the notation of Chapter 5, the data set $X_1, \ldots, X_{10}$ is a simple (i.e., without-replacement) random sample of size 10 from a finite population of size $N = \theta$ (the "population" here is the set of all tickets sold). An analog estimator of θ is obtained by noting that θ is the population maximum, so that the sample analog is then the *sample* maximum,

$$M = \max(X_1, \ldots, X_{10}) \tag{6.1.3}$$

Use of analog estimators is basically an *ad hoc* approach. It would be nice to have formal methods available that could be used to generate estimators in general situations. Two commonly used such methods are moment-matching (MM) and maximum likelihood (ML):

Moment-matching (or "The Method of Moments")

Recall from Section 4.3 that the kth moment of a random variable X is defined to be $E(X^k)$. The sample version of this is

$$\frac{1}{n}\sum_{i=1}^{n} X_i^k \tag{6.1.4}$$

Suppose we have r population quantities $\theta_1, \ldots, \theta_r$ to be estimated; we wish to find estimators $T_1, \ldots, T_r$. First, we find expressions for the first r population moments; these expressions will involve $\theta_1, \ldots, \theta_r$. We substitute T_i for θ_i ($i = 1, \ldots, r$) in these expressions, equate the r resulting expressions to the corresponding r sample moments, and then solve for the T_i.

Example 6.1.4

Consider Example 6.1.3. The first population moment is EX. Since X takes on the values $1, \ldots, \theta$ with probability $1/\theta$ each,

$$EX = \sum_{x=1}^{\theta} x\left(\frac{1}{\theta}\right)$$

$$= \frac{\theta + 1}{2} \tag{6.1.5}$$

Thus we equate $(T + 1)/2$ to the first sample moment $\overline{X}$, and solve for T. This gives

$$T = 2\overline{X} - 1 \tag{6.1.6}$$

which is very different from (but not necessarily better or worse than) the estimator M in Equation (6.1.3).

Example 6.1.5

Consider again the problems of estimating the population mean and variance, μ and σ^2; here $r = 2$. First write down expressions for the first two population moments, in terms of the quantities that we wish to estimate, μ and σ^2:

$$EX = \mu$$

$$E(X^2) = \sigma^2 + \mu^2 \tag{6.1.7}$$

---

where we have used the fact that for any random variable W having finite variance

$$\mathrm{Var}(W) = E(W^2) - (EW)^2$$

[This is from Equation (3.2.16); this relation will be useful throughout this chapter, and should be reviewed before continuing.] Then substitute T_1 and T_2 for μ and σ^2, and equate to the sample moments:

$$\overline{X} = T_1$$

$$\frac{1}{n}\sum_{i=1}^{n} X_i^2 = T_2 + T_1^2$$

After solving for T_1 and T_2, we find that the method of moments produces the same estimators for μ and σ^2 as obtained through the analog method; that is, T_1 turns out to be $\overline{X}$ and T_2 turns out to be

$$\frac{1}{n}\sum_{i=1}^{n} X_i^2 - \overline{X}^2$$

which can easily be shown to be algebraically the same as S^2 from Equation (6.1.2):

$$\begin{aligned}\sum_{i=1}^{n}(X_i - \overline{X})^2 &= \sum_{i=1}^{n}(X_i^2 - 2\overline{X}X_i + \overline{X}^2)\\ &= \sum_{i=1}^{n} X_i^2 - 2\overline{X}\sum_{i=1}^{n} X_i + n\overline{X}^2\\ &= \sum_{i=1}^{n} X_i^2 - 2n\overline{X}^2 + n\overline{X}^2\\ &= \sum_{i=1}^{n} X_i^2 - n\overline{X}^2\end{aligned}$$

so that

$$\frac{1}{n}\sum_{i=1}^{n} X_i^2 - \overline{X}^2 = S^2$$

The Method of Maximum Likelihood (ML)

This is a completely different approach (although, as with the method of moments above, different approaches do not always produce different results). To illustrate the method, consider the following example.

Example 6.1.6

Suppose we wish to estimate the probability π of heads for a certain coin that is known to be unbalanced, based on the results of tossing the coin three times. Suppose we obtain a head on the first and third tosses, and a tail on the second. The probability of this result is $\pi^2(1 - \pi)$. Now, keep in mind that we do not know the value of π; our job is to make a reasonable guess as to the value, based on this observed data.

Would you be willing to guess that π has the value of 0.1? You shouldn't: If the probability of heads for the coin were 0.1, it would be very unlikely to get two heads among three tosses; in fact, the probability of this happening (in the order we observed, head-tail-head) for a coin having 0.1 probability of heads is only $(0.1)^2(1 - 0.1)^1 = 0.009$. Thus, although events of low probability do of course occur sometimes, it would be hard to believe that our coin has $\pi = 0.1$.

For this reason, it would seem more reasonable to take as our guess for π a value under which it *would* be fairly likely to get two heads out of the first three tosses, that is, a value of π that makes

$$\pi^2(1 - \pi) \tag{6.1.8}$$

reasonably big. In fact, we might push this philosophy to the extreme and guess π to be the value which *maximizes* expression (6.1.8). This is, then, a simple calculus problem: Set

$$0 = \frac{d}{d\pi} \pi^2(1 - \pi) \tag{6.1.9}$$

and then solve for π. The answer turns out to be 2/3, so we guess π to be 2/3 (which also happens in this case to be the estimate obtained using the analog method: we estimate the population proportion of heads, π, by the sample proportion of heads, 2/3).

Thus the ML method guesses π to be the value that would have made the observed sample data (two heads out of three) most likely to occur. By the way, it is extremely important to note that we should not interpret this to mean that 2/3 is the "most likely" or "most probable" value of π. The quantity π is not a random variable; it is a fixed (although unknown) number, so that it is nonsense to talk about the probability that π takes on one value or another.

The ML method can also be used with continuous random variables. In this case we cannot work with the probability of the observed sample data as we did above, since that probability will be zero for continuous data. However, recall that for such data the density is analogous to probability. Thus, instead of maximizing the probability of the observed data, we maximize the **likelihood**, which is formed using densities in place of probability mass functions.

To see how the likelihood function is defined, note that expression (6.1.8) has the form

$$P(X_1=a, X_2=b, X_3=c) = P(X_1=a)P(X_2=b)P(X_3=c) \tag{6.1.10}$$

where a, b, and c are equal to either 1 or 0, depending on whether the ith toss results in heads or tails; for our data, $a = 1$, $b = 0$, and $c = 1$. By analogy, for continuous data we can define the likelihood function to be the product of the density function values at each observed data point.

Example 6.1.7

Suppose we have a sample of size n from the density family $f_X(t) = rt^{r-1}(0 < t < 1)$, where r is the unknown parameter value to be estimated. Let the n observed data values be denoted by $a_1, a_2, \ldots, a_n$, where a_1 is the value we observe for X_1, and so on. Then the likelihood function is

$$f_X(a_1) \ldots f_X(a_n) = r^n(a_1 \ldots a_n)^{r-1} \tag{6.1.11}$$

In most cases, the calculus is much easier if we maximize the *logarithm* of the likelihood, rather than the likelihood itself; since $\ln(u)$ is an increasing function of u, these two maximization problems are equivalent. The log likelihood is

$$n \ln r + (r-1)\sum_{i=1}^{n} \ln(a_i) \tag{6.1.12}$$

Taking d/dr and setting the result to zero, we find that the maximizing value of r is

$$\frac{-1}{\frac{1}{n}\sum_{i=1}^{n} \ln(a_i)} \tag{6.1.13}$$

The value computed in Equation (6.1.13) is then our estimate for r, which we will denote by $\hat{r}$.

For example, suppose X is the proportional content of a certain mineral in rocks in a quarry, and that geologists have found in the past that the density of X tends to be of the form rt^{r-1}. Suppose we wish to estimate the value of r for this particular quarry, and we collect a sample of $n = 3$ rocks, finding that the percentages of the mineral in the three rocks are 27%, 62%, and 55% ($X_1 = 0.27$, $X_2 = 0.62$, and $X_3 = 0.55$). Then we find, from Equation (6.1.13), that $\hat{r} = 1.26$.

6.2 CRITERIA FOR GOOD ESTIMATION

In Section 6.1, we discussed several classes of estimators: analog, MM, and ML. There are many other classes of estimators as well. In a given problem, how can we compare the merits of competing estimators?

Of course, the answer to this question depends on one's criterion for measuring the quality of an estimator. The purpose of this section is to informally introduce some of the usual criteria used in practice. As always, one can discuss new concepts much more clearly by keeping a concrete example close at hand. The example used here will be the lottery setting from Section 6.1.3.

Example 6.2.1

Recall the two estimators for the parameter θ developed in Examples 6.1.3 and 6.1.4; denote them by W_1 and W_2, so that

$$W_1 = \max(X_1, \ldots, X_{10})$$

and

$$W_2 = 2\overline{X} - 1 = 2\left(\frac{X_1 + \cdots + X_{10}}{10}\right) - 1$$

Which estimator is better?

Recall from Section 5.2 that the sample observations are random variables. Thus estimators, which are functions of the sample data, are random variables too. This is a very important point to keep in mind, since it implies that any criteria we use must be based on some probabilistic quantity—probabilities, expected values, variances, and so on. For example, consider W_1 and W_2 above. Neither estimator is *uniformly* better than the other: In some samples, W_1 will be the value closer to θ, while in other samples W_2 will be the more accurate estimator (unfortunately, we will not know whether the particular sample *we* have is in the first category or the second, since we do not know the value of θ!). Thus we must phrase our comparison of the two estimators probabilistically, perhaps stating that one estimator is "usually" closer to θ, or that one estimator has a smaller estimation error "on the average." Many specific forms of such comparisons are possible.

(Before continuing, let us agree to make the simplifying assumption that the tickets in Example 6.1.3 are chosen *with* replacement; this will make $X_1, \ldots, X_{10}$ independent. Of course, the tickets are actually drawn *without* replacement, but if θ is reasonably large, then mathematical results obtained from this assumption will be very close approximations to the true values.)

For example, let us first try a criterion based on probability: Let us deem the better estimator to be that estimator T for which $P(|T-\theta| \leq 2)$ is larger, that is, the estimator which has a greater chance of being within 2 units of θ. This quantity can easily be computed mathematically in the case of W_1:

$$\begin{aligned} P(|W_1 - \theta| \leq 2) &= P(W_1 \geq \theta - 2) \\ &= 1 - P(W_1 < \theta - 2) \\ &= 1 - P(\max(X_1, \ldots, X_{10}) < \theta - 2) \\ &= 1 - P(X_1 < \theta - 2 \text{ and } \ldots \text{ and } X_{10} < \theta - 2) \\ &= 1 - P(X_1 < \theta - 2) \ldots P(X_{10} < \theta - 2) \\ &= 1 - \left(\frac{\theta - 3}{\theta}\right)^{10} \end{aligned} \qquad \textbf{(6.2.1)}$$

Unfortunately, as pointed out above, we do not know the value of θ, and thus cannot evaluate the expression in Equation (6.2.1) in general. Thus our analysis of W_1 will have to be based on a "what if" point of view. For example, we can ask ourselves, "What if the true value of θ were actually 50? If that were the case, what chance would W_1 have of coming within 2 units of the true value 50?" We can then ask the question for other potential values which we think θ might have, thus getting an idea of the performance of W_1 in the range of possible settings that we feel we may encounter. Substituting 50 for θ in Equation (6.2.1), we see that W_1 would have only a 46% chance of being in error by less than 2 units—a little disappointing, although with a sample size of 10, we should not expect any estimator to have really good accuracy.

Let's see whether W_2 is any better. The exact distribution of W_2 would be far too complicated to try to compute. However, recall that $\overline{X}$ is approximately normally distributed (a fact established at the end of Section 5.3). Since $W_2 = 2\overline{X} - 1$, the Linearity Property of the Normal Family implies that W_2 has an approximately normal distribution. To exploit this fact, we will need to know the mean and variance of W_2. From Equation (6.1.5), we know that

$$E\,X_i = \frac{\theta + 1}{2} \qquad \textbf{(6.2.2a)}$$

and from Equations (3.2.16) and (6.1.5), and the relation

$$\sum_{j=1}^{n} j^2 = \frac{2n^3 + 3n^2 + n}{6}$$

we have

$$\begin{aligned}\mathrm{Var}(X_i) &= E(X_i^2) - (EX_i)^2 \\ &= \sum_{j=1}^{\theta} j^2\left(\frac{1}{\theta}\right) - \left(\frac{\theta+1}{2}\right)^2 \\ &= \frac{\theta^2 - 1}{12}\end{aligned} \quad \textbf{(6.2.2b)}$$

Now we can use the Central Limit Theorem. Still considering what would happen if the actual value of θ were 50, we would have

$$E\,X_i = \frac{50+1}{2} = 25.5$$

and

$$\mathrm{Var}(X_i) = \frac{50^2 - 1}{12} = 208.25$$

Thus from the discussion summary "Large-Sample Distribution of the Sample Mean" at the end of Section 5.3, we have that $\overline{X}$ is approximately normally distributed with mean 25.5 and variance $208.25/10 = 20.825$.

Then, by the Linearity Property of the Normal Family, W_2 would also have an approximately normal distribution, with mean $2(25.5) - 1 = 50$, and variance $(2^2)20.825 = 83.3$. Thus

$$P(|W_2 - \theta| \leq 2) \approx P\left(|Z| \leq \frac{2}{\sqrt{83.3}}\right) = 0.1742 \quad \textbf{(6.2.3)}$$

so W_2 has even less chance than W_1 of coming within 2 units of the right answer. W_1 seems to be the winner, at least on this criterion.

Again, many other criteria are possible; the choice of criterion is up to the user. For example, one could look at $p = P(|W_1 - \theta| < |W_2 - \theta|)$. There was such a difference between W_1 and W_2 under our first criterion that we would guess that the probability p here is substantially larger than 0.5. Let us see if this is true.

In contrast to the techniques used above, the evaluation of p is amenable to neither exact nor approximate mathematical solutions; thus, we turn to simulation. While we are at this, we might as well compare W_1 and W_2 in another way, by finding $e_1 = E|W_1 - \theta|$ and $e_2 = E|W_2 - \theta|$:

Program 6.2.1

```
program Ex62(input,output);

   const N = 10;

      Theta = 50;

   var Rep,I,Max,Count,XI,Seed: integer;
       A1,A2,Tot1,Tot2,W1,W2,Sum: real;

begin
   Count := 0; Tot1 := 0.0; Tot2 := 0.0;  Seed := 9999;
   for Rep := 1 to 1000 do
      begin
      Sum := 0.0;  Max := 0;
      for I := 1 to N do
         begin
         XI := Theta * random(Seed) + 1;
         Sum := Sum + XI;
         if XI > Max then Max := XI
         end;
      W1 := Max;  W2 := 2.0 * Sum/N - 1.0;
      A1 := abs(W1 - Theta);  A2:=abs(W2 - Theta);
      if A1 < A2 then Count := Count + 1;
      Tot1 := Tot1 + A1;  Tot2 := Tot2 + A2
      end;
   writeln('p = ',Count/1000.0);
   writeln('e1 and e2 = ',Tot1/1000.0,Tot2/1000.0)
end.
```

The results were $p = 0.58$, $e_1 = 5.97$, and $e_2 = 7.90$. Thus, although the value of p is perhaps not as much greater than 0.50 as we might have guessed, on the whole W_1 seems to be a much better estimator than W_2.

Another criterion which is sometime used to evaluate estimators is that of **bias**. The bias of an estimator $\hat{\theta}$ of a quantity θ is defined to be $E(\hat{\theta}) - \theta$, that is, the average amount by which $\hat{\theta}$ overestimates or underestimates θ.

Example 6.2.2

Recall the estimator S^2 for σ^2 in Example 6.1.2. The bias of S^2 may be found as follows: (A side benefit of the computation below will be to give the reader practice in use of the properties of mean and variance.)

$$E(S^2) = E\left[\frac{1}{n}\sum_{i=1}^{n}(X_i - \overline{X})^2\right]$$
$$= \frac{1}{n}\sum_{i=1}^{n} E\left[(X_i - \overline{X})^2\right]$$
$$= E\left[(X_1 - \overline{X})^2\right] \qquad \textbf{(6.2.4)}$$

since

$$E\left[(X_1 - \overline{X})^2\right] = E\left[(X_2 - \overline{X})^2\right] = \cdots = E\left[(X_n - \overline{X})^2\right]$$

from symmetry.

Now

$$E\left[(X_1 - \overline{X})^2\right] = E\left(X_1^2 - 2X_1\overline{X} + \overline{X}^2\right)$$
$$= E(X_1^2) - 2E(X_1\overline{X}) + E(\overline{X}^2) \qquad \textbf{(6.2.5)}$$

since the expected value of the sum of several random variables is the sum of the individual expected values [Equation (4.2.4)].

Let us simplify Equation (6.2.5) term by term:

From Equation (3.2.16) applied to X_1, we see that the first term in Equation (6.2.5) is equal to $\mu^2 + \sigma^2$.

The second term is

$$E\left(X_1\frac{X_1 + \cdots + X_n}{n}\right) = \frac{1}{n}\left[E(X_1^2 + X_1X_2 + \cdots + X_1X_n\right]$$
$$= \frac{1}{n}\left[E(X_1^2) + E(X_1X_2) + \cdots + E(X_1X_n)\right]$$
$$= \frac{1}{n}[(\mu^2 + \sigma^2) + (n-1)\mu^2] \qquad \textbf{(6.2.6)}$$

again by Equation (3.2.16), and by Equation (4.2.17).

For the third term in Equation (6.2.5), apply Equations (3.2.16), (5.3.2), and (5.3.3) to the random variable $\overline{X}$, yielding

$$E(\overline{X}^2) = \mu^2 + \frac{\sigma^2}{n}$$

After substituting in Equation (6.2.4), we find that

$$E(S^2) = \frac{n-1}{n}\sigma^2 \qquad \textbf{(6.2.7)}$$

Thus (on the average) S^2 tends to underestimate σ^2 slightly, by a proportion of $1/n$. We say that S^2 is a **biased** estimator of σ^2.

The amount of bias is actually minor in this case (compare this *estimator* bias to the *sampling* bias discussed in Section 5.1). However, it turns out that we can easily remove the bias, by a slight adjustment, so we might as well do so:

From Equation (6.2.7) [using Equation (3.2.15)], we see that

$$E\left(\frac{n}{n-1}S^2\right) = \sigma^2 \tag{6.2.8}$$

so

$$\frac{n}{n-1}S^2 = \frac{1}{n-1}\sum_{i=1}^{n}(X_i - \overline{X})^2 \tag{6.2.9}$$

is an **unbiased** estimator of σ^2.

From here on, we will use Equation (6.2.9) as our definition of S^2, instead of Equation (6.1.2). However, the reader should not conclude from this that one should insist on unbiasedness in choosing an estimator (in fact, it is not always possible to get unbiasedness anyway; see Exercise 6.31). Although many standard unbiased estimators in statistics, including S^2 in Equation (6.2.9), have been retained for historical reasons, the modern point of view is to keep bias small but not necessarily zero. In this way, other important estimation criteria may be taken into account as well.

Using formal criteria for estimator quality is also useful in evaluating the effectiveness of specialized sampling schemes.

Example 6.2.3

Suppose a certain population with mean μ is divided into five subpopulations; for example, a city may be divided into five districts. Suppose for simplicity that all the subpopulations are of the same size. Consider two possible sampling plans:

I. Take a random sample of size 10 from the overall population, without regard to subpopulation.

II. Take a random sample of size 2 from each of the five subpopulations.

Intuitively, Plan II sounds better, more "representative," since it guarantees that we will have the same number of observations from all subpopulations. Let us investigate whether it really is better and, if so, see how much better.

Let X_i, $i = 1,2,\ldots,10$, be the observations collected under Plan I, and let Y_{ij}, $i = 1,2,\ldots,5$, $j = 1,2$, be the data gathered under Plan II (where Y_{ij} denotes the jth observation sampled from Subpopulation i). Our estimators then would be

$$\overline{X} = \frac{1}{10}\sum_{i=1}^{10} X_i$$

and

$$\overline{Y} = \frac{1}{10} \sum_{i=1}^{5} \sum_{j=1}^{2} Y_{ij}$$

Since we have an ordinary random sample, we know from Section 5.3 that $\overline{X}$ has mean μ and variance σ^2, where σ^2 is the overall population variance. Let us relate this latter quantity to the subpopulation means and variances, as follows.

Let Subpopulation i have mean μ_i ($i = 1, 2, 3, 4, 5$), and suppose that each subpopulation has variance ω^2. (Again, keep in mind that we do not know the values of these, but they do exist, and we can use them as a basis for our comparison between $\overline{X}$ and $\overline{Y}$.) Let X be a value chosen at random from the overall population, and let I denote the number of the subpopulation from which it happens to be drawn. Then using the discrete version of Equation (4.1.17), we have

$$\begin{aligned}
\sigma^2 &= \mathrm{Var}(X) \\
&= \sum_{i=1}^{5} \mathrm{Var}(X|I = i) p_I(i) + \sum_{i=1}^{5} \left[E(X|I = i) - EX\right]^2 p_I(i) \\
&= \sum_{i=1}^{5} \omega^2 \left(\frac{1}{5}\right) + \sum_{i=1}^{5} (\mu_i - \mu)^2 \left(\frac{1}{5}\right)
\end{aligned}$$

so

$$\begin{aligned}
\mathrm{Var}(\overline{X}) &= \frac{1}{10} \mathrm{Var}(X) \\
&= \frac{1}{10}\omega^2 + \frac{1}{50} \sum_{i=1}^{5} (\mu_i - \mu)^2
\end{aligned}$$

Now we can compute $E(\overline{Y})$ and $\mathrm{Var}(\overline{Y})$:

$$\begin{aligned}
E(\overline{Y}) &= E\left[\frac{1}{10} \sum_{i=1}^{5} \sum_{j=1}^{2} Y_{ij}\right] \\
&= \frac{1}{10} \sum_{i=1}^{5} \sum_{j=1}^{2} EY_{ij} \\
&= \frac{1}{10} \sum_{i=1}^{5} \sum_{j=1}^{2} \mu_i \\
&= \frac{1}{5} \sum_{i=1}^{5} \mu_i \\
&= \mu
\end{aligned}$$

with the last equality stemming from the fact that all the subpopulations are of equal size, so that the overall population mean μ is an unweighted average of the subpopulation means.

Now

$$\begin{aligned}
\operatorname{Var}(\overline{Y}) &= \operatorname{Var}\left[\frac{1}{10}\sum_{i=1}^{5}\sum_{j=1}^{2} Y_{ij}\right] \\
&= \frac{1}{100}\sum_{i=1}^{5}\sum_{j=1}^{2} \operatorname{Var}(Y_{ij}) \\
&= \frac{1}{100}\sum_{i=1}^{5}\sum_{j=1}^{2} \omega^2
\end{aligned}$$

so

$$\operatorname{Var}(\overline{Y}) = \frac{1}{10}\omega^2$$

We have found that

$$\operatorname{Var}(\overline{X}) = \frac{1}{10}\omega^2 + \frac{1}{50}\sum_{i=1}^{5}(\mu_i - \mu)^2$$

and

$$\operatorname{Var}(\overline{Y}) = \frac{1}{10}\omega^2$$

Thus $\operatorname{Var}(\overline{Y}) < \operatorname{Var}(\overline{X})$, that is (recall the definition of variance),

$$E[(\overline{Y} - E\overline{Y})^2] < E[(\overline{X} - E\overline{X})^2]$$

or

$$E[(\overline{Y} - \mu)^2] < E[(\overline{X} - \mu)^2]$$

This last inequality says that on the average, $\overline{Y}$ is closer than $\overline{X}$ to μ, confirming our intuitive feeling that $\overline{Y}$ is the better estimator. Note, moreover, that the larger the differences among the subpopulation means μ_i, the more of an advantage $\overline{Y}$ has over $\overline{X}$.

6.3 CONFIDENCE INTERVALS

So far we have been concerned with **point estimators**; one can discuss **interval estimators** as well. For example, in addition to making a statement such as, "We estimate the value of θ to be about 12.8," it is customary (and more informative) to say something like, "We estimate θ to be between 11.9 and 14.6," although we will see below that such statements must be carefully qualified.

Example 6.3.1

Suppose we are using $\overline{X}$ to estimate the mean μ of a normal distribution that is known to have variance 9; suppose also that the sample size is $n = 100$. We can derive an interval estimate as follows: From Example 5.3.2, we know that $\overline{X}$ has a normal distribution with mean μ and variance 9/100. Thus probabilities of the form

$$P\left(-t < \frac{\overline{X} - \mu}{0.3} < t\right) \tag{6.3.1}$$

can be found from Table 2 in Appendix B. In particular, we can find a value of t that makes the probability in Equation (6.3.1) equal to 0.95. A glance at Table 2 in Appendix B shows that $t = 1.96$. Thus

$$\begin{aligned} 0.95 &= P\left(-1.96 < \frac{\overline{X} - \mu}{0.3} < 1.96\right) \\ &= P(-0.59 < \overline{X} - \mu < 0.59) \\ &= P(-\overline{X} - 0.59 < -\mu < -\overline{X} + 0.59) \\ &= P(\overline{X} - 0.59 < \mu < \overline{X} + 0.59) \end{aligned} \tag{6.3.2}$$

Equation (6.3.2) says that the interval

$$(\overline{X}-0.59,\ \overline{X}+0.59) \tag{6.3.3}$$

has probability 0.95 of containing μ. It is vital to understand this statement well. Recall that the meaning of probability stems from the long-run frequency of occurrence of a particular event, in infinitely many repetitions of the experiment. In the context here, our "experiment" is to sample 100 individuals from a given population. To repeat the experiment thus would mean obtaining a new sample of 100 individuals. The point is that in the context of repeated sampling, the quantity $\overline{X}$ in Equation (6.3.3) changes with each sample of 100 individuals. On the other hand, μ is a *population* quantity, and thus always has the same value (even though we do not know that value). Thus, as we change from one sample to another, the interval (6.3.3) wanders around, sometimes containing the fixed value μ, and sometimes missing it. In repeated sampling, 95% of the intervals (6.3.3) will contain μ.

The interval in Equation (6.3.3) is called a **95% confidence interval for μ.** We say that we are "95% confident" that μ is contained in the interval. What does this mean, especially in view of the fact that we take only *one* sample, rather than infinitely many? Suppose that after we take our sample of 100 individuals, we find that $\overline{X}=24.06$. The interval (6.3.3) then becomes (23.47, 24.65). We do not know whether μ is in this interval, but we are 95% "confident" that it is, *in the following sense*: Of all intervals (6.3.3) obtained from all possible samples of 100 individuals, 95% of these intervals do contain μ, so in this sense we are 95% sure that *our* particular interval contains μ.

This situation is similar to the following one: Suppose your friend tosses a fair coin in the adjoining room, letting it fall the floor. He then returns to you and asks you for your assessment of the probability that the result of the toss was heads. The coin is still lying on the floor in the adjoining room, and your friend knows which side is facing up; thus *to him*, the probability of heads is either 0 or 1, depending on whether tails or heads is facing up. However, *from your point of view*, the probability that the coin has heads facing up is 0.5, because if you do this experiment many times, the outcome will be heads in 50% of them.

The case is the same for a confidence interval as for the coin. The population mean μ either is or is not in the interval you form from your sample, so if you knew μ, the probability that the interval contains μ—from the point of view of such knowledge—would be either 0 or 1. But you do not know the value of μ, so your assessment of your confidence that your interval contains μ is 0.95, in the sense that 95% of all such intervals do contain μ.

By tracing through the above derivation with more general values, we can find a more general formula. Suppose that we take a sample of size n from a normal distribution with unknown mean μ and known variance σ^2. Then a $100(1-\alpha)\%$ confidence interval for μ is

$$\left[\overline{X} - z\left(\frac{\alpha}{2}\right)\frac{\sigma}{\sqrt{n}},\ \overline{X} + z\left(\frac{\alpha}{2}\right)\frac{\sigma}{\sqrt{n}}\right] \qquad \textbf{(6.3.4)}$$

where $z(t)$ denotes a value to the right of which there is area t under the $N(0,1)$ density curve; that is, $z(t)$ satisfies $P[Z > z(t)] = t$. (Note that our original example used $\alpha = 0.05$.)

However, the expression (6.3.4) needs to be generalized even more. First, the assumption that we are sampling from a normal population is often not true. However, note that the above derivation did not directly utilize the fact that the observations X_i were normally distributed. All that was used was the normality of $\overline{X}$—but the Central Limit Theorem (see Section 5.3) says that this is true approximately, even if the distribution of X_i is not even approximately normal. Thus the normality assumption is not really crucial here, and the interval (6.3.4) will still hold approximately if the sampled population is not normal.

Second, it is rare that we know the value of σ, at least in situations in which we do not know the value of μ. This problem has an easy solution too: From Equation (6.2.9), we have an *estimator* S^2 for the population variance σ^2, so we can substitute S for σ.

$$\left[\overline{X} - z\left(\frac{\alpha}{2}\right)\frac{S}{\sqrt{n}}, \overline{X} + z\left(\frac{\alpha}{2}\right)\frac{S}{\sqrt{n}}\right] \tag{6.3.5}$$

Thus interval (6.3.5) is an *approximate* $100(1-\alpha)\%$ confidence interval for μ, valid (at least approximately) for any sampled population having finite variance, whether or not the population distribution is normal.

Example 6.3.2

In the reliability literature, there is a famous data set consisting of "up" times of 220 airline air conditioning units. The values of $\overline{X}$ and S for this data set are 93.35 and 105.77, respectively. Using interval (6.3.5), we find that we are 95% confident that the mean lifetime of *all* units, that is, not just the ones sampled, is between 79.69 and 107.01 hours.

If we do know that the sampled population has a normal distribution, interval (6.3.5) can be made exact. To do this, we need a new parametric family of distributions, known as the **Student-*t*** family. It was mentioned at the end of Section 3.3 that a Student-t distribution with d degrees of freedom is defined to be that of the random variable

$$Q = \frac{Z}{\sqrt{Y/d}}$$

where Z has a $N(0,1)$ distribution, Y has a chi-square distribution with d degrees of freedom, and Z and Y are independent random variables. In our setting here, we set

$$Z = \frac{\overline{X} - \mu}{\sigma}$$

$$Y = \frac{(n-1)S^2}{\sigma^2}$$

and $d = n - 1$. Using advanced methods that are beyond the scope of this book, it can be shown that Y and Z as defined above are independent, and that Y has a chi-square distribution with $n - 1$ degrees of freedom (keep in mind, though, that this is only true in the context being considered here, in which the sampled population has a normal distribution). Thus, since we already know that Z has a $N(0,1)$ distribution, it follows that Q has a Student-t distribution with $n - 1$ degrees of freedom. (We will denote this distribution by T_{n-1}.)

Then by duplicating the steps in Equation (6.3.2), we find that the interval

$$\left[\overline{X} - t\left(\frac{\alpha}{2}, n-1\right)\frac{S}{\sqrt{n}}, \overline{X} + t\left(\frac{\alpha}{2}, n-1\right)\frac{S}{\sqrt{n}}\right] \tag{6.3.6}$$

is an *exact* $100(1 - \alpha)\%$ confidence interval for μ. Here, $t(\alpha/2, n-1)$ is analogous to $z(\alpha/2)$, but for T_{n-1} instead of $N(0,1)$. These values are tabulated in Table 3 in Appendix B.

The shape of the t-distribution is very similar to that of the normal, with the limiting shape actually being $N(0,1)$ as d approaches infinity. Thus interval (6.3.6) is really not much different from interval (6.3.5), except for very small n. For example, $t(0.05, 10)$ is equal to 1.81, not very much different from the 1.65 value for $z(0.05)$, and the corresponding t-value for $d = 25$ is already down to 1.71. Thus the usefulness of interval (6.3.6) is not as great as it may first seem. For large samples, (6.3.6) is essentially the same as (6.3.5), so that (6.3.6) is not especially useful in those cases. Interval (6.3.6) would be of some use in very small samples, since the approximation in interval (6.3.5) may not be very accurate then. However, even in those cases, we can only use interval (6.3.6) if we know that the sampled population is close to normal, and this is very difficult to determine with small samples.

By the way, the accuracy of the approximation (6.3.5) is not something about which we can only make guesses; we can use simulation to investigate its accuracy. For example, suppose the sampled population has an exponential distribution, say with mean $\mu = 4.0$, and that we have a random sample of size 25. Of course, in situations in which we use confidence intervals, we do not know the value of μ. But here we are asking a "what if" type of question, namely, if we use interval (6.3.5) as a confidence interval for μ with *approximate* confidence level 0.95, what is the *true* confidence level, that is, P[the interval (6.3.5) contains μ]?

To find this probability mathematically would be an extremely difficult task, so it is an excellent example of the great utility of simulation. Writing a simulation program for finding this probability is quite simple.

Program 6.3.1

```
program Ci(input,output);

   const Mu = 4.0;
```

```
      N = 25;

   var XBar,S,BiasCorrection,Mult: real;
       Rep,Count,Seed: integer;
       X: array[1..N] of real;

function Expon: real;
      var U: real;
   begin
      U := random(Seed);
      if U < 0.00001 then U := 0.00001;
      Expon := -Mu*ln(U)
   end;

procedure GenXI;
      var I: integer;
   begin
      for I := 1 to N do
         X[I] := Expon
   end;

   procedure FindXBarAndS;
      var I: integer;      SumX,SumX2,S2: real;
   begin
      SumX := 0.0;  SumX2 := 0.0;
      for I := 1 to N do
         begin
         SumX := SumX + X[I];
         SumX2 := SumX2 + sqr(X[I])
         end;
      XBar := SumX/N;
      S2 := SumX2 - sqr(XBar);  {biased version, see material
                             following Example 6.1.5}
      S2 := BiasCorrection*S2;
      S := sqrt(S2)
   end;

begin
   Seed := 9999;
   BiasCorrection := N/(N - 1.0);
   Mult := 1.96/sqrt(N);
   Count := 0;  {this will be a count of the number of
                             intervals which contain Mu}
   for Rep := 1 to 1000 do
   begin
      GenXI;  {generate the sample}
      FindXBarAndS;
      {now check to see if the interval contains Mu}
```

```
      if abs(XBar - Mu) < Mult*S then
         Count := Count + 1
      end;
   writeln('the true confidence level is ',Count/1000)
end.
```

Running this program produced the output 0.917, so the alleged 95% confidence interval is actually a 92% confidence interval. Thus interval (6.3.5) was a fairly good approximation.

Since interval (6.3.5) is a valid approximation for any distribution, it holds in particular for the case in which the observations X_i each take on only the values 0 and 1. In this case,

$$\mu = EX = 0 \cdot P(X = 0) + 1 \cdot P(X = 1) = P(X = 1)$$

so that we are estimating some proportion π, where π is the proportion of the time that X takes on the value 1. For example, suppose we wish to estimate the proportion π of people in a certain region who are in favor of building a new dam nearby. We take a random sample of n of the citizens in the region, and ask each one whether he or she is in favor of the dam. The response of the i-th person sampled is denoted by X_i, and we might use $X_i = 1$ to indicate "in favor," and $X_i = 0$ to indicate "opposed."

We then use the sample data $X_1, \ldots, X_n$ to estimate the population proportion π. Our estimator $\hat{\pi}$ of π is the proportion of X_i in our sample which were equal to 1. But this is equivalent to

$$\frac{X_1 + \cdots + X_n}{n} \tag{6.3.7}$$

so $\hat{\pi}$ is actually $\overline{X}$. Thus since $\mu = \pi$ and $\overline{X} = \hat{\pi}$, we are set up to use interval (6.3.5).

In practice, the original version of S is used, that is, the one in Equation (6.1.2) rather than the one in Equation (6.2.9). It can be shown (Exercise 6.16) that S^2 then is equal to $\hat{\pi}(1 - \hat{\pi})/n$. Thus the interval (6.3.5) for the case of estimating a proportion is

$$\left[\hat{\pi} - z\left(\frac{\alpha}{2}\right)\sqrt{\frac{\hat{\pi}(1-\hat{\pi})}{n}},\ \hat{\pi} + z\left(\frac{\alpha}{2}\right)\sqrt{\frac{\hat{\pi}(1-\hat{\pi})}{n}}\right] \tag{6.3.8}$$

Example 6.3.3

A bus company wishes to estimate the proportion π of runs in which there are fewer than 10 passengers. A random sample of 25 runs produces 6 in which there are fewer than 10 people riding the bus.

Thus we estimate π to be $6/25 = 0.24$. However, since $n = 25$ is such a small sample size, we know that this estimate may contain considerable error. Using the interval (6.3.8), we can get some indication of the magnitude of this error, by saying that we are 95% confident that the true value of π is somewhere between 0.08 and 0.40. This is a very big range, again due to the fact that n is so small; thus, the consumer of this information can see that a larger sample is needed.

There are also two-sample versions of the foregoing material. They are derived in the same way as those above. Here are the most commonly used ones:

Some Two-Sample Confidence Intervals

a. Suppose we have independent samples of sizes n_1 and n_2 from populations having means μ_1 and μ_2 and standard deviations σ_1 and σ_2. Let the sample estimates of means and standard deviations be denoted by $\overline{X}_i$ and S_i. Then an approximate $100(1-\alpha)\%$ confidence interval for $\mu_1 - \mu_2$ is

$$\overline{X}_1 - \overline{X}_2 \pm z\left(\frac{\alpha}{2}\right)\sqrt{\frac{S_1^2}{n_1} + \frac{S_2^2}{n_2}} \qquad \textbf{(6.3.9)}$$

b. In (a), if both sampled populations are normally distributed, *and* we have $\sigma_1 = \sigma_2$, then an *exact* $100(1-\alpha)\%$ confidence interval for $\mu_1 - \mu_2$ is

$$\left[\overline{X}_1 - \overline{X}_2 - t\left(\frac{\alpha}{2}, k\right)\frac{S}{\sqrt{m}},\ \overline{X}_1 - \overline{X}_2 + t\left(\frac{\alpha}{2}, k\right)\frac{S}{\sqrt{m}}\right] \qquad \textbf{(6.3.10)}$$

where $k = n_1 + n_2 - 2$, $m = 1/n_1 + 1/n_2$, and S^2 is a combined estimator of the common variance $\sigma_1^2 = \sigma_2^2$:

$$S^2 = \frac{(n_1 - 1)S_1^2 + (n_2 - 1)S_2^2}{k} \qquad \textbf{(6.3.11)}$$

c. The two-sample extension of interval (6.3.8) is

$$\left[\hat{\pi}_1 - \hat{\pi}_2 - z\left(\frac{\alpha}{2}\right)s,\ \hat{\pi}_1 - \hat{\pi}_2 + z\left(\frac{\alpha}{2}\right)s\right] \qquad \textbf{(6.3.12)}$$

where

$$s = \sqrt{\frac{\hat{\pi}_1(1-\hat{\pi}_1)}{n_1} + \frac{\hat{\pi}_2(1-\hat{\pi}_2)}{n_2}}$$

Example 6.3.4

Suppose the manager of a timeshare system is pondering which of two text editors, E1 or E2, to implement on her system. E2 has many nice features which are said to help the programmers in their code development activities; it is hoped that E2 would result in substantial time savings for them. However, E2 uses more computing resources, in the form of more frequent disk accesses.

The manager decides to conduct an experiment. She asks 20 programmers to develop a certain program using E1, and 20 other programmers to write the same program using E2. Let X_i denote the time (in hours) needed for the *i*th programmer in the first group to get the program working correctly, and let Y_i be the corresponding value for the second group ($i = 1,2,\ldots,20$). Let μ_1 represent the population mean for E1, that is, the mean time over *all* programmers using that timeshare system, not just the ones who participate in the experiment. Similarly, let μ_2 denote the mean for E2. Suppose we find that

$$\overline{X}=6.2,\ S_1 = 1.4$$

$$\overline{Y}=5.3,\ S_2 = 2.6$$

Of course, we are actually interested in μ_1 and μ_2, rather than $\overline{X}$ and $\overline{Y}$; the latter two quantities only serve as estimates for the former two. Using the expression (6.3.9) to get a 95% confidence interval for $\mu_1 - \mu_2$, we have

$$6.2-5.3\pm1.96\sqrt{\frac{1.4^2}{20}+\frac{2.6^2}{20}}$$

which yields the interval $(-0.4, 2.2)$.

So, we find that we are 95% confident that the difference $\mu_1 - \mu_2$ lies between -0.4 and 2.2. This suggests that although E2 might be substantially better than E1 (2.2 hours faster on the average), it might actually be even a little slower than E1, by perhaps 0.4 hours. Again, we see that a larger sample is needed for more accurate results.

It should be emphasized here that in (a) above, the two samples must be independent. This is not always the case.

Example 6.3.5

In Example 6.3.4, suppose the design of the experiment were slightly different, with 10 programmers instead of 20. Each programmer uses E1 to develop Program 1, and E2 to develop Program 2. Let X_{ij} be the completion time for the ith programmer on the jth program ($i = 1, \ldots, 10$; $j = 1, 2$). Then the samples are definitely not independent. For example, suppose Programmer 3 has a reputation in the company as being an unusually fast worker. Thus there will be a tendency for *both* X_{i1} and X_{i2} to be smaller than most of the other observations in our sample. In other words, those values of i for which X_{i1} is relatively small probably have X_{i2} relatively small too, so X_{i1} and X_{i2} are positively correlated.

In cases such as this, we can still form a valid confidence interval for $\mu_1 - \mu_2$ by looking at *differences*. The problem then reverts to the one-sample setting—we have a single sample of differences from the "population" of all such differences. Thus we compute the following quantities:

$$d_i = X_{i1} - X_{i2}$$

$$\bar{d} = \frac{1}{n} \sum_{i=1}^{n} d_i$$

$$S_d^2 = \frac{1}{n-1} \sum_{i=1}^{n} (d_i - \bar{d})^2 \qquad \textbf{(6.3.13)}$$

and then use interval (6.3.5) to compute a confidence interval for $\mu_d = \mu_1 - \mu_2$, with $\bar{X} = \bar{d}$ and $S = S_d$.

Finally, it should also be mentioned that one can form one-sided confidence intervals. The derivations of these intervals are parallel to those above. For example, in Equation (6.3.2),

$$0.95 = P\left(-1.96 < \frac{\bar{X} - \mu}{0.3} < 1.96\right)$$

can be replaced by

$$0.95 = P\left(-1.65 < \frac{\bar{X} - \mu}{0.3}\right)$$

with the result that $(-\infty, \bar{X} + 0.49)$ is a 95% one-sided confidence interval for μ; in other words, we are 95% confident that μ is at most $\bar{X} + 0.49$. We could derive a lower bound similarly.

Example 6.3.6

Recall Example 6.3.4 above. One can derive a one-sided version of interval (6.3.9) for an upper bound for $\mu_1 - \mu_2$ (Exercise 6.12):

$$\left[-\infty, \overline{X}_1 - \overline{X}_2 + z(\alpha)\sqrt{\frac{S_1^2}{n_1} + \frac{S_2^2}{n_2}}\right] \tag{6.3.14}$$

We then find that a 95%-confident upper bound for $\mu_1 - \mu_2$ is 2.0, rather than the 2.2 value we found in Example 6.3.4. It is clear that what has happened is that we have obtained a sharper upper bound for $\mu_1 - \mu_2$, at the expense of relinquishing our right to calculate a lower bound. This may be quite reasonable in the setting considered here; for example, a very small upper bound may indicate that the nice features of E2 are not good enough to compensate for the drain on computing resources caused.

6.4 NONPARAMETRIC DENSITY ESTIMATION: HISTOGRAMS

[Before beginning this section, it is suggested that the reader review three concepts: (a) the intuitive ideas behind densities and cumulative distribution functions; (b) the concept from calculus of a derivative being the tangent to a curve, and thus the limit of secant slopes on that curve; (c) the analog-estimation idea in Section 6.1, in which we estimate *population* quantities from *sample* quantities. Most readers are apt to find this section difficult reading, but it is an "investment" that will pay excellent dividends, both for its direct practical value, and also for its use in review of previous concepts.]

We have previously presented measures of location and dispersion as descriptive tools for a distribution. However, these two values cannot completely capture all of the information in a density f. Often the *shape* of a density is of interest. For example, suppose we are studying the memory requirements of jobs submitted to a given computer system. The mean and standard deviation of this distribution give us some information, but knowledge of the density would allow us to ask many other questions, such as: Are most of the values concentrated near the mean? What is the effective range of the distribution? Is there more than one peak to the density (possibly indicating that we should study two subpopulations separately)?

Another very important reason to be able to estimate densities is that of selection of a parametric family for the distribution we are studying. We estimate the density nonparametrically first, and then see if the shape seems to fit one of the well-known parametric families.

Thus, just as we estimate such quantities as population means and variances from sample data, we also wish to estimate population densities from such data. Note first that if we do know that a density belongs to a certain parametric family (perhaps from previous experience with this kind of data), estimation of the density is easy: We simply estimate the parameters, and then "plug in" to the density formula.

Example 6.4.1

Suppose we know that f belongs to the normal family, that is,

$$f(t) = \frac{1}{\sqrt{2\pi\sigma}} e^{-0.5(t-\mu/\sigma)^2} \tag{6.4.1}$$

Suppose from our sample data we find that $\overline{X} = 22.4$ and $S=5.3$. Then our estimate of the density would be

$$\hat{f}(t) = \frac{1}{\sqrt{2\pi(5.3)}} e^{-0.5(t-22.4/5.3)^2} \tag{6.4.2}$$

However, in this section we are concerned with the case in which we do not know a parametric family for the given density. Estimation in this situation is considerably more complicated.

The simplest form of density estimation is known as a **histogram**. To define this, we will first define a related estimator.

Since $f(t) = (d/dt)F(t)$, we have, for small h,

$$f(t) \approx \frac{F(t+h) - F(t)}{h}$$

Another approximation, similar in spirit but better for our purposes, is

$$f(t) \approx \frac{1}{h}\left[F\left(t + \frac{h}{2}\right) - F\left(t - \frac{h}{2}\right)\right] \tag{6.4.3}$$

Thus, since $F(w) = P(X \le w)$,

$$f(t) \approx \frac{P(t - h/2 < X \le t + h/2)}{h}$$

so that a reasonable estimate of $f(t)$ is

$$\hat{f}(t) = \frac{1}{h}\left[\frac{C(t - h/2,\ t + h/2)}{n}\right] \tag{6.4.4}$$

where we are using $C(a, b)$ to denote the count of sample observations X_i that fall into the interval $(a, b]$.

In practice, Equation (6.4.4) requires too much computation, since it must be evaluated separately for each value of t. (Recall that since the density f is a *function*, it consists of infinitely many values, one $f(t)$ for each t. Thus, in theory, infinitely many values—or at least a very large number—must be estimated.) In the histogram method, only a small number of these values are computed, in the following way:

An interval (l, r) is chosen that is large enough to include all the sample observations X_i, such that $r - l$ is an even multiple of h. Typically, we choose l to be

$$\min\{X_i : i = 1, \ldots, n\}$$

and r to be a bit larger than

$$\max\{X_i : i = 1, \ldots, n\}$$

The exact details of our choice of $[l, r]$ do not matter much. Then $[l, r]$ is divided up into k intervals of length h, where $k = (r - l)/h$.

Let t_i be the midpoint of the i-th interval:

$$t_i = l + (i - 1)h + \frac{h}{2} \tag{6.4.5}$$

Then to avoid computing Equation (6.4.4) for a very large number of values of t in $[l, r]$, we make a further approximation by assuming that the value of $f(t)$ is constant throughout each interval (note that this approximation becomes more and more accurate as h becomes smaller and smaller). Thus we make $\hat{f}(t)$ flat through each interval: For all t in the ith interval, we set $\hat{f}(t) = \hat{f}(t_i)$. In other words, to compute $\hat{f}(t)$, we first find the corresponding t_i. Then we evaluate Equation (6.4.4) with $t = t_i$. A formal description of this is

$$\hat{f}(t) = \frac{1}{h}\frac{C(u - h/2, u + h/2)}{n}$$

where u is the midpoint of whatever interval t happens to be in, or

$$u = a + \left\lfloor \frac{t - a}{h} \right\rfloor + \frac{h}{2}$$

where $\lfloor s \rfloor$ means the greatest integer less than or equal to s; for example, $\lfloor 3.7 \rfloor = 3$. All this may sound complicated, but it becomes much clearer after one does the computations once.

Example 6.4.2

Recall the air conditioner data from Example 6.3.2. Figure 6.1a shows a histogram estimate for the population density of air conditioner lifetimes. Here we have used $h = 104.0$, with $l = 0.0$ and $r = 520.0$, where the value of h was chosen so that the number of subintervals, $(r - l)/h$, would be an integer. In order to make the graph a little more descriptive, we have "connected the dots" at the points $(t_i, \hat{f}(t_i))$, resulting in Figure 6.1b. The general shape of the graph suggests that it may be reasonable to model this density as an exponential distribution.

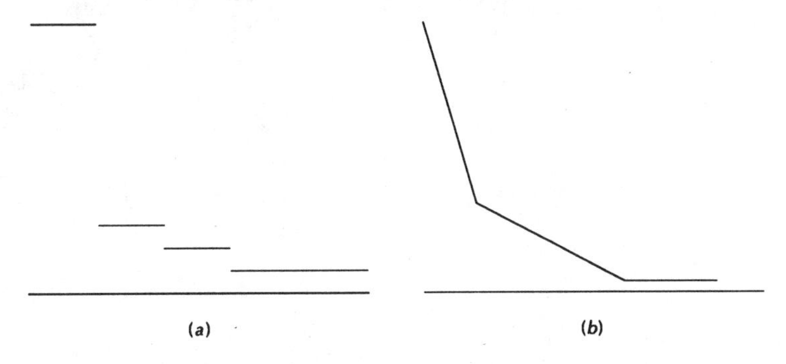

Figure 6.1 Histogram estimate for Example 6.4.2, $h = 104$.

It was mentioned above that the exact details of the choice of $[l, r]$ are not too important. However, the choice of the value of h is tremendously important. From Equation (6.4.3) and our knowledge of calculus, we can see that h should be kept fairly small: We wish to estimate $f(t)$, which is the slope of the tangent line of F at t. Equation (6.4.3) gives the slope of a secant line, which will be close to the slope of the tangent only if h is small. If h is large, the graph for $\hat{f}$ will be too flat. For example, if we set $h = r - l$, then the entire graph of $\hat{f}$ will be a horizontal line, suggesting a uniform distribution on $[l, r]$!

On the other hand, we can not make h *too* small, either. In fact, if we were to make h smaller than $\min\{|X_i - t| : i = 1, ..., n\}$, then $C(t - h/2, t + h/2)$ would be 0, and thus $\hat{f}(t)$ would be 0 too! In fact, for very small h, $\hat{f}(t)$ would be 0 for most values of t. In general, use of too small an h will make the graph of $\hat{f}$ look too ragged. For example, in Figure 6.2a (and the connected version, Figure 6.2b), we have set h to 11.56. Intuitively, we would feel that the true density f is much smoother than this.

The graphs corresponding to an intermediate value, $h = 34.67$, appear in Figures 6.3a and 6.3b. Given the amount of oscillation present, it would seem that a value of h between this one and 104.0 might be appropriate.

Whatever the "proper" value for h in this example, the graphs above do demonstrate that in choosing a value of h, there is a tradeoff. If we choose too small an h, then artificial oscillations will appear in the graph, while choosing too large an h may hide the real oscillations. Unfortunately, statistical researchers have not as yet found practical methods for determining a good value for h. There are formulas for the optimal value of h, which depend strongly on the sample size n (h should be smaller for larger n), and on the density f itself. Since we do not know f (if we did, we would not wish to estimate it!), the dependence of the formulas on f make the usefulness of these formulas quite unclear. However, as a rough rule of thumb, one will usually produce good results by choosing h as small as possible, subject to the constraint that the resulting $\hat{f}$ should be fairly smooth.

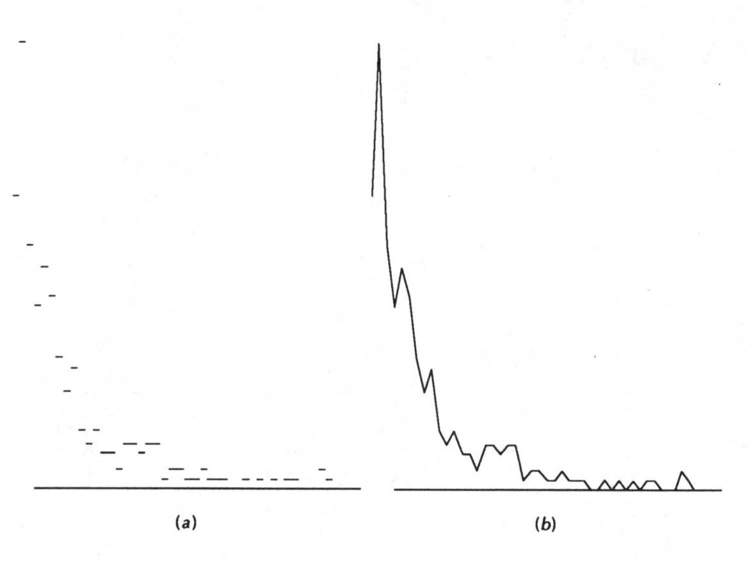

Figure 6.2 Ragged graphs of f.

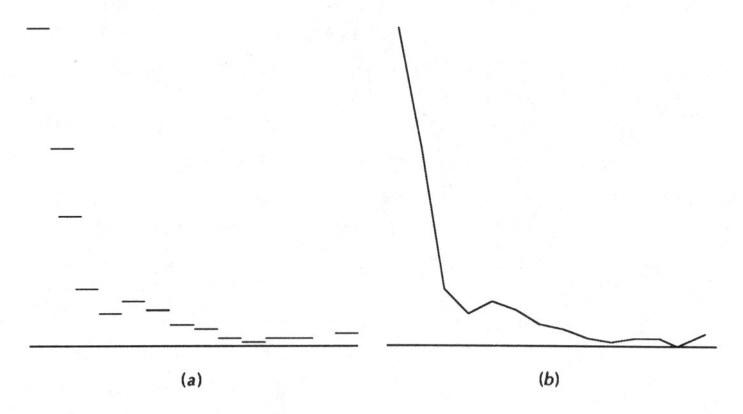

Figure 6.3 Graphs with $h=34.67$.

6.5 APPLICATIONS OF ESTIMATION THEORY TO SIMULATION METHODOLOGY

In all of the simulation programs that we have run so far, we have ignored the issue of *how long* to run the program. For example, look at Program 2.4.3, which gives the general format for a simulation program that finds a probability $P(A)$. How large should the value of TotReps be?

It is clear that TotReps should be made as large as possible, with perfect accuracy in the reported value of $P(A)$ coming only if TotReps is infinite; with finite values, the reported value of $P(A)$ will only be an approximation. Of course, an infinite value for TotReps is impossible, and even finite (but large) values of TotReps are infeasible. For example, if we are paying for computer time, we can only afford limited values of TotReps. However, even if the computer time is free, we may not have the patience to use a value of TotReps which requires, say, three weeks of computer time!

A related question is this: For a given value of TotReps, how accurate is the resulting approximation that the program reports for $P(A)$?

These questions are easily answered with the material learned earlier in this chapter. Let us consider the last question first (i.e., the one in the last paragraph), concerned with the accuracy of our approximation for $P(A)$.

Denote the exact value of $P(A)$ by π, and TotReps by n. Note that the n repetitions of the "for" loop in Program 2.4.3 simulate n independent repetitions of our experiment. In each repetition, the event A either occurs or does not occur, with probabilities π and $1-\pi$, respectively. The *reported* value of $P(A)$, which we will now denote by $\hat{\pi}$, is then simply an *estimate* of π, based on n repetitions of a sampling experiment. We can actually form a confidence interval for the exact $P(A)$ using the interval (6.3.8).

Example 6.5.1

Look back at Program 2.4.6. There TotReps was equal to 1000, and the value of $P(A)$ reported by the program was 0.331. Thus, an approximate 95% confidence interval for the exact value of $P(A)$ is

$$\left(0.331 - 1.96\sqrt{\frac{(0.331)(0.669)}{1000}},\ 0.331 + 1.96\sqrt{\frac{(0.331)(0.669)}{1000}}\right)$$

Using this expression, we are 95% confident that the true value of $P(A)$ is between 0.302 and 0.360.

Furthermore, this reasoning can also be used to decide how large to make TotReps before we run the program. We know that the half-width (i.e., radius) of the confidence interval will be $z(\alpha/2)[\hat{\pi}(1-\hat{\pi})/n]^{1/2}$. Of course, we will not know $\hat{\pi}$ before running the program, but often we have a rough idea as to the value of $P(A)$ beforehand, and we can use this as a guess as to the value that $\hat{\pi}$ will have. Then we can simply choose TotReps to be

$$\frac{z^2(\alpha/2)g(1-g)}{r^2}$$

where g is the guessed value for $P(A)$, and r is the maximum value of the radius that we can tolerate. Or, we can use 0.5 as a conservative guess for $P(A)$, for this purpose; see Exercise 6.9.

Similarly, suppose we are using simulation to find an expected value. In this case we can use the interval expression (6.3.5).

Example 6.5.2

Consider Example 3.2.11. Approximations for $E(T)$ and $\mathrm{Var}(T)$ were 17.50 and 128.91, based on TotReps equal to 100. Here, μ and σ^2 from Section 6.3 will take the form of the exact values of $E(T)$ and $\mathrm{Var}(T)$. Thus, we have

$$\overline{X} = 17.50$$

$$S^2 = \frac{100}{99}(128.91) = 130.0$$

[The factor 100/99 is due to the fact that we are using Equation (6.2.9), instead of Equation (6.1.2), for S^2.] Using the interval (6.3.5), we find that we are 95% confident that the exact value of $E(T)$ is between 15.28 and 19.72. This is a rather broad range. For better accuracy, we may wish to run the program for a longer time, say with TotReps equal to 1000.

As before, when we estimated a probability to find a suitable value for TotReps, we can use an estimated S in planning the value of TotReps before running the experiment. However, guessing a value for S is usually more difficult.

If the value of TotReps needed to achieve a desired level of accuracy is infeasible, we may wish to try **variance-reduction methods**. The goal of these methods is to increase accuracy without increasing TotReps. As a sample of what is involved, let us look at Example 6.5.2 above. Here $\overline{X}$, the reported simulation estimate of $E(T)$, has variance σ^2/n (which of course must also be estimated, by S^2/n). This formula is quite familiar, from Equation (5.3.3). (It would be very helpful for the reader to review that derivation before reading further.) It assumes independence of the observations X_i, which in our present context means independence between separate repetitions within the "for" loop in our simulation program. This assumption is justified, so the calculations in Example 6.5.2 are valid. However, we may wish to *intentionally* introduce some kind of dependence among the X_i. If this is done well, we may be able to achieve negative values for $\mathrm{Cov}(X_i, X_j)$ in Equation (4.2.6), which will reduce $\mathrm{Var}(\overline{X})$, that is, produce a more accurate estimate of $E(T)$, since the radius of the confidence interval is a multiple of this variance.

Although variance-reduction methods can be extremely successful in reducing the value of TotReps needed for a given level of accuracy, achieving this success usually is highly dependent on special aspects of the process being simulated. Thus, such methods are beyond the scope of this book. The interested reader is referred to the "Further Reading" section at the end of this chapter.

6.6 LINEAR STATISTICAL MODELS

The subject matter of this section forms an introduction to one of the most widely used statistical tools. Although it is impossible to do justice to this area in any less space than a full-length book, the material is so useful that we are including a very short introduction here. The reader is urged to take a course covering this subject in depth.

A large class of statistical models concern linear relationships between several variables. The formulation is as follows: We have **predictor variables** $X^{(1)}, \ldots, X^{(r)}$ and a **response variable** Y. The predictors may be either random variables or fixed constants, while the response variable is random. We are interested in the mean value of the response variable, *given a specified set of values for the predictor variables*. In other words, we are interested in the **regression function**

$$\mu(s_1, \ldots, s_r) = E\left(Y \mid X^{(1)} = s_1, \ldots, X^{(r)} = s_r\right) \qquad \textbf{(6.6.1)}$$

Such models are useful in a variety of applications, which may be loosely organized into the following categories:

- Prediction
- Effect measurement
- Control
- Adjustment for bias

Below are several examples of the Prediction category, followed by one example each of the other three application types.

Example 6.6.1 (Prediction)

Suppose we are developing a computer operating system, in which we anticipate very heavy usage of a Pascal compiler. Suppose we wish to schedule compile jobs according to their anticipated compile time, with the operating system giving priority to those which it thinks will require less compilation time.

To discuss this, let us briefly review what a compiler does. A user will first use a text editor to create a disk file containing the Pascal version of a program; this file is called the **source file**. The user will then submit a command to the operating system, requesting that the source file be **compiled**, which means that the compiler will read the source file, translate the Pascal program to a version in the computer's machine language, and then store the machine language program in another disk file, the **object file**.

Some compilers also will have a number of options that the user may request. One of these is known as **code optimization**, which refers to special attempts by the compiler to produce object code having shorter execution time. This requires considerable extra compilation time.

Of course, there is no way to predict perfectly how long a given source file will take to compile. However, given the right predictor variables, we may be able to achieve an accuracy level sufficient for scheduling purposes (Exercise 6.25). For example, we might try to predict compilation time Y from the number of lines of source code $X^{(1)}$, the number of variables declared in "var" statements, $X^{(2)}$, and an indicator variable $X^{(3)}$, for which the value 1 indicates that the user has requested code optimization, while the value 0 indicates otherwise.

Example 6.6.2 (Prediction)

Suppose a civil engineering company is bidding on a river bridge construction contract. The company may be required to give an approximate figure for the length of time needed to complete the project. Thus, they might wish to predict Y, the time needed, from the predictors $X^{(1)}$, the length of the bridge; $X^{(2)}$, the maximum height of the bridge above the water; and $X^{(3)}$, the maximum depth of the river. Again, we probably will not be able to make perfect predictions, but we hope to be able to achieve a level of accuracy which is sufficient for our purposes.

Example 6.6.3 (Prediction)

Machine recognition of speech has a variety of applications. One example is that in which a factory worker can control a machine verbally, freeing the worker's hands for other tasks to be done at the same time, such as loading a heavy object into the machine. Another example is that of a "dictating typewriter"; instead of inputting text through a keyboard, one can simply dictate the text verbally.

Consider the recognition of a particular sound, such as the short "e" sound in the word "get." Here Y would be an indicator variable for which the value 1 would indicate that the sound uttered was that of a short "e," while 0 would indicate that the sound was something else. The predictors $X^{(j)}$ would then be various measurements on the graphical wave form recorded from the utterance.

Note that since a variable W that takes on only the values 0 and 1 has $EW = P(W = 1)$, Equation (6.6.1) in this case reduces to

$$\mu(s_1, \ldots, s_r) = P(Y = 1 | X^{(1)} = s_1, \ldots, X^{(r)} = s_r)$$

In a simple analysis, we might adopt the rule that we will guess the utterance to be a short "e" sound if this probability is larger than 0.5. More complex analyses would be based on the relative costs of misclassification, a subject that we leave for the speech recognition textbooks and papers.

Note again that we can not expect perfect accuracy in prediction. However, we may not need it. In the dictating typewriter case, for instance, the machine's guess as to which word(s) you said will appear on a screen, so that you can indicate errors. As long as the errors are not too frequent, the time and money saved through bypassing the keyboard input will far outweigh the small effort needed to correct errors.

Example 6.6.4 (Effect Measurement)

Traffic engineers are very concerned about the factors associated with accidents. For example, when maximum speed limits were reduced to 55 miles per hour in the 1970s, one benefit was supposed to be a reduction in accident rates. Suppose we wish to formally investigate the effects of speed on accidents. We might let Y be the number of accidents per year on a given road, and let $X^{(1)}$ denote the speed limit, $X^{(2)}$ the volume of traffic, $X^{(3)}$ the number of rainy days per year at that site, and $X^{(4)}$ the number of road lights per mile. If we can find the relationship expressing the mean value of Y as a function of $X^{(1)}$, $X^{(2)}$, $X^{(3)}$, and $X^{(4)}$, then we may be able to obtain insight into the role of speed in accident rates. For example, we can compare the effects on mean number of accidents per year of

a. lowering the speed limit by 5 miles per hour, versus

b. increasing the number of road lights by 25%.

Example 6.6.5 (Control)

Suppose we are in the business of manufacturing foodstuffs, and we have developed a new soft drink. We are trying to decide what the concentration of sugar should be in the new drink. Let Y be a consumer rating of the flavor of the drink, and let $X^{(1)}$ denote the concentration of sugar. Then we wish to find the value of s_1 at which the peak of the graph of $\mu(s_1)$ occurs. (Actually, "control" operations are very common in engineering, especially in the fields of mechanical and electrical engineering. These applications do use the material in this section, including the least-squares estimators presented below, but the complexity of this area makes inclusion in this text infeasible.)

Example 6.6.6 (Adjustment for Bias)

Suppose a large manufacturing corporation has just installed a number of units of a new machine. An industrial engineer employed by the company is investigating which of two methods should be used to teach the use of the new equipment. The company has two plants. The investigator decides to try Method A on a sample of employees at the first plant, and Method B on a sample from the second plant. However, she is worried about potential bias that may result from this kind of sampling. For example, suppose that, on the average, the employees at the first plant are more experienced than those of the second plant. A more experienced worker might be able to learn new equipment more quickly. If so, this might produce a bias that would falsely make Method A look better. As will be discussed later, estimation of the function in Equation (6.6.1) may be able to solve this bias problem, with Y denoting the length of time needed to learn the use of the machine, $X^{(1)}$ denoting the number of years of experience of the employee, and $X^{(2)}$ taking on the values 1 and 0, denoting Group A versus Group B.

With these examples as motivation, let us consider how we might estimate the function (6.6.1) from sample data. First note that without any parametric structure imposed on the form of this function, we again have an infinite-parameter estimation problem, just as in Section 6.4: We must estimate the value of function (6.6.1) for each of the possibly infinitely many r-tuples $(s_1, \ldots, s_r)$. On the other hand, use of an appropriate parametric model would change this to a finite-parameter problem, again as discussed in Section 6.4.

We will discuss both nonparametric and parametric approaches, with emphasis on the latter. In order to facilitate the discussion, let us concentrate for the moment on the case $r = 1$, and write $X^{(1)}$ simply as X. To further clarify the concepts, let us treat a specific example, namely Example 6.6.2 (without $X^{(2)}$ and $X^{(3)}$). Our sample observations would consist of data from n bridges constructed by this company in the past. Denote the sample data points by $(X_1, Y_1), \ldots, (X_n, Y_n)$, where, for example, X_3 and Y_3 are the span length and construction time of the third bridge in our sample.

Suppose that we wish to bid on a project calling for the construction of a bridge of length 500 feet. Thus, we wish to predict Y, knowing that $X = 500$. A reasonable predictor would be the mean Y value among *all* bridges of length 500, $\mu(500)$ (see Exercise 6.15). Thus, we must consider methods for estimating $\mu(500)$.

Let us consider nonparametric analysis first. There probably will not be any bridges in our sample of length exactly 500 feet, and even if there are some, there will be very few—too few for a good estimate. So, let us look at bridges in our sample of length *near* 500 feet; that is, as in Section 6.4, let us base our analysis on intervals. For example, we could choose the interval (475, 525), and take as our estimate of $\mu(500)$ the average of all the sample observations Y_i for which X_i falls into the interval (475, 525):

$$\hat{\mu}(500) = \frac{1}{m} \sum_{X_i \epsilon (475,525)} Y_i$$

where m is the number of observations X_i such that

$$X_i \epsilon (475, 525)$$

and $\hat{\mu}(t)$ denotes our estimator of $\mu(t)$.

This type of analysis is very simple, but it has the same type of problem we encountered in Section 6.4: It is not at all clear how to choose the interval size. For example, why not use (490, 510) or (460, 540)? As will be seen in the following heuristic argument, there is a tradeoff between bias and variance: Suppose the density of X is asymmetric around 500, with more weight to the right of 500 than to the left. Then there are apt to be more X_i to the right of 500, and since $\mu(s)$ will probably be an increasing function of s in this particular application, the result will be that our estimate of $\mu(500)$ will be biased upward. The magnitude of this bias might be rather small if our interval is small, since the density of X will be approximately constant in a small interval; thus it is tempting to choose a small interval. On the other hand, if we make the interval *too* small, there may be very few X_i in it, maybe even none. This creates a variance problem, as indicated by the inverse dependence of the variance of a sample mean on the sample size [recall Equation (5.3.3)].

As in the case of nonparametric density estimation, nonparametric regression analysis has been the subject of considerable attention in the statistical research community in recent years. It has been in use for many years, particularly in pattern recognition applications, but only recently has there been a major effort to find methods for determining such quantities as the interval size above. A completely reliable method for choosing the interval size has not yet been developed, although some proposals have been made that appear to have some promise.

Now let us consider parametric estimation of $\mu(500)$, and the values of the function $\mu(s)$ for other values of s. The simplest and most common model is a linear one:

$$\mu(s) = \beta_0 + \beta_1 s \tag{6.6.2}$$

Here we are assuming that the graph of $\mu(s)$ takes the form of a straight line. This then becomes a two-parameter estimation problem, so that estimation of the infinitely many values $\mu(s)$ (as s varies) simply requires estimation of the two parameters β_0 (intercept) and β_1 (slope).

Of course, we should make sure that this linear model is reasonable first, for example by plotting the data points $(X_1, Y_1), \ldots, (X_n, Y_n)$ and seeing whether they seem to form a "cloud" centered around a straight line. Nonparametric estimation might also be used, estimating $\mu(s)$ at a number of s values, to get a rough idea as to whether the graph of $\mu(s)$ is approximately a straight line. The general area of model selection is treated extensively in the references at the end of this chapter.

The next question is how to estimate β_0 and β_1. Let us first consider the idea of "analog" estimators from Section 6.1, in the context in which the predictor X is random. From Exercise 6.15 (adapted to conditional expectation), we know that $\mu(s)$ minimizes

$$E\,[(Y - c)^2 \mid X = s] \tag{6.6.3}$$

over all choices of c, so Equation (6.6.2) implies that β_0 and β_1 minimize

$$E\,[(Y - b_0 - b_1X)^2 \mid X = s] \tag{6.6.4}$$

over all choices of b_0 and b_1. [Recall the discussion near Equation (4.2.10).]

It can be shown that this remains true if we "uncondition": β_0 and β_1 minimize

$$E\,[(Y - b_0 - b_1X)^2] \tag{6.6.5}$$

over all choices of b_0 and b_1. [This is a consequence of Equations (4.1.15).]

The sample analog of expression (6.6.5) is

$$\frac{1}{n}\sum_{i=1}^{n}(Y_i - b_0 - b_1X_i)^2 \tag{6.6.6}$$

so the concept of analog estimators would suggest estimating β_0 and β_1 by whatever values of b_0 and b_1 minimize expression (6.6.6). Another interpretation of this would be that we are choosing b_0 and b_1 so as to best "predict" the Y_i by the X_i.

In the case of general r, our model for $\mu(s_1, \ldots, s_r)$ is

$$\mu(s_1, \ldots, s_r) = \beta_0 + \beta_1 s_1 + \cdots + \beta_r s_r \quad \textbf{(6.6.7)}$$

Denote the data for the i-th unit in our sample (e.g., the ith bridge) by $X_i^{(1)}$, $\ldots$, $X_i^{(r)}$, Y_i. Then we estimate $\beta_0, \beta_1, \ldots, \beta_r$ by the values of $b_0, b_1, \ldots,$ b_r that minimize

$$\sum_{i=1}^{n} [Y_i - b_0 - b_1 X_i^{(1)} + \cdots + b_r X_i^{(r)}]^2 \quad \textbf{(6.6.8)}$$

Of course, the $1/n$ factor in Equation (6.6.6) had no effect on the minimization, so we did not need to include it here.

This minimization process is a straightforward calculus problem. Setting to zero the partial derivatives of the expression (6.6.8) with respect to the b_j, we obtain a system of $r + 1$ equations (the details of the differentiation are omitted here). Since a system of linear equations is much easier to handle in matrix form, let us use this mode. The matrix equation turns out to be

$$X^t X \beta = X^t Y \quad \textbf{(6.6.9)}$$

where:

$$X = \begin{bmatrix} 1 & X_1^{(1)} & \cdots & X_1^{(r)} \\ \cdot & \cdot & \cdot & \cdot \\ \cdot & \cdot & \cdot & \cdot \\ \cdot & \cdot & \cdot & \cdot \\ 1 & X_n^{(1)} & \cdots & X_n^{(r)} \end{bmatrix} \quad \textbf{(6.6.10a)}$$

$$\beta = \begin{bmatrix} \beta_0 \\ \cdot \\ \cdot \\ \cdot \\ \beta_r \end{bmatrix} \quad \textbf{(6.6.10b)}$$

and

$$Y = \begin{bmatrix} Y_1 \\ \cdot \\ \cdot \\ \cdot \\ Y_n \end{bmatrix} \quad \textbf{(6.6.10c)}$$

For illustration, suppose that in Example 6.6.2, our sample data are taken from 50 bridges constructed in the past, and that the first of these bridges was of length 228 feet and was 50 feet above the river, with the maximum depth of the river 11 feet; also, the construction time was 312 days. Then the first row of the matrix X would be

$$1 \quad 228 \quad 50 \quad 11$$

and Y_1 would be 312.

Thus, from Equation (6.6.9) we can see that if the matrix

$$(X^t X)^{-1} \tag{6.6.11}$$

exists, then the vector of estimates $\hat{\beta}_j$ is

$$\hat{\beta} = \begin{bmatrix} \hat{\beta}_0 \\ \cdot \\ \cdot \\ \cdot \\ \hat{\beta}_r \end{bmatrix} = (X^t X)^{-1} X^t Y \tag{6.6.12}$$

The estimator $\hat{\beta}$ derived above has some reasonable statistical properties, including unbiasedness (Exercise 6.21). However, it is not necessarily the best estimator of β. Before discussing this, recall from Section 6.2 that the word "best" here depends on what criterion we are using. We will use the criterion of *minimum variance* among unbiased estimators which are linear in Y, that is, of the form CY for some matrix C.

It turns out that there is a very simple condition under which the solution to Equation (6.6.9) is best in this sense. To describe this condition, define the variance function analogously to the mean function in Equation (6.6.1):

$$\sigma^2(s_1, \ldots, s_r) = \text{Var}\,(Y \mid X^{(1)} = s_1, \ldots, X^{(r)} = s_r) \tag{6.6.13}$$

Then, using advanced methods beyond the scope of this text, the solution to Equation (6.6.9) can be shown to be best if this variance function is a constant; that is, if

$$\sigma^2(s_1, \ldots, s_r) = \omega^2 \tag{6.6.14}$$

for some value ω. For instance, in Example 6.6.2, this condition would mean that the variance in construction time among 500 foot bridges is the same as the variance in time among 2000 foot bridges. This is probably not true; it seems intuitively clear that the larger construction jobs should have more variation in project completion time. However, it may be approximately true.

If Equation (6.6.14) does *not* hold, then a slightly modified version of Equation (6.6.8) must be minimized to produce the best estimators:

$$\sum_{i=1}^{n} w_i(Y_i - b_0 - b_1 X_i^{(1)} + \cdots + b_r X_i^{(r)})^2 \tag{6.6.15}$$

where the weight w_i is the reciprocal of the (conditional) variance of Y_i. Unfortunately, this information is often not available in practice.

We now assume that Equation (6.6.14) does hold, and also that the conditional distribution of Y, given $X^{(1)}, \ldots, X^{(r)}$, is normal. Also, we assume that the observations $Y_1, \ldots, Y_n$ are (conditionally) independent. Then it can be shown that $\hat{\beta}$ is the maximum likelihood estimator of β, and that confidence intervals can be formed as follows: A $100(1-\alpha)\%$ confidence interval for the quantity $\lambda_0\beta_0 + \lambda_1\beta_1 + \cdots + \lambda_r\beta_r$ is

$$\lambda_0\hat{\beta}_0 + \lambda_1\hat{\beta}_1 + \cdots + \lambda_r\hat{\beta}_r \pm t\left(\frac{\alpha}{2}, n-r-1\right) S\sqrt{\lambda^t(X^tX)^{-1}\lambda} \tag{6.6.16}$$

where

$$S^2 = \frac{1}{n-r-1}(Y-X\hat{\beta})^t(Y-X\hat{\beta}) \tag{6.6.17}$$

and

$$\lambda = \begin{bmatrix} \lambda_0 \\ \lambda_1 \\ \cdot \\ \cdot \\ \cdot \\ \lambda_r \end{bmatrix}$$

and the vector λ is any set of values chosen by the analyst. The expression $t(\alpha\,2, n-r-1)$ involves the same function as used in Equation (6.3.6).

The equal-variance and normality assumptions do not have to hold exactly in order to get fairly accurate intervals. This is true especially with respect to the normality assumption, due again to the Central Limit Theorem, as in Section 6.3. The effects of violations of these assumptions are explored in Exercise 6.22.

The most common use of Equation (6.6.16) is with the λ vector consisting of a 1 in λ_j and 0s in all other components. Equation (6.6.16) then yields a confidence interval for β_j:

$$\hat{\beta}_j \pm t\left(\frac{\alpha}{2}, n-r-1\right) S\sqrt{c^{j+1,j+1}} \tag{6.6.18}$$

where $c^{k,l}$ denotes the element in the kth row and lth column of $(X^tX)^{-1}$.

However, other λ vectors are useful too. For instance, in Example 6.6.4, we may wish to form a confidence interval for the quantity

$$b = -5.00\beta_1 + 0.25d\beta_4$$

where d is the current number of road lights per mile. This quantity is the expected mean effect of making both changes proposed in that example, that is, decreasing the speed limit by 5 miles an hour and increasing the number of road lights by 25%. We could form the confidence interval for

b by taking λ to be

$$\begin{bmatrix} 0.00 \\ -5.00 \\ 0.00 \\ 0.00 \\ 0.25d \end{bmatrix}$$

in Equation (6.6.16).

It should be mentioned that although the predictors $X^{(j)}$ and the response variable Y are often in the form of a measurement, such as the length of a bridge, there are other common forms too. One such form is that of an **indicator variable** (sometimes called a **dummy variable**). An indicator variable takes on the values 1 and 0, indicating the presence or absence of some particular trait. In fact, we have already seen the use of such variables: In Examples 6.6.1, 6.6.3, and 6.6.6, the variables $X^{(3)}$, Y, and $X^{(2)}$ were indicator variables, with the traits being code optimization, a short "e" sound, and Group A, respectively.

In some applications, *all* of the predictor variables $X^{(j)}$ are of the indicator type. In such cases, the calculations above, in particular Equations (6.6.12) and (6.6.16), have traditionally been referred to as the **analysis of variance** (ANOVA).

Example 6.6.7

Suppose we have three different brands of gasoline, denoted Brands 1, 2, and 3, that we wish to compare with respect to the levels of pollutants they emit when burned. We sample $3k$ cars at random, and randomly assign k of the cars to each brand. Let Y denote the pollutant level, and let $X^{(1)}$, $X^{(2)}$, and $X^{(3)}$ be the indicator variables for Brands 1, 2, and 3. For example, $X^{(2)}$ will take on the values 1 and 0, 1 indicating Brand 2, and 0 indicating that the brand is not 2.

Let ν_i denote the mean pollutant level for Brand i, and let ν denote the overall mean among the three brands,

$$\nu = \frac{\nu_1 + \nu_2 + \nu_3}{3}$$

Then Equation (6.6.7) holds with $\beta_0 = \nu$, and $\beta_i = \nu_i - \nu$. If we assume that the variance of the pollutant level is the same for the three brands, then, Equation (6.6.14) holds too, so that we can use the interval (6.6.16); for example, we can form a confidence interval for $\nu_1 - \nu_2$ by setting $\lambda = (0, 1, -1)^t$.

Let W_{ij} denote the pollutant measurement from the jth car assigned to Brand i. Two-dimensional subscript notation is very natural here, but in order to use the framework in Equations (6.6.10), we need one-dimensional subscripts. Thus, let us rename W_{ij} as $Y_{(i-1)k+j}$. In other words, the cars assigned to Brand 1 become Observations

1 to k, those using Brand 2 become Observations $k + 1$ to $2k$, and Observations $2k + 1$ to $3k$ are those using Brand 3.

Then we can set up the matrix X and the vector Y accordingly. For example, suppose $k = 5$, and the fourth car assigned to Brand 2 had a pollutant level of 12.7. Then the $(2 - 1)5 + 4 = 9$th element in Y would be 12.7, and the ninth row of X would be

1 0 1 0

However, there is a basic problem with this: Note that since

$$X^{(1)} + X^{(2)} + X^{(3)} \equiv 1 \quad \textbf{(6.6.19)}$$

then in any given row of the X matrix, we would have the sum of the second, third, and fourth elements equal to the first element. In other words, the sum of the second, third, and fourth columns of the matrix X is equal to the first column. This will imply that the matrix inverse in Equation (6.6.11) does not exist (the reader should verify this with a simple example). This should not be really surprising, since Equation (6.6.19) shows that one of our three predictors variables is redundant (e.g., if we know the values of $X^{(1)}$ and $X^{(2)}$, then we know the value of $X^{(3)}$); this redundancy is then reflected in one of the equations in the system (6.6.9) being redundant, resulting in the nonexistence of the matrix inverse.

This problem occurs in all ANOVA applications. There are a number of ways to solve it:

a. In some simple cases, including the one above, formulas for $\hat{\beta}$ and the other quantities have already been worked out directly from the set of linear equations summarized by the matrix equation (6.6.9), so one can bypass the matrix inversion.

b. It is possible to substitute what is called a **generalized inverse** for matrix inversion. We will not discuss this process here.

c. Since the inversion problem stems ultimately from the fact that the columns of X are linearly dependent, we can remove one of them. For example, the above discussion showed that $X^{(3)}$ is redundant, so we could remove it from our model, and the matrix X would have only three columns, instead of four. Note that if we do this, the interpretation of the parameters β_i changes. Parameter β_0 is now ν_3, and for $i = 1, 2$, we would have $\beta_i = \nu_i - \nu_3$. Then, for example, if we wished to form a confidence interval for $\nu_1 - \nu_2$, we would do so for $\beta_1 - \beta_2$.

In addition to indicator variables, another important category of nonmeasurement variables is that of **product variables**. For instance, suppose in Example 6.6.2 we had reason to believe that the effect of span length on construction time is quadratic, rather than linear. Then we could change r

from 3 to 4 in Equation (6.6.1), setting

$$X^{(4)} = [X^{(1)}]^2$$

in which our product variable consists of the product of one of the original predictors with itself.

By the way, this would not be considered a "nonlinear" model: Although it is true that we would be representing the mean of Y as a nonlinear function of $X^{(1)}$, the point is that it still would be linear with respect to the parameters β_i, so we would still have a linear system (6.6.9). Truly nonlinear regression models are very useful, though. For example, in applications in which the response variable Y is of the indicator type, such as in Example 6.6.3, a very common model is that of the **logistic form**:

$$\mu(s_1, \ldots, s_r) = P(Y = 1 | X^{(1)} = s_1, \ldots, X^{(r)} = s_r) = \frac{1}{1 + \exp[-(\beta_0 + \beta_1 s_1 + \cdots + \beta_r s_r)]} \quad \textbf{(6.6.20)}$$

This model has the advantage that its value always lies between 0 and 1, very appropriate since in this application, $\mu(s_1, \ldots, s_r)$ is a probability.

Product variables also arise very frequently as **interaction terms**. To illustrate the use of these, consider Example 6.6.6. In this example, Equation (6.6.7) says

$$\text{Mean (learning time} \mid \text{experience, method)} = \beta_0 + \beta_1(\text{experience}) + \beta_2 U$$

where U is 1 for Method A, 0 for Method B. This model implicitly assumes that the difference in mean learning time between the two methods is the same at all levels of experience; for example, the difference between the two methods for employees having two years of experience is the same as the difference for ten-year veterans of the trade.

If the differences between the two methods are independent of experience, then the above model may be fine (this is called the **analysis of covariance**). However, the differences might *not* be uniform with respect to experience. It may be that employees with long years of experience are able to learn a new machine quickly using *any* training method, while the choice of training method may make a big difference when training employees who have little experience.

If it is felt that such *interaction* exists between length of experience and training method, we could change r from 2 to 3 in (6.6.1), setting

$$X^{(3)} = X^{(1)} X^{(2)}$$

Again, we must carefully note the meanings of the resulting parameters. For example, the difference in mean learning time between the two methods among ten-year veterans may be expressed as $\beta_2 + 10\beta_3$.

Product variables representing interaction effects are also very common in ANOVA models.

Example 6.6.8

Consider a more sophisticated version of Example 6.6.7, in which we sample in such a way as to take into account the brand of car. For the sake of simplicity, assume that we look at only two brands of cars, Brands A and B, and that we take measurements on five cars at each of the six combinations of brands of gasoline and cars (thus $n = 30$). This is called a "two-way" ANOVA, alluding to the fact that we are studying the effects of two factors—gasoline brand and car brand—on pollutant level. The original version of Example 6.6.7 was a one-way ANOVA.

Suppose we define $X^{(3)}$ as the indicator variable for Brand A. (The original $X^{(3)}$, the indicator for Brand 3 gasoline, has been deleted from our model, as suggested in the above discussion of nonexistence of the matrix inverse.) Then the model (6.6.1) is

$$E(Y \mid X^{(1)}, X^{(2)}, X^{(3)}) = \beta_0 + \beta_1 X^{(1)} + \beta_2 X^{(2)} + \beta_3 X^{(3)} \qquad \textbf{(6.6.21)}$$

We will leave it to the reader to verify that this is a "no-interaction" model; that is, it tacitly assumes that the differences among gasolines are the same for each brand of car. If this assumption is not justified, we can add product variables to the model:

$$X^{(4)} = X^{(1)} X^{(3)}, \quad X^{(5)} = X^{(2)} X^{(3)} \qquad \textbf{(6.6.22)}$$

Again, the reader should consider carefully the implications of this, for example, how Equation (6.6.16) might be put to use here.

Again, as mentioned at the outset of this chapter, the area of linear models is one of the most useful of all fields in statistics, and one of the richest in ideas. What we have discussed here can only be considered a very brief introduction. You are urged to explore the references listed below.

FURTHER READING

Peter Bickel and Kjell Doksum, *Mathematical Statistics: Basic Ideas and Selected Topics*, Holden-Day, 1977.

Paul Bratley, Bennett Fox, and Linus Schrage, *A Guide to Simulation*, Springer-Verlag, 1983.

John Chambers, *Computational Methods for Data Analysis*, Wiley, 1977.

William Cochran, *Sampling Techniques*, Wiley, 1977.

D. R. Cox and D. V. Hinkley, *Theoretical Statistics*, Chapman and Hall, 1974. At the level of Bickel and Doksum, but less formally presented. Includes some "statistical philosophy."

N. R. Draper and H. Smith, *Applied Regression Analysis*, Wiley, 1981.

Richard Duda and Peter Hart, *Pattern Classification and Scene Analysis*, Wiley, 1973.

Lloyd Fisher and John McDonald, *Fixed Effects Analysis of Variance*, Academic Press, 1978. Treats general linear models, including regression and analysis of variance.

Richard Johnson and Dean Wichern, *Applied Multivariate Statistical Analysis*, Prentice-Hall, 1982.

William Kennedy and James Gentle, *Statistical Computing*, Marcel Dekker, 1980.

J. H. Maindonald, *Statistical Computation*, Wiley, 1984.

L. R. Rabiner and R. W. Schafer, *Digital Processing of Speech Signals*, Prentice-Hall, 1978.

Reuven Rubinstein, *Simulation and the Monte Carlo Method*, Wiley, 1981.

G. A. F. Seber, *Linear Regression Analysis*, Wiley, 1977.

B. W. Silverman, *Nonparametric Density Estimation*, Marcel Dekker, 1986.

J. T. Tou and R. C. Gonzalez, *Pattern Recognition Principles*, Addison-Wesley, 1974.

Sanford Weisberg, *Applied Linear Regression*, second edition, Wiley, 1985.

EXERCISES

6.1 (S) In Example 6.2.1 concerning the estimation of the number of lottery tickets sold, the Central Limit Theorem was used to find that an approximate value for $P(|W_2 - \theta| \leq 2)$ is 0.1742. Use simulation to determine the accuracy of this approximation.

6.2 Consider Example 6.1.7, in which the MLE of r was found for a random sample of size n from the density rt^{r-1}.

- **a.** (M) Find the Method of Moments Estimator of r, W_{MM}.
- **b.** (S) Develop a Pascal function X(R,Seed) that will simulate a random variable X having the density rt^{r-1}. (Use the inverse c.d.f. method from Section 3.4.)
- **c.** (S) Denote the Maximum Likelihood Estimator of r by W_{ML}. Compare the accuracies of W_{MM} and W_{ML}, for the case in which the true value of r is 2, by computing $P(|W_{ML} - r| < c)$ and $P(|W_{MM} - r| < c)$ for various values of c of your choice.
- **d.** (M) Use the Central Limit Theorem to do the comparison in Part (c) above.

6.3 (M) In Example 6.2.3, suppose $\mu_i = 10i$, for $i = 1, 2, 3, 4, 5$, and $\omega = 8$. Also, suppose Plan I takes a sample of size 100, and Plan II samples 20 from each of the five subpopulations. Use the Central Limit Theorem to find $P(|\overline{Y} - \mu| < c)$ and $P(|\overline{X} - \mu| < c)$ for various values of c of your choice.

6.4 (M) In Example 6.3.2 concerning airline air conditioners, the confidence interval turned out to be rather wide, with a radius of $(107.01 - 79.69)/2 = 13.66$. This was with a sample of 220 units. Approximately how many additional units would need to be sampled in order to get an interval with a radius of 10.0?

6.5 (M) Consider Example 6.5.1, which analyzed the output of the simulation in Program 2.4.6. That program used a TotReps value of 1000, producing a confidence interval for $P(A)$ of radius $(0.360 - 0.302)/2 = 0.029$. Suppose we had used TotReps = 5000. Approximately what radius would have resulted?

6.6 Consider the raffle problem in Example 6.1.3. Continue to assume independence of $X, \ldots, X_{10}$, as discussed in Example 6.2.1.

a. (M) Show that the analog estimator T in Equation (6.1.3) is also the Maximum Likelihood Estimator. (Note: You will not be able to do the maximization through use of derivatives, since the likelihood function is not differentiable. Instead, try to maximize directly. "Practice" with the case $n = 2$ first, before going to the actual value of $n = 10$. Make sure to keep in mind the values that will make the likelihood function 0.)

b. (M) Show that $[cT, \infty)$ is an approximate $100(1-\alpha)\%$ confidence interval for θ, where

$$c = \frac{1}{(1 - \alpha)^{1/10}}$$

c. (S) Use simulation to determine the accuracy of the approximation in Part (b) above. That is, find out how close the true probability of coverage comes to $1-\alpha$. Look at the case $\theta = 50$ and $\alpha = 0.05$.

6.7 S) Consider Program 6.3.1. Rerun the program using Equation (6.3.6) instead of Equation (6.3.5), to see how well an analyst would do if he or she assumed a normal population when it was actually exponential.

6.8 (S) Run an analog of Program 6.3.1 for the case in which the observations X_i are all either 0 or 1, say for $\pi = 0.4$.

6.9 Consider the material following Example 6.5.1, regarding the choice of TotReps.

a. (M) Show that the largest value of the quantity $\pi(1 - \pi)$ occurs for $\pi = 0.5$, and state why this value was termed a "conservative guess."

b. (M) Suppose we are planning a simulation experiment in which we will be estimating a probability π, which we believe will be at most 0.02. How large should we make TotReps, if we desire the radius of our confidence interval for π to be around 0.004?

6.10 (M) Suppose we are planning to collect a random sample from a population having unknown mean μ and known standard deviation 3.0. How large should we choose the sample size n so that there will only be about a 2% chance that the sample mean will overestimate μ by more than 0.5 unit?

6.11 (M) Suppose we have sampled some industrial process, using measurements X and Y from two machines. The standard deviation of the measurement from the second machine is known to be 3 times that of the first. However, both measurements are unbiased; that is, $EX = EY = \gamma$, where γ is the quantity to be measured. We will use a combined estimate $T = aX + bY$.

a. What relationship should a and b have, for T to be unbiased?

b. Determine a and b so that T will have minimum variance, under the constraint that it be unbiased.

6.12 (M) Derive the confidence intervals (6.3.9) and (6.3.14).

6.13 (M) An urn contains ω white marbles and one black marble. We wish to estimate ω using the single observation X, where X is the number of times we must sample marbles (without replacement) until we draw a black one. Thus, X can take on the values $1, 2, \ldots, \omega + 1$. Find the Method of Moments Estimator of ω based on X.

6.14 (M,S) Recall Examples 3.1.3 and 3.2.2, involving sampling of straight pieces of wire from a bag. Assume that the longer wires have a greater chance of being sampled, as before. Suppose someone wishes to estimate the mean length μ of all wires in the bag. He or she samples n wires with replacement from the bag, resulting in measured lengths $L_1, \ldots, L_n$, and estimates μ by

$$\bar{L} = \frac{L_1 + \cdots + L_n}{n}$$

As mentioned in Section 5.1, $\bar{L}$ will then be a biased estimator of μ (although it will be an unbiased estimator of EL). Note that $\mu = 4.00$, although of course this is unknown to the person doing the sampling.

a. Find the bias of $\bar{L}$ as an estimator of μ.

b. Suppose $n = 100$. Use the Central Limit Theorem to find the approximate probability that $\bar{L}$ comes within c units of the true value of μ, for various values of c of your choice. Check your answers, using simulation.

6.15 (M) Suppose we wish to predict a random variable W, knowing only its density f_W.

a. Suppose that we use as our criterion of prediction accuracy the mean squared prediction error $E[(W - c)^2]$. Show that the optimal prediction value under this criterion is $c = EW$.

b. Suppose instead that we use as our criterion the mean absolute error $E[|W - c|]$. Show that the optimal prediction value c then will be the median of W.

6.16 (M) Show that if all observations X_i in a sample take on only the values 0 and 1, then

$$S^2 = \frac{\hat{\pi}(1 - \hat{\pi})}{n}$$

where, in this context, S^2 is defined as in Equation (6.1.2).

6.17 (M) Rederive Equation (6.2.7), by writing $\sum_{i=1}^{n}(X_i - \bar{X})^2$ as

$$\sum_{i=1}^{n} [(X_i - \mu) - (\bar{X} - \mu)]^2$$

6.18 (S) Let us investigate the question of how large to make the subinterval size h in a histogram. We can do this by sampling from a population with *known* density f, and examining the histograms that result from the use of various values of h. Of course, as mentioned before, if we know f, there is no point in estimating it. However, we can ask the question, *if an analyst did not know this* f *and estimated it from a sample, what value of* h *would give the best results?*

One way to sample from a known density is to use simulation, which we will do here.

a. Generate a sample of size $n = 100$ from an exponential density having mean 10.0. Draw histograms for various values of h. Indicate which one gives the most accurate estimate of f, that is, which one most resembles an exponential density.

b. Do (a) again for a sample of size $n = 1000$. The optimal value of h should be smaller here.

6.19 (S) In Section 4.4, it was mentioned that the Central Limit Theorem does not say that the density of the quantity W_n converges to that of $N(0, 1)$; the only claim made by the theorem is that the cumulative distribution function F_{W_n} converges to the c.d.f. of $N(0, 1)$. However, there is a theorem that says convergence in the density sense does occur too. The mathematical conditions for this will not be set forth here, but instead we will explore this issue through simulation. Let us estimate the density of W_{10}. To do this, generate 1000 samples of size 10 from an exponential distribution having mean, say, 4.0. This gives 1000 values for W_{10}, from which you can get a histogram estimate of the density of that random variable.

6.20 (S) An important concept in statistical applications is that of **multiple inference**. To illustrate this concept, suppose we take independent random samples, say of size 25, from three separate populations having unknown means μ_1, μ_2, and μ_3, and suppose we wish to form 95% confidence intervals for all possible differences, $\mu_1 - \mu_2$, $\mu_1 - \mu_3$, and $\mu_2 - \mu_3$, using the three sample means $\overline{X}_1$, $\overline{X}_2$, and $\overline{X}_3$. For simplicity, let us suppose that we know that the populations are normal with equal variances, so that we can use Equation (6.3.10) to form exact confidence intervals. The problem is this: For each interval that we make, we are 95% confident that the interval is correct, that is, that the true value being estimated is contained by the interval ("confident" in the sense explained in Section 6.3). However, it is clear that we are *not* 95% confident that all three intervals are correct. In fact, if the three intervals were independent, the *overall* confidence level would then be $(0.95)^3$, or about only 86%. On the other hand, the three intervals are not independent; for example, the ones for $\mu_1 - \mu_2$ and $\mu_1 - \mu_3$ are not independent, since they both involve $\overline{X}_1$.

a. (M) Using Bonferroni's Inequality (Exercise 2.17), show that the true overall confidence level is at least 85%.

b. (S) Find the true overall confidence level.

6.21 (M) In Section 6.6, consider the case in which $r = 1$ and the predictor X is a random variable, that is, X and Y have a joint distribution.

a. Use Equation (6.6.9) to derive the following explicit expressions for $\hat{\beta}_0$ and $\hat{\beta}_1$:

$$\hat{\beta}_1 = \frac{\sum_{i=1}^{n}(X_i - \overline{X})(Y_i - \overline{Y})}{\sum_{i=1}^{n}(X_i - \overline{X})^2}$$

$$\hat{\beta}_0 = \overline{Y} - \hat{\beta}_1 \overline{X}$$

(*Hint:* First review the algebraic manipulations near the end of Example 6.1.5.)

b. Show that $\hat{\beta}_0$ and $\hat{\beta}_1$ are unbiased. *Hint:* First condition on the X values, and use Equation (4.1.15).

c. Show that $\hat{\beta}_0$ and $\hat{\beta}_1$ are **consistent**, that is, that

$$\lim_{n \to \infty} \hat{\beta}_j = \beta_j$$

for $j = 0, 1$.

6.22 (S) Suppose $X^{(1)}$ has a $U(0, 10)$ distribution, and that we will be using the material in Section 6.6 to investigate the relationship between $X^{(1)}$ and another random variable Y.

a. Suppose (unknown to the one collecting and analyzing the sample data) the conditional distribution of Y, given $X^{(1)} = s$, is $U(s - 5, s + 5)$. Explain why Equations (6.6.7) and (6.6.14) hold here, and determine the values of β_0, β_1, and ω^2. Suppose Equation (6.6.18) is used to form a 95% confidence interval for β_1. Use simulation to find the true confidence level, for $n = 10$, 50, and 200, and comment on the "robustness" of linear model theory with regard to the normality assumption.

b. Suppose the conditional distribution of Y, given $X^{(1)} = s$, is exponential with mean s. Explain why Equation (6.6.7) holds here, and give the values of β_0 and β_1. Explain why Equation (6.6.14) does *not* hold here. Suppose Equation (6.6.18) is used to form a 95% confidence interval for β_1. Use simulation to find the true confidence level for $n = 100$, and comment on the "robustness" of linear model theory with regard to the equal-variance assumption.

6.23 You have the opportunity to buy a certain kind of electronic component very inexpensively, but you are concerned about the general proportion of defectives, π. To investigate this, you keep testing these components until you find one that does not work. Let N be the number that you must inspect in this way.

a. (M) Determine the probability mass function p_N.

b. (M) Find the ML estimator of π.

c. (M) Show that $(1 - \alpha^{1/N}, 1)$ is an approximate $100(1 - \alpha)\%$ confidence interval for π. [*Hint*: First note that

$$P(N > k) = (1 - \pi)^k$$

Make use of logarithms in algebraic manipulations needed to find $P(1 - \alpha^{1/N} \leq \pi)$.] Note that the Central Limit Theorem-based intervals do not work here; why not?

d. (S) For $\pi = 0.10$, use simulation to find the exact confidence level of the interval in Part (c).

6.24 (M) Suppose a television manufacturer is measuring the degree of radiation to which a viewer is exposed. It is reasonable to assume that this is inversely related to the square of the distance from the viewer to the TV set. Suppose measurements are taken on 12 TV sets, according to the following scheme: The distances examined are 3 feet, 6 feet, and 9 feet, and at each distance, four of the televisions are measured for radiation.

a. Suggest a linear model for this setting , and write down the numerical X matrix for it.

b. Make up data for the Y vector, and find a 95% confidence interval for the mean radiation level at a distance of 5 feet.

c. Suppose we were to take measurements on each TV at each distance, giving a total of three measurements for each TV. Comment on the effect of this on the independence assumption made in the linear model.

d. Suppose we are comparing two different kinds of televisions. Suggest changes to the above sampling scheme, and models to use for the analysis. State precisely how you would use these models to compare the two kinds of televisions.

6.25 (S) Consider Example 6.6.1, simplified by deleting the predictors $X^{(2)}$ and $X^{(3)}$. Suppose that the distribution of $X^{(1)}$ is $N(200, 2500)$, and that the conditional distribution of Y given $X^{(1)} = s$ is $N(30 + 20s, 100)$, with all of these means and variances known to the analyst. "Shortest Job First" scheduling should minimize mean job completion time (waiting time plus actual compilation time) among all jobs. However, if we use Equation (6.6.7) to predict Y for a job when we are deciding when to schedule the job, we will occasionally make mistakes. Suppose we are scheduling k jobs. As suggested in Example 6.6.1, we can schedule them according to their predicted Y values. This ordering might not be identical to that based on the actual Y values. Find the mean job completion time if ordering is done on the basis of predicted Y values, and the mean time that we would obtain if we knew the actual Y values and scheduled the jobs on that basis. Compare, and comment on whether the lack of knowledge of the true Y values hurts us much. Do this problem for several different values of k.

6.26 (M) Show that the logistic model (6.6.20) is implied under either of the following settings, and express the parameters β_j in terms of the specified parameters [i.e., in Part (a), express the β_j in terms of the μ_i and Σ, and in (b), in terms of the p_{ij}]:

a. The conditional distribution of $X^{(1)}, \ldots, X^{(r)}$ given $Y = i$ is r-variate normal, with mean vector μ_i and covariance matrix Σ (with *no* subscript i), $i = 0, 1$.

b. Conditional on $Y = i$, $X^{(1)}, \ldots, X^{(r)}$ are independent random variables, with $P(X^{(j)} = 1) = p_{ij}$ and $P(X^{(j)} = 0) = 1 - p_{ij}$, $i = 0, 1$.

6.27 (S) An analog estimator of the correlation $\rho(X, Y)$ (Section 4.2) is

$$r = \frac{1/n \sum_{i=1}^{n} (X_i - \overline{X})(Y_i - \overline{Y})}{\sqrt{(1/n) \sum_{i=1}^{n} (X_i - \overline{X})^2} \sqrt{(1/n) \sum_{i=1}^{n} (Y_i - \overline{Y})^2}}$$

This estimator is biased upward if $\rho > 0$. The bias can be rather substantial for small n, although the bias goes to zero as n goes to infinity. Show this using simulation, using sample sizes of 5, 10, 20, and 50, in the following setting: X has a $N(0, 1)$ distribution, and the conditional distribution of Y given $X = s$ is $N(s, 1)$. Note that you will need to find ρ first; Equations (4.1.15) and (4.1.17) will be helpful in this regard.

6.28 (S) We have just invented a new, improved battery. The battery lifetime has an exponential distribution whose new mean μ is longer than the old value of 100. To investigate how much better the new type of battery is, we adopt the following scheme: We will test 10 batteries at first. If the resulting $\overline{X}$ value is over 100, we will use this value as our estimate of μ. On the other hand, if $\overline{X} < 100$, we will decide that our sample was "not representative," and throw it out. We will then take a new sample of 10 batteries, and in fact will continue to take and throw out 10-battery samples until one of them has $\overline{X} > 100$. This $\overline{X}$ will then be our estimate of μ. Obviously this procedure results in a biased estimator of the mean battery lifetime. Find the amount of the bias, if in fact $\mu = 102$.

6.29 (M) Consider Example 6.6.8, which analyzes brands of gasoline and cars. Suppose we have measurements on only two cars for each gasoline brand/automobile brand combination, so that $n = 12$.

Suppose the data collected are those displayed below, as follows: There are three rows and two columns, corresponding to the three brands of gasoline and two brands of cars. The two observations for each gasoline brand/automobile brand combination are listed together, with a comma after the first. For example, for the third brand of gasoline and second brand of car, the two measurements were 21 and 23.

25, 24	28, 27
22, 22	25, 26
20, 21	21, 23

a. First, suppose we include the "interaction terms" in Equation (6.6.22) in our analysis, so that Equation (6.6.21) becomes

$$E(Y \mid X^{(1)}, X^{(2)}, X^{(3)}, X^{(4)}, X^{(5)}) = \beta_0 + \beta_1 X^{(1)} + \beta_2 X^{(2)} + \beta_3 X^{(3)} + \beta_4 X^{(4)} + \beta_5 X^{(5)}$$

Compute the vector $\hat{\beta}$ in Equation (6.6.12). Find a 95% confidence interval for the population mean difference between gasoline Brands 1 and 2, for automobile Brand 1. Also, find 95% confidence intervals for β_4 and β_5, and comment on whether the interaction terms seem necessary.

b. Recalculate the first confidence interval in (a), assuming a no-interaction model [regardless of your answer in (a) concerning the necessity of the interaction terms].

6.30 (M) In Example 6.3.4, suppose there are three text editors, E1, E2, and E3, and that the manager asks 25 programmers to use each of the three editors for a period of time long enough to establish a preference for one of them. Let X_{ij} be 1 or 0, depending on whether Programmer i decides that Ej is the best editor. For example, if Programmer 4 feels that E3 is the best editor, we would have $X_{43} = 1$ and $X_{41} = X_{42} = 0$. Suppose the manager wants to find a 95% confidence interval for

$\pi_1 - \pi_2$, where π_j is the population proportion of all programmers who prefer Ej. Since X_{i1}, X_{i2}, and X_{i3} are not independent, we can not use Equation (6.3.12) here. The object of this exercise is thus to derive a formula appropriate for this setting, as follows. Let

$$\hat{\pi}_j = \frac{1}{25} \sum_{i=1}^{25} X_{ij}$$

be the estimate of π_j, $j = 1, 2, 3$. The vector

$$T = \begin{bmatrix} \sum_{i=1}^{n} X_{i1} \\ \sum_{i=1}^{n} X_{i2} \\ \sum_{i=1}^{n} X_{i3} \end{bmatrix}$$

is multinomial, and thus has an approximate multivariate normal distribution (recall Example 4.5.4). Now use the linearity property of that distribution on the vector

$$\hat{\pi} = \begin{bmatrix} \hat{\pi}_1 \\ \hat{\pi}_2 \\ \hat{\pi}_3 \end{bmatrix}$$

Then follow a line similar to that of Examples 4.5.2 and 4.5.5.

6.31 (M) Suppose we wish to estimate $1/\pi$, where π is the probability of heads for a certain coin, based on a single toss X ($X = 0$ or 1, for tails or heads). Show that there does not exist an unbiased estimator of $1/\pi$, that is, there is no function $g(X)$ such that $E\,[g(X)] = 1/\pi$ for all π in (0, 1].

CHAPTER 7

Analysis of Sample Data: Making Decisions

7.1 SIGNIFICANCE TESTS FOR MEANS AND PROPORTIONS

The subject matter for this section is best introduced by way of example:

Example 7.1.1

A manufacturer of a certain machine component has found that 4% of all components produced are defective. Recently, there has been a sudden increase in demand for the parts, so most workers have been working overtime. It is felt that perhaps this has resulted in an increase in the proportion of defectives. Out of the last 500 components produced, 30 (or 6.0%) have been defective. Should we conclude from this that the proportion of defectives has in fact increased?

Let π be the proportion of defectives produced under current conditions; note that π is the day-in, day-out defective rate, not the proportion in our *sample* of 500 (which we will denote by $\hat{\pi}$). We have two **hypotheses** here. The hypothesis that nothing has changed is called the **null hypothesis**, and is denoted by H_0. The **alternative hypothesis**, denoted by H_A, is that the proportion has increased:

$$H_0: \pi = 0.04$$
$$H_A: \pi > 0.04 \qquad \textbf{(7.1.1)}$$

We wish to choose one of these two hypotheses as representing the true state of our manufacturing process.

We can approach this question as follows. We take the "innocent until proven guilty" attitude, adopting the convention that we will believe H_0 until we receive strong evidence to the contrary, that is, strong evidence in support of H_A. In this case, "strong evidence" would be getting many more than the (0.04)(500) = 20 defectives we would "expect" to get in our sample of 500 if H_0 were true. The question here is whether 30, the number of defectives that our sample turned out to have, should be considered "many more" than 20.

Let X denote the number of defectives out of 500. Keep in mind that, even if H_0 were true, it still would be possible to get a very large value of X, even $X = 500$. As an analogy, suppose we wish to test the hypothesis that a certain coin has exactly 50% chance of coming up heads. Even if the coin were perfectly balanced, it would still be possible to get 500 heads out of 500 tosses of that coin. Of course, the key point is that this would be extremely unlikely. If we were to get 500 heads out of 500 tosses, we would conclude that the coin was *not* balanced. However, we could not be absolutely sure of this, since it would be *possible* to get 500 heads out of 500 with a balanced coin, but this event would be so unlikely for a balanced coin that we choose to believe instead that the coin is not balanced. The point is that all we can do is make our best guess, with the understanding that we still could be wrong.

Relating this coin-tossing example to the question of the defective components, recall that we said that we would believe H_0 (i.e., *accept* H_0) unless X came out much larger than 20, in which case we would believe H_A (i.e., *reject* H_0). Thus we will reject H_0 if and only if $X > c$, where c is some suitably chosen cutoff value.

What value should be chosen for c? This question can be answered by noting that we said above that we will reject H_0 if the event $\{X > c\}$ would be extremely unlikely if H_0 were true. A commonly-used standard for "extremely unlikely" is a probability of only 5%; that is, we choose c so that, under H_0, there would be only a 0.05 probability that we would get more than c defectives in our sample of 500.

We can thus calculate c, using the Central Limit Theorem, as in Example 4.4.3. Since we are asking what would happen if H_0 were true, the following calculations all use $\pi = 0.04$: X is approximately normally distributed with mean (500)(0.04) = 20, and variance (500)(0.04)(0.96) = 19.2. Thus

$$0.05 = P(X > c) = P\left(\frac{X-20}{\sqrt{19.2}} > \frac{c-20}{\sqrt{19.2}}\right) \tag{7.1.2}$$

From the $N(0,1)$ in Table 2 in Appendix B we see that 0.05 =

$P(Z > 1.65)$, for any $N(0,1)$ random variable Z. Thus we can set $1.65=(c-20)/\sqrt{19.2}$, and find that $c=27.2$, which we will round off to 27. [By the way, for better accuracy, we really should have used the "correction for continuity" for the normal approximation to the binomial distribution, Exercise 4.29. We have not done so here, so that we could simplify the presentation.]

Thus, we have set a policy that we will reject H_0 only if there are more than 27 defectives in our sample of 500. The rationale for this is that it would be "extremely unlikely" (5% chance) to get this many defectives if π were only 0.04. Thus, if we get 30 defectives, this would be considered sufficiently strong evidence to reject H_0.

The 5% value used above is called the **significance level** of our hypothesis test; it is usually denoted by α, and it is also called the probability of a **Type I error**. The latter term stems from the fact that if H_0 were actually true, there would be probability α that we make the wrong decision, that is, reject H_0. Although 5% is the most commonly used value, it is rather arbitrarily chosen, and each analyst can certainly choose whatever value is most meaningful in the context of the analyst's investigation. Perhaps a value of 1% or 10% may make more sense in some cases. For example, for $\alpha = 0.01$, we would have $2.33 = (c-20)/\sqrt{19.2}$ in Equation (7.1.2), which would give a cutoff point of 30 instead of 27. Thus we would *not* reject H_0 in this case; we would demand stronger evidence than that provided by the value $X = 30$.

How should α be chosen? One criterion involves β, the probability of a **Type II error**. This type of error means that we **accept** H_0 in a situation in which H_A is true. If β turns out to be too large, we may wish to make α a little larger, so that we may reduce the value of β. For example, suppose in Example 7.1.1 that the true state of nature is that $\pi = 0.08$. This would represent a severe deterioration in the quality of our manufacturing process, with the rate of defectives actually doubling. If this were the case, we would want to know about it.

Let us investigate what the chances are that we *would* know about it, that is, that the result of our hypothesis test would be to reject H_0: Keeping in mind that in this situation X would have mean $(500)(0.08) = 40$ and variance $(500)(0.08)(0.92) = 36.8$, rather than the values used in Equation (7.1.2), we have, for $\alpha = 0.05$,

$$\begin{aligned}\beta &= P(X \leq 27)\\ &= P\left[\frac{X-40}{\sqrt{36.8}} \leq \frac{27-40}{\sqrt{36.8}}\right]\\ &= P(Z < -2.14)\\ &= 0.0162 \end{aligned} \tag{7.1.3}$$

This is good; if the defective rate has actually doubled, we will have less than a 2% chance of falsely deciding that π is still at the 0.04 level.

Thus, in this example we are fortunate enough to have both α and β reasonably small. It is vitally important for the reader to recognize that this is a consequence of the fact that we have a fairly large sample ($n=500$). Consider what would have happened if n had been 50, for example (you should verify the following calculations). Using the "standard" α value of 0.05, the cutoff point c would be 4, and then the β value would be 0.50. Thus, if the defective rate had doubled, we would have a 50% chance of falsely deciding that π had not increased. This is too large a probability, so we may have to allow α to increase somewhat, unless we are able to increase n. All of these aspects must be considered in the planning stages of an investigation, especially so that the analyst can determine a reasonable value for the sample size n.

Example 7.1.2

Suppose the analyst testing Equation (7.1.1) wishes to have $\alpha = 0.05$ and $\beta = 0.10$. What is the minimum sample size that will allow us to meet these criteria? We need to find n (and c) satisfying

$$0.05 = P(X > c) = P\left(\frac{X - n(0.04)}{\sqrt{n(0.04)(0.96)}} > \frac{c - n(0.04)}{\sqrt{n(0.04)(0.96)}}\right)$$

and

$$0.10 = P(X < c) = P\left(\frac{X - n(0.08)}{\sqrt{n(0.08)(0.92)}} < \frac{c - n(0.08)}{\sqrt{n(0.08)(0.92)}}\right)$$

Since $0.05 = P(Z > 1.65)$ and $0.10 = P(Z < -1.28)$ for Z which has a $N(0,1)$ distribution, we set

$$1.65 = \frac{c - n(0.04)}{\sqrt{n(0.04)(0.96)}}$$

and

$$-1.28 = \frac{c - n(0.08)}{\sqrt{n(0.08)(0.92)}}$$

Solving these two equations, we find that $n=289$ (and $c=17$). Thus a sample of 289 components is large enough for our needs.

In planning an investigation, the analyst must choose two of the three quantities n, α, and β. This then determines the value of the third quantity.

Note also that we really should not be talking about *the* value of β. Actually, there are infinitely many values for β, one for each value of π. Thus it is more appropriate to write $\beta(\pi)$, e.g. we calculated $\beta(0.08)$ above. Of course, it should be kept in mind that we do not know the true value of π (otherwise, the sample and the test would of course be completely unnecessary). We are only asking "what if" questions: "What if the true value of π were actually 0.08? This would be a severe increase in our defective rate. Would we be able to detect this increase?"

Because of questions like this, we usually discuss $1 - \beta$ instead of β itself, and call this quantity the **power** of the test. In the current example, involving π, we would write power (π). The connotation of the word "power" is that our test has strong ability to reject H_0. In the situation discussed above, for example, we would like power(0.08) to be large; that is, if the defective rate actually has doubled, we would like to have a large probability of detecting the increase.

For fixed α, the power of a test is highly affected by both the sample size n and the amount of departure of the true parameter value from H_0, in this case the size of $\pi - 0.04$. You should carefully review the above example, and convince yourself of these facts.

Let us look once more at the original setting that we considered above, in which n was 500 and α was 0.05. Since our observed value of X, 30, was larger than the cutoff point of 27, we rejected H_0. In fact, with $X = 30$, we would have rejected H_0 even with an α value smaller than 0.05. To see this, work backwards in Equation (7.1.2), by replacing c by 29, and solve for a probability that will replace the 0.05. We find that this new probability is

$$P\left(Z > \frac{29 - 20}{\sqrt{19.2}}\right) = 0.02$$

Thus, if we had originally chosen $\alpha = 0.02$, we would have had a cutoff point of 29, and $X = 30$ would have just barely resulted in rejecting H_0 [since we used the strict $>$ in Equation (7.1.2), rather than $\geq$]. So, for our observed value of $X = 30$, the smallest α for which we would reject H_0 is 0.02. This "minimum α under which we reject H_0" value will be called the **observed significance level** (OSL) here. (It often is called the **p-value**.)

In reporting the results of a hypothesis test, the OSL value is considered more informative than merely reporting whether H_0 was rejected or not. It allows each reader of such a report to set his or her own personal value of α; the reader simply compares his or her own α to the OSL, and rejects H_0 if and only if the OSL is smaller.

The OSL is also reported for another reason: It gives us some idea as to "how strongly" we rejected H_0 (if we did reject it). A very small OSL suggests that, not only was H_0 false, but also it was "not even close" to be being true. *However,* this conclusion does not necessarily follow, and may lead to quite incorrect analysis. This will be discussed further in Section 7.3.

The form of H_A, which uses the $>$ symbol, gives rise to the description of H_A as a **one-sided alternate** (of course, the use of the $<$ symbol would also be described in this way). It is also possible to have a **two-sided alternate** H_A:

$$\begin{aligned} H_0&: \pi = 0.04 \\ H_A&: \pi \neq 0.04 \end{aligned} \tag{7.1.4}$$

In Example 7.1.1, we were reasonably sure that the increase in overtime work had either a harmful effect ($\pi > 0.04$) or no effect ($\pi = 0.04$), but definitely not a beneficial effect ($\pi < 0.04$). If we had not been so sure of this, Equation (7.1.4) would have been more appropriate than Equation (7.1.1).

In the two-sided case, the probability $P(X > c)$ in Equation (7.1.2) would have been replaced by $P(X < c_1 \text{ or } X > c_2)$, since either an unusually low value of X (i.e., one much lower than 20) or an unusually high value (much higher than 20) would be evidence that H_0 is false. In practice, the values c_1 and c_2 are taken to be symmetric with respect to the mean of X under H_0, giving, in this case, $c_1 = 20 - d$ and $c_2 = 20 + d$ for some d. Now Equation (7.1.2) would start out as

$$0.05 = P(X < 20 - d \text{ or } X > 20 + d)$$

The reader should verify that the result of this is $d = 8$, so that we would reject H_0 if we get either fewer than 12 defectives in our sample of 500, or more than 28.

Although the idea of one-sided and two-sided alternate hypotheses is related to the concept of one-sided and two-sided confidence intervals in Section 6.3, there is one crucial difference: In order to validly set up a one-sided hypothesis such as in Equation (7.1.1), you must be reasonably sure that the excluded case [in Equation (7.1.1), we excluded the case $\pi < 0.04$] is not possible. On the other hand, one-sided confidence intervals are quite valid without this provision.

Two-sample tests can be formed in exactly the same way as shown above, as can tests concerning means. The key step is to use the (exact or approximate) normality of the observed summary statistic, such as X in Example 7.1.1.

Example 7.1.3

Recall Example 6.3.4, concerning a comparison of two competing text editors. Suppose we are fairly sure that editor E2 is at least as good as, if not better, than E1, in terms of program development time. The reader should verify that we can test

$$H_0: \mu_1 = \mu_2$$
$$H_A: \mu_1 > \mu_2 \qquad \textbf{(7.1.5)}$$

with $\alpha = 0.05$ by rejecting H_0 if and only if

$$\frac{\bar{X} - \bar{Y}}{\sqrt{\frac{S_1^2}{20} + \frac{S_2^2}{20}}} > 1.65 \qquad \textbf{(7.1.6)}$$

Note that under H_0 we have $E(\bar{X} - \bar{Y}) = 0$, so that the numerator in Equation (7.1.6) represents $\bar{X} - \bar{Y} - 0$, just as we have $X - 20$ in the numerator in Equation (7.1.2). Since our data set yields a value of only 1.36 for the left side of Equation (7.1.6), a value not greater than 1.65, we do not reject H_0. The data suggest that H_A may be true, but the evidence is not strong enough to meet our criterion of $\alpha = 0.05$.

Of course, there is no particular reason for choosing this value of α; this is just the one most frequently used in practice. It may be that in this and many other cases in practice, a hypothesis test is not appropriate anyway. The reason for this is that we are not merely interested in whether E2 is better than E1—which is all Equation (7.1.5) is concerned with—but rather we are interested in *how much* better E2 is than E1. After all, E2 adds a considerable burden to the work of our system disk unit, slowing down user response time. If E2 is only a little better than E1, we may not wish to use E2. This discussion will be continued in Section 7.3.

7.2 SOME OTHER KINDS OF TESTS

There is a tremendous variety of hypothesis tests which have been devised and are in common use. In this section, we present some of the most frequently encountered ones.

Let us look at tests concerning variances first, a natural step to follow our tests concerning means in Section 7.1. Suppose we have a random sample $Y_1, \ldots, Y_n$ from a population having variance σ^2. In the past, we measured the variance to be σ_0^2, but we suspect that it may have increased recently. We may therefore wish to test

$$H_0: \sigma^2 = \sigma_0^2$$
$$H_A: \sigma^2 > \sigma_0^2 \qquad \textbf{(7.2.1)}$$

The obvious statistic to use in our test is the sample variance

$$S^2 = \frac{1}{n-1} \sum_{i=1}^{n} (Y_i - \bar{Y})^2$$

where of course

$$\bar{Y} = \frac{1}{n}\sum_{i=1}^{n} Y_i$$

As mentioned previously, in Section 6.3, it can be shown that *if the sampled population has a normal distribution*, then

$$\frac{(n-1)S^2}{\sigma^2} \tag{7.2.2}$$

has a chi-square distribution with $n-1$ degrees of freedom (this distribution was defined near the end of Section 3.3). Thus, under H_0, the quantity

$$W = \frac{(n-1)S^2}{\sigma_0^2} \tag{7.2.3}$$

has this distribution, and thus, we can reject H_0 if W is greater than $\chi^2(\alpha; n-1)$, the upper-α cutoff point of this distribution.

Example 7.2.1

Recall Example 7.1.1. Suppose each manufactured component has a certain measurement Y, with mean μ. Using the methods of Section 7.1, we could test to see if there was a change in μ following the increase in overtime hours. Suppose we do so, and decide that μ has not changed. This still may not be an indication that there have been no ill effects from the increase in overtime. For example, the variance σ^2 may have increased, and if the customers depend on a consistent (i.e., small σ) value for Y, an increased variance may be very undesirable.

Suppose we collect a sample of 50 components, and find that $S^2 = 24.2$. The old value of σ^2, before the increase in overtime, was 15.6. Suppose also that we are satisfied that the sampled population is normal (we may have used the methods of Section 6.4 to decide this). The value of Equation (7.2.3) is then 76.0, which is greater than the 67.5 cutoff point (for $\alpha = 0.05$) in the chi-square table in Table 4 in Appendix B.

Thus we would conclude that the variance has increased. Note, however, that this does *not* imply that the variance has increased *enough* for us to be worried. We should estimate the amount of the increase. One can form a confidence interval for the variance; you should verify, using Equation (7.2.2), that a 95% confidence interval for σ^2 is

$$\left(\frac{(n-1)S^2}{a_2}, \frac{(n-1)S^2}{a_1}\right) \tag{7.2.4}$$

where $P(a_1 < T) = P(T < a_2) = 0.025$ for a random variable T having a chi-square distribution with $n - 1$ degrees of freedom; that is, $a_1 = \chi^2(0.025; n-1)$ and $a_2 = \chi^2(0.975; n-1)$. (See Exercise 7.1.)

With our data, we get a confidence interval for σ^2 of (16.6, 36.6). Note that this interval shows that the increase in σ^2 might actually have been very small, perhaps an increase only from 15.6 to 16.6. A small increase such as this may not justify abandoning our overtime policy. On the other hand, the interval is very wide, indicating that the increase in σ^2 might be either small *or* large. We would need a larger sample to get a shorter interval, and thus get a better idea as to whether the increase was in fact small. We can not make a well-informed decision with the data we have available here.

It is very important to note that the use of the chi-square distribution here assumes that the sampled population is normal. In Section 6.3, Equation (6.3.6), we encountered another situation in which this assumption was made. However, in that case, the normality assumption was not really very important, except in small samples. The reason for this is that for large samples, Equation (6.3.6) is almost identical to Equation (6.3.5), which is valid *without* the normality assumption.

But what about the accuracy of the chi-square test here, in the case that the normality assumption does not hold? Again, we can investigate this using simulation: Suppose, for example, the sampled population has a $U(0,a)$ distribution, and H_0 is true, that is, $\sigma^2 = 15.6$. (You should verify that this implies that $a = 13.68$ and $\mu = 6.84$.) Suppose the analyst mistakenly believes the sampled population to be normal (or does not even try to check this). Will the test used above work correctly, that is, will it have 0.05 probability of rejecting H_0? Let us see.

Program 7.2.1

```
program ChiSqr(input,output);

   var I,Seed,NSam: integer;
       S2,Count: real;

   function random(var Seed:integer): real;
   begin
      Seed := (25173*Seed + 13849) mod 16384;
      random := Seed/16384.0
   end;

procedure GenS2;
   var X,T: real;
       I: integer;
```

```
begin
   T := 0.0;   S2:= 0.0;
   for I := 1 to 50 do
      begin
      X := 13.68*random(Seed);
      T := T + X;
      S2 := S2 + sqr(X)
      end;
   S2 := S2 - sqr(T)/50.0 {numerator of (7.2.3)}
   end;

begin
   writeln('enter NSam');
   readln(NSam);
   Seed := 9999;
   Count := 0.0;
   for I := 1 to NSam do
      begin
      GenS2;
      if S2/15.6 > 67.5 then Count := Count + 1.0
      end;
   writeln('true alpha is ',Count/NSam)
end.
```

Running the program with NSam equal to 10,000, we find that the true α value is 0.003, rather than the alleged value of 0.05. Thus, α is wrong by a factor of more than 10.

The result of this program suggests that the chi-square test for variance is very sensitive to departures from the normality assumption. This is in fact generally true for inferences (i.e., tests and confidence intervals) involving variances and quantities related to variances, such as covariances and correlations. If an inference procedure for one of these quantities assumes sampling from a normal population, then it is probably not trustworthy. The sampled populations must be very close to normal in order for inferences of this type to be valid. On the other hand, inferences for means (Section 7.1) and generalized means (like the linear models studied in Section 6.6) *are* fairly accurate even when the sampled populations are not normal. We say that tests on variances are **nonrobust** to the normality assumption, while tests on means are **robust**.

Another important category of tests is that concerning **contingency tables**. These tables arise quite frequently in a wide variety of applications. In fact, the **chi-square test**, which we will present here, has been described as one of the twenty most important scientific developments since 1900. (See "20 Discoveries that Shaped Our Lives," special issue of *Science 84*, November 1984.)

Contingency tables occur in a situation in which we have defined k categories of some kind. We sample n individuals from some population and see how many fall into each of the k categories. Let N_i be the number in Category i, among the *n sampled* individuals, and let p_i be the *population* proportion of individuals in that category. (Note that $N_1 + \cdots + N_k = n$ and $p_1 + \cdots + p_k = 1$.) Usually we wish to test some hypothesis concerning the proportions p_i.

Example 7.2.2

Suppose we develop a new random number generator, for $U(0,1)$ random variables, to be used in place of the function "random" presented in Section 3.4. We wish to evaluate the accuracy of the new generator, which we will name Rnd. Although there are many sophisticated methods used for such purposes, an elementary approach might be through the following hypothesis test.

Let $X_1, \ldots, X_{1000}$ be the result of 1000 calls to Rnd, and let $N_1, \ldots, N_{20}$ be the numbers of generated X_i that fall in the intervals $[0.00,0.05)$, $[0.05,0.10), \ldots, [0.95,1.00)$. Ideally, we should have $p_1 = \cdots = p_{20} = 0.05$. Thus, we can test the hypothesis

$$H_0\colon p_1 = \cdots = p_{20} = 0.05 \tag{7.2.5}$$

with the alternate hypothesis H_A being the negation of H_0: H_A states that at least one of the probabilities p_i is *not* equal to 0.05.

As we have seen, the form of any hypothesis test depends on the distribution of some quantity under H_0. This quantity is called the **test statistic**. For example, the quantity (7.2.3) was the test statistic above for the hypothesis (7.2.1). In the case of a contingency table, the test statistic must be some function of the category counts $N_1, \ldots, N_k$. That function turns out to be as follows.

Chi-Square Test for Contingency Tables

Let e_i denote the expected value of N_i under H_0, and let

$$W = \sum_{i=1}^{k} \frac{(N_i - e_i)^2}{e_i} \tag{7.2.6}$$

Also, let $d = (k-1) - m$, where m is the number of algebraically independent p_i under H_0. This will be clarified below. Then, under H_0, W has an approximate chi-square distribution, with d

degrees of freedom, and a test of H_0 having significance level α should reject H_0 if and only if $W > \chi^2(\alpha; d)$.

The meaning of the phrase "algebraically independent" in the definition of m is best explained by first stating that under H_A, there are $k-1$ algebraically independent p_i. What this means is that once $k-1$ of the p_i have been specified, for example, $p_1, \ldots, p_{k-1}$, then the kth one, p_k (in our example) is then determined by the constraint that the k numbers must sum to 1. On the other hand, under H_0 in Example 7.2.2, the p_i are *all* completely determined, each having the value 0.05. Thus $m=0$ and $d=k-1=19$ in that setting.

We can now perform the test in Example 7.2.2. First, let us determine the e_i. For convenience, consider e_1; all of the others will be the same. We have generated 1000 numbers, and under H_0, each of these numbers has probability 0.05 of falling into Category 1, [0.00, 0.05). Thus, N_1 has a binomial distribution with $n=1000$ and $p=0.05$, so $e_1=E(N_1)=np=50$. Using $\alpha=0.01$, for example, we then find that H_0 should be rejected if and only if

$$\frac{1}{50}\sum_{i=1}^{20}(N_i-50)^2 > \chi^2(0.01; 19) = 36.19$$

Again, however, we should be careful not to rely solely on the test results in making our decision, our decision in this case being whether to use Rnd in our future simulation work. No random number generator is perfect [see the remarks surrounding Equation (3.4.1)]. Therefore, H_0 is automatically false in a technical sense, although it may be "true" on the practical level, that is, the imperfections in Rnd may be so small that for all practical purposes H_0 should be considered "true." A problem then arises from the fact that our test is deciding whether H_0 is true in the *technical* sense, not in the *practical* sense. With a large enough sample, our test will detect even very small discrepancies from H_0. We should not simplistically decide that Rnd is an unacceptable generator merely because we reject H_0. This is an especially important consideration in simulation settings, in which it is usually very easy to generate huge samples.

One of the most common applications of contingency tables is a **test of independence**:

Example 7.2.3

A city is considering forming a special bus/carpool lane on one of its busiest freeways, and conducts a survey of regular drivers on this freeway. Each surveyed driver is asked two questions:

1. Is most of your use of this freeway for traveling to work?
2. Would you favor formation of a special bus/carpool lane?

Since in this setting there are two data items on each sampled individual, it is natural to use double subscripts instead of single ones as we did above. For example, the observations X_i above will be replaced by X_{ij}, where X_{ij} is the response of the *i*th driver to Question *j* (1 for a "yes" answer, 2 for "no"). Similarly, although we could label our four category counts N_1, N_2, N_3, and N_4 (letting $k = 4$), it is clearer to use double subscripts: N_{st} will denote the number of drivers who respond *s* to Question 1 and *t* to Question 2 (s, t = 1, 2). For example, N_{21} is the number of drivers who do *not* use the freeway primarily for work travel, but who *do* favor having a bus/carpool lane.

We may wish to test the hypothesis that responses to Questions 1 and 2 are independent, that is, that X_{i1} and X_{i2} are independent random variables. Let $p_{st} = P(X_{i1} = s,\ X_{i2} = t)$, so that, for example, p_{11} is the population proportion of drivers who use the freeway primarily to go to and from work, and who are in favor of the special lane.

Note that the marginal probabilities can be obtained from the p_{st} as follows (you may wish to review the material on marginal distributions in Section 4.1 before continuing): Let $p_{s*} = P(X_{i1} = s)$ and $p_{*t} = P(X_{i2} = t)$, for example, p_{*1} is the population proportion of drivers in favor of the special lane, regardless of whether or not they use the freeway for travel to work. Then $p_{s*} = p_{s1} + p_{s2}$, and $p_{*t} = p_{1t} + p_{2t}$. (The * notation indicates summation over the given subscript.)

Then the hypothesis of independence can be stated as

$$H_0\colon p_{st} = p_{s*}p_{*t}\ (s = 1, \ldots, r;\ t = 1, \ldots, c) \tag{7.2.7}$$

for a general table having *r* rows and *c* columns (here $r = c = 2$). H_A is again the negation of H_0.

The test statistic in this case will be the same as that of Equation (7.2.6), except for the labeling change from single to double subscripts:

$$W = \sum_{s=1}^{r} \sum_{t=1}^{c} \frac{(N_{st} - e_{st})^2}{e_{st}} \tag{7.2.8}$$

However, a problem does arise here in that it is not so clear as before how to compute the quantities e_{st}, since the probabilities p_{st} are not specified under H_0. The natural solution is to estimate the probabilities, and then compute corresponding estimates of the quantities e_{st}:

$$\hat{p}_{s*} = \frac{1}{n}\sum_{t=1}^{c} N_{st}$$

$$\hat{p}_{*t} = \frac{1}{n}\sum_{s=1}^{r} N_{st} \tag{7.2.9}$$

Under H_0, we then use these estimates to in turn estimate p_{st} as

$$\hat{p}_{st} = \hat{p}_{s*}\hat{p}_{*t} \tag{7.2.10}$$

We are now in a position to estimate the quantities e_{st}. Since $e_{st} = np_{st}$, we can form the estimate

$$\hat{e}_{st} = n\hat{p}_{s*}\hat{p}_{*t} \tag{7.2.11}$$

and then evaluate W by using Equation (7.2.8).

The only remaining question for this testing procedure is the number of degrees of freedom in the chi-square distribution, that is, the quantity d following Equation (7.2.6). Of course, we still use the same formula, $d=(k-1)-m$ (keep in mind that we are still in the context of that formula; the only thing we have changed is labeling). Recall that k here is the number of categories, which in this case is rc. What about m? To answer this question, note that under H_0, the parameters p_{st} are determined by the marginal probabilities $p_{1*}, \ldots, p_{(r-1)*}$ and $p_{*1}, \ldots, p_{*(c-1)}$ (p_{r*} is not needed, since it is determined by $p_{1*}, \ldots, p_{(r-1)*}$, by subtraction from 1, and similarly for p_{*c}). So $m = r - 1 + c - 1$, and

$$d = rc - r - c + 1 = (r - 1)(c - 1)$$

Another category of tests that we will introduce here is that of **nonparametric**, or **distribution-free** tests. These tests have as their goal avoidance of assumptions that the distributions of the sampled populations belong to some specified parametric family (e.g., normal). The chi-square test for variance presented at the beginning of this section is considered to be a **distribution-dependent** test, since it relies on the assumption of a normal population.

The simple example we will present here is called the **sign test**. It is used in the same kind of situation as in Example 7.1.3: We are comparing two populations, using a sample of size n from each. We wish to compare the populations with respect to some measure of location. In Example 7.1.3, we did this for means, while the sign test is based on a median.

Let X and Y denote observations drawn at random from the two populations. In Example 7.1.3, we were testing

$$H_0\colon EX = EY \tag{7.2.12}$$

which of course is equivalent to

$$H_0\colon E(X - Y) = 0 \tag{7.2.13}$$

The sign test is analogous to Equation (7.2.13), using a median instead of a mean:

$$H_0\text{: median}(X - Y) = 0 \tag{7.2.14}$$

The alternative, H_A, can be either one-sided or two-sided. Here we will use a one-sided alternative, as in Example 7.1.3:

$$H_A\text{: median}(X - Y) > 0 \tag{7.2.15}$$

The test itself is quite simple, being based on the binomial distribution. By the way, note that this does not contradict the distribution-free property of our test. The phrase "distribution-free" refers to the lack of an assumption concerning the distributions in the sampled populations. It does not prevent us from knowing the distribution of our test statistic, which is in fact vital to any test.

Let X_i and Y_i denote the ith observations from the two samples. Under H_0, the events $\{X_i - Y_i > 0\}$ and $\{X_i - Y_i < 0\}$ each have probability 0.5, for continuous distributions. Thus, under H_0, the random variable

$$N = \text{the number of indices } i \text{ for which } X_i - Y_i > 0$$

has a binomial distribution with parameters n and 0.5, so that

$$P(N = k) = \binom{n}{k} 0.5^k 0.5^{n-k} = \frac{\binom{n}{k}}{2^n} \tag{7.2.16}$$

We will suspect H_0 is false if N is much larger than $n/2$ [but not if it is much smaller than $n/2$, since H_A has the one-sided form (7.2.15)]. Our rejection rule will be that we will reject H_0 if N is greater than or equal to some value r.

Let us think about what value of r would be good in Example 7.1.3. For example, consider $r = 14$. This would result in a significance level α of

$$\sum_{k=14}^{20} \frac{\binom{20}{k}}{2^{20}} = 0.0577 \tag{7.2.17}$$

If we desire a smaller α than this, then we must choose a larger value than 14 for r.

Is the use of the sign test appropriate in a case such as Example 7.1.3? Recall that the whole point of nonparametric tests is to avoid making distributional assumptions concerning the sampled populations. However, in the "z-test" used in that example, given by Equation (7.1.6), there were no distributional assumptions made. We *could* have assumed normal distributions, and then used the Student-t distribution for the test statistic in Equation (7.1.6). This would have given us a test that would have been exact rather than approximate, but of course this exactness would have depended on the normality assumption. However, the z-test does *not* make any distributional assumptions about the sampled population. Therefore, the z-test really should be considered distribution-free, so the sign test does not have an advantage in this respect.

Both the z-test and the sign test do make assumptions of independence. In the z-test, we assume that the observation pairs (X_i, Y_i) are independent over all i [e.g., the pair (X_2, Y_2) is independent of the pair (X_5, Y_5)], and also that X_i and Y_i are independent of each other. The sign test only makes the first of these two independence assumptions, and would appear to have an advantage in this respect. However, if necessary, the z-test can be used even if the second assumption does not hold; we simply look at the differences $X_i - Y_i$, and test H_0: $\mu_d = 0$, as in Example 6.3.5.

So far in this discussion, the sign test does not look like a good alternative procedure to the z-test. In fact, one can show, using the advanced theory of mathematical statistics, that if the population of differences $X - Y$ is normal, then the sign test is only 64% as efficient as the z-test. This means that the z-test can achieve the same power as the sign test with only 64% as many observations.

However, the sign test does have one important advantage, in that it is robust to **outliers**, which are observations of extreme size, quite possibly errors. Recall that the chi-square test for variance in the beginning of this section was found to be nonrobust to the normality assumption. A t-test in Example 7.1.3, which assumes normality, *would* be robust to the normality assumption, since for large n, the t-test and z-test are almost identical, and the z-test does not assume normality. However, the t-test and z-test are *not* robust to outliers. For example, suppose in that example X_4 was accidentally recorded in minutes, instead of hours. This 60-fold error in the value of X_4 would result in very substantial errors in $\overline{X}$ and in S_1; thus, the t-test and z-test may give very misleading advice in such a situation. On the other hand, while there would be some resulting effect on the sign test, the damage would be very limited; even if X_4 were to have a million-fold error, the error in the value of N would be at most one count. This could be an invaluable advantage for the sign test over the z-test.

Errors in sampled data are much more common than most people might guess. This aspect of nonparametric statistics should always be kept in mind, even if distributional assumptions of parametric tests seem justified.

The last category of tests we will cover here consists of tests to be used in the linear models introduced in Section 6.6. (The reader should review the notation used in that section before continuing. Also, as in that section, we remind the reader that the material below is simply an introduction. More detailed information is available in the references given earlier.) Here we will make the same normality, equal-variance, and independence assumptions that were used in Equation (6.6.16), although we mention again that this analysis is somewhat robust to the first two of these assumptions, especially that of normality.

First, consider the hypothesis

$$H_0\colon \beta_i = 0 \tag{7.2.18}$$

This may be tested by using the fact that the quantity

$$\frac{\hat{\beta}_i - \beta_i}{S\sqrt{c^{i+1,i+1}}} \tag{7.2.19}$$

has a t-distribution with $n - r - 1$ degrees of freedom. Therefore, a test of Equation (7.2.18) can be performed by computing the expression (7.2.19) with β_i set to 0. If, for example, the alternative hypothesis is

$$H_A\colon \beta_i > 0 \tag{7.2.20}$$

then we would reject H_0 if the quantity (7.2.19) turns out to be larger than $t(\alpha, n - r - 1)$.

More complicated tests in linear models are also possible, in which we may test hypotheses of the form

$$\begin{aligned} H_0\colon \Lambda\beta &= 0 \\ H_A\colon \Lambda\beta &\neq 0 \end{aligned} \tag{7.2.21}$$

where Λ is a matrix with $r+1$ columns and s rows; here the value of s, and the matrix Λ, are chosen by the analyst. We are doing a simultaneous test that s different linear combinations of β are all equal to 0.

To test Equation (7.2.21), we need some notation. First, set

$$RSS = (Y - X\hat{\beta})^t(Y - X\hat{\beta})$$

Then, recalculate $\hat{\beta}$ according to H_0, which means to minimize Equation (6.6.8) under the constraint $\Lambda\beta = 0$; call the minimizing value $\tilde{\beta}$. Then define

$$RSS_0 = (Y - X\tilde{\beta})^t(Y - X\tilde{\beta})$$

Then under H_0, the quantity

$$\frac{(RSS_0 - RSS)/s}{RSS/(n - r - 1)} \tag{7.2.22}$$

has an F distribution with $(s, n-r-1)$ degrees of freedom (this distribution was defined near the end of Section 3.3). H_0 is rejected if this quantity exceeds the upper-α level of this F distribution.

To make these rather abstract ideas precise, let us look at an example.

Example 7.2.4

Consider Example 6.6.8, comparing the no-interaction and interaction models. We may wish to test the hypothesis of no interaction,

$$H_0: \beta_4 = \beta_5 = 0 \quad \textbf{(7.2.23)}$$

which corresponds to Equation (7.2.21) with

$$\Lambda = \begin{bmatrix} 00010 \\ 00001 \end{bmatrix}$$

We calculate $\hat{\beta}$ by fitting the five-predictor model, and thus form RSS above. Then we fit the three-predictor model, and determine RSS_0. Then we compute Equation (7.2.22), with $r=5$ and $s=2$.

Once again, we must be very careful to keep the concept of power in mind here. It turns out that the test for interaction has relatively low power (Exercise 7.3). Therefore, if we accept H_0, we should not immediately conclude that interaction does not exist, nor should we conclude that interaction effects are small.

7.3 ROLE OF HYPOTHESIS TESTING IN PRACTICAL DECISION MAKING

In Sections 7.1 and 7.2, we introduced the concept of hypothesis testing in a number of examples, but we included cautions against its misuse. Since the use of hypothesis testing is extremely widespread (indeed, many analysts regard it as the ultimate goal of their analyses), it is important to discuss what problems can occur. This is the subject of this section.

For instance, after developing a test for the validity of the random number generator Rnd in Example 7.2.2, we suggested that a simple test might be misleading. The reason we gave was that any random number generator will have some imperfections, which will be detected—if a large enough sample size is used—in the hypothesis test. If we base our decision as to whether to use Rnd solely on the results of the test, then no random number generator could possibly be declared "correct," if a large sample size is used. In other words, the test could reject *any* generator, even though many fine ones exist.

You might (justifiably) find this development to be somewhat ironic: It seems intuitively clear that a larger sample is desirable, and indeed this has been stressed in this chapter and in Chapter 6. Yet we see now that in some settings, the larger the sample, the more the test is capable of misleading us.

To investigate this issue more thoroughly, let us use the concept of **power** of a hypothesis test, introduced in Section 7.1. (You may wish to review this concept before continuing.) Consider Example 7.1.3. There we decided to reject H_0 if Equation (7.1.6) were to hold. In order to see the effect of sample size, replace the values 20 in that equation by a general sample size n, so that we can compute the limiting power as n goes to infinity. Recall first that power depends on the precise details of *how* H_0 is false. In our present context, this means specifying a value for the difference $\mu_1 - \mu_2$. Consider what would happen if the actual values of the two means were $\mu_1 = 6.27$ and $\mu_2 = 6.22$, so that the difference would be 0.05.

Let

$$d_n = \sqrt{\frac{S_1^2}{n} + \frac{S_2^2}{n}}$$

Then

$$\begin{aligned} \lim_{n\to\infty} \text{power}(0.05) &= \lim_{n\to\infty} P\left[\frac{\overline{X} - \overline{Y}}{d_n} > 1.65\right] \\ &= \lim_{n\to\infty} P\left[\frac{\overline{X} - \overline{Y} - 0.05}{d_n} > 1.65 - \frac{0.05}{d_n}\right] \\ &= \lim_{n\to\infty} P\left[Z > 1.65 - \frac{0.05}{d_n}\right] \end{aligned} \tag{7.3.1}$$

where Z has an approximate (and in the limit, exact) $N(0,1)$ distribution.

As n goes to infinity, the sample variances S_1^2 and S_2^2 go to the population variances σ_1^2 and σ_2^2, so d_n goes to zero. Then

$$\lim_{n\to\infty} P\left[Z > 1.65 - \frac{0.05}{d_n}\right] = P(Z > -\infty) = 1 \tag{7.3.2}$$

Thus

$$\lim_{n\to\infty} \text{power}(0.05) = 1 \tag{7.3.3}$$

so that the power will be very high (i.e., near 1) for sufficiently large sample sizes n. We should also note here that the high power will result in an OSL (observed significance level) that is very close to 0. This fact is demonstrated in Exercise 7.10.

At first glance, the high power that our test has for large n would seem to be very good. It implies that if H_0 is false, we have an excellent chance of discovering that this is the case. In fact, much of formal statistical theory is concerned with finding a test of maximum power, consistent with the positive connotation of the word "power." However, as with the word "power" in other contexts, hypothesis testing power can be abused, with harmful effects.

The problem is that the test might be answering the wrong question. For instance, we pointed out above that in the evaluation of the random number generator Rnd in Example 7.2.2, we were *not* interested in whether

Rnd generates numbers whose distribution is *exactly* $U(0,1)$. We know before even collecting any data that Rnd could not be exactly $U(0,1)$, since no generator could be exactly correct. Instead, we are interested in whether the inevitable inaccuracy in Rnd is sufficiently small for us to use Rnd in simulation programs. *Yet, the test is not answering this question; it is answering the question of whether Rnd is an exact U(0,1) generator.* For a large enough sample size, the answer to *this* question will be "no," even if Rnd works well enough for practical use.

There are similar problems with the test in Example 7.1.3. The test is answering the question as to whether the two competing text editors are *exactly* of equal effectiveness, but this is not really what we wish to determine. We instead wish to determine whether the efficiency advantage of E2 over E1 is substantial enough to overcome certain drawbacks of E2. If the two editors are *approximately* of equal effectiveness, for example, having means of 6.27 and 6.22 as above, then other considerations will be much more important when we decide which editor to use. For example, as pointed out at the end of Section 7.1, our decision may be heavily influenced by the relative amounts of system resources consumed by the two editors. If Editor E2 places a heavier burden on the system disk, E2's slight advantage of 0.05 (6.27 versus 6.22 in our example) would certainly not justify its use. Here again, the test is not answering the real question in which we are interested.

The problem described above is made worse by the fact that it is customary to use the word "significant" in conjunction with rejecting H_0. For example, if one were to reject H_0 in Example 7.1.3, the customary way to phrase this rejection is that Editor E2 had been found to be "significantly better" than E1. This term has evolved over the years from the original term, "statistically significant," which meant that the observed difference in sample means was not just due to some random accident of sampling. In other words, the term means that we believe that there is indeed some nonzero difference between E2 and E1. *However,* this definitely does not automatically mean that there is an *important* difference, even though the word "significant" carries such a connotation. Thus, there is great potential here for misinterpreting the results of a test.

Note that of course we have the opposite problem with small samples. Here the power is too *low*. In Example 7.1.3, the sample size of 20 might not be large enough for our test to have a good chance of detecting a fairly substantial difference in efficiency between the two editors. Thus in a situation in which our test has low power, we have the opposite problem from that which might occur for the high-power cases: Instead of having the danger that the test may declare an unimportant difference "significant," we have the danger that the investigator will accept H_0 and report something like, "There is no significant difference between E2 and E1." Such a declaration may be highly misleading, since there may well be an important difference, which was not detected due to lack of power in the test.

Thus, if one does a hypothesis test, it is of crucial importance to calculate the power at various points. Based on the examples given above, we see that we should check for two possibly dangerous situations:

a. The test has too much power at points very near H_0.

b. The test has too little power at points far enough away from H_0 that we would want to be able to detect such departures.

Note that (a) is particularly important if we reject H_0, while (b) is particularly important in the event that we accept H_0.

As illustration of this, in Example 7.1.3, we could calculate the power for the cases $\mu_1-\mu_2=0.05$, $\mu_1-\mu_2=0.20$, $\mu_1-\mu_2=1.00$, and so on. We should make sure that the test has sufficient power to detect substantial departures from H_0, but at the same time we should check to see whether the power is so high that the test could mislead us by pouncing on a tiny, unimportant deviation from H_0.

Due to the possible dangers of hypothesis testing, many statisticians use confidence intervals as their main statistical tool in the decision-making process. Confidence intervals tend to be much more informative than hypothesis tests.

Again, let us use Example 7.1.3 for illustration. We can compute a confidence interval for the difference $\mu_1-\mu_2$, as in Example 6.3.4. Suppose this interval turned out to be (0.03,0.07). Then we would be fairly confident that the difference in efficiency between the two editors was very small, and we should weigh other considerations in deciding which one to use, as discussed above. On the other hand, suppose the interval turned out to be (3.03,3.07). In this situation, we would be fairly confident that there is a very large difference between the two editors, one that might outweigh the fact that E2 consumes a much larger amount of system resources.

Note, though, that both of the intervals in the last paragraph were very narrow. What if the interval we get is not so narrow, such as (0.03, 3.07)? This would mean that, as far as we could tell from our sample data, the difference between the two editors might be either very small or very substantial. In other words, we would not know much at all! The fact that the interval was so wide would signal us that our sample size was just too small; we would need to collect more data before we could make an informed decision as to whether there was a sizable difference between the two editors.

Keep in mind, by the way, that we are not particularly interested in whether the interval actually contains the value 0. The question of whether the interval contains 0 is relevant to the hypothesis (7.1.5), but not necessarily to the question of *practical* interest to us. If the only way that we use the interval is to check whether it contains 0, then what we are doing reduces to a hypothesis test, and we are defeating the entire purpose of forming the interval.

This approach can be applied to the other examples as well. Consider the test for independence in Example 7.2.3. Again, the test given there asks a question in terms of some condition holding *exactly*, in this case exact independence of the two factors. But if Equation (7.2.7) comes very close to being satisfied, we may well consider the two factors independent, for all practical purposes. Thus, the test might mislead us into making the wrong decision (again, "wrong" from the practical point of view, rather than the theoretical one). We can instead find confidence intervals for the cell probabilities p_{st}, and assess how close these probabilities come to satisfying Equation (7.2.7).

In other words, what we are saying here is that one should make a decision based on a careful, thorough look at both the sample data and one's external knowledge of the problem (e.g., disk usage of Editor E2), and that the data side of this analysis is more fully described through confidence intervals than through tests. All the various factors should be weighed before a decision is made.

In some cases, the "weighing process" can be very difficult. For instance, in Example 7.2.4, we are trying to decide whether a no-interaction model is sufficiently accurate for future use. We might decide to choose the no-interaction model if the intervals for β_4 and β_5 indicate that these two parameters are small, that is, that the no-interaction model is a close approximation. However, the question of what our definition of "small" should be does not have an obvious answer. This makes the decision-making process a bit less crisp than we would prefer. Nevertheless, we should not be fooled by the easy psychological satisfaction that we would get from using a hypothesis test to make the decision for us. As emphasized above, the test is simply not designed to assess whether the no-interaction model is a close enough approximation for practical purposes; the test can only tell us whether the no-interaction model is *exactly* correct, to infinitely many decimal places, and this is usually not the question of interest.

In summary, hypothesis testing should be done very carefully in the decision-making process, with power calculations playing a vital role in the analysis, supplemented by the use of confidence intervals. The reader is urged to read further in this area. The author highly recommends Chapter 29 ("A Closer Look at Tests of Significance") in the text *Statistics*, by Freedman, Pisani, and Purves (Norton, 1978). Also, for a discussion of this problem in an engineering context, see the editorial, "Don't Do Hypothesis Testing," in the *IEEE Transactions on Reliability*, April 1976.

FURTHER READING

Peter Bickel and Kjell Doksum, *Mathematical Statistics: Basic Ideas and Selected Topics*, Holden-Day, 1977.

D.R. Cox and D.V. Hinkley, *Theoretical Statistics*, Chapman and Hall, 1974. At the level of Bickel and Doksum, but less formally presented. Includes some "statistical philosophy."

N.R. Draper and H. Smith, *Applied Regression Analysis*, Wiley, 1981.

Lloyd Fisher and John McDonald, *Fixed Effects Analysis of Variance*, Academic Press, 1978. Treats general linear models, including regression and analysis of variance.

Richard Johnson and Dean Wichern, *Applied Multivariate Statistical Analysis*, Prentice-Hall, 1982.

G.A.F. Seber, *Linear Regression Analysis*, Wiley, 1977.

Sanford Weisberg, *Applied Linear Regression*, second edition, Wiley, 1985.

EXERCISES

7.1 (M) Verify that the interval in Equation (7.2.4) is a 95% confidence interval for σ^2, taking as given the statement about the distribution of S^2. State formulas for one-sided confidence intervals.

7.2 (S) Modify Program 7.2.1 to investigate the robustness of the chi-square test for variance in the case that the underlying population has an exponential distribution.

7.3 (M, S) Perform the test in Example 7.2.4, using the data in Exercise 6.29. Then use simulation to find the power of the test, in the following setting: $\beta_0 = 25, \beta_1 = \beta_2 = \beta_3 = 5$, and $\beta_4 = \beta_5 = \tau$. Try various values of τ, for example, 1, 2, 5, and 10. (Note: The simulation will be extremely slow if a matrix inversion is repeated many times. Find the matrix inverses needed by hand first, and incorporate the resulting expressions into your program.)

7.4 (S) Consider Example 7.1.1, with the following change in procedure: We will sample 500 items, as before, and then perform the test. However, if the result of the test is to accept H_0, then we sample an additional 250 items, and then do a new test on the combined sample of 750. Suppose we set α to 0.10 in both cases.

a. The overall α of this procedure should actually be higher than 0.10. Explain why, and find the true value.

b. On the other hand, overall β values should be lower than those that the calculations in Section 7.1 indicate; for example, for $\beta(0.08)$ with $\alpha = 0.05$, the true value should be lower than the 0.0162 which we calculated. Explain why this should be the case, and find the true value.

7.5 Consider the test in Example 7.2.3.

a. (M) Suppose we have the following data:

$N_{11}=20,\ N_{12}=52,\ N_{21}=6,\ N_{22}=21$

Perform the test, using $\alpha=0.10$, and find the approximate OSL value.

b. (S) Investigate the accuracy of the chi-square approximation in the test procedure, for $n=20$, 40, 60, and 120 (invent your own values for p_{st}).

c. (S) Find the power of the test, for $n=20$, 40, 60, and 120 (invent your own values for p_{st}).

7.6 Consider Example 7.1.3, but with

$$H_A\colon \mu_1 \neq \mu_2$$

for when we do not have *a priori* knowledge that a particular one of the text editors is better. Suppose someone reacts to the discussion in Section 7.3 by saying something like, "I don't care *how much* different E1 and E2 are. I just want to know which one is better. Thus hypothesis testing is OK for my purposes." Either defend or criticize such a statement. A defense might be based on giving an example in which there are absolutely no external factors to consider, such as the heavy disk usage of E2. A criticism might be based on a claim that such examples (in which there are no external factors to consider) occur rarely in practice; another critical argument might mention the effects of possible sampling bias.

7.7 (M) In Example 7.1.1, suppose our procedure is to sample items until we get the first defective, recording N, the number of correct items sampled. Devise a test based on this sampling scheme, with $\alpha = 0.05$, and find the power of this test for the case $\pi = 0.08$.

7.8 (M, S) Suppose in Example 7.1.2 we wish to have $\alpha = 0.01$ and $\beta = 0.05$. Following the approach of that example, find the sample size needed, and verify the answer you find by using simulation.

7.9 (M, S) Review the material on tests based on chi-square and F distributions in Examples 7.2.1 and 7.2.4, and the definitions of those distributions in Section 3.3. Then propose an F-test for the hypothesis $H_0\colon \sigma_1^2 = \sigma_2^2$, where σ_1^2 and σ_2^2 are the variances of two populations from which two independent samples have been drawn. Then investigate the robustness of this test, in a manner similar to that of Program 7.2.1.

7.10 Consider the effects of large n, discussed at the beginning of Section 7.3 and near Equation (7.3.1). It was stated there that

$$\lim_{n\to\infty} \text{OSL} = 0$$

the point then being that a small OSL value does not necessarily imply that the difference $\mu_1 - \mu_2$ is large.

a. (M) Show that the limit stated above is zero.

b. (S) "Watch" that convergence of OSL to 0, in the following way: Generate data (X_1, Y_1), (X_2, Y_2), (X_3, Y_3), . . . , assuming normal populations, $\sigma_1^2 = \sigma_2^2 = 5$, and $\mu_1 = 6.27$, $\mu_2 = 6.22$. Calculate the OSL value based on the first 25 observations, then on the first 50, the first 75, . . . , and finally the first 500. Note the convergence of OSL to 0.

7.11 (M) Suppose we wish to test the hypothesis

$$H_0\colon \nu = 32.0$$
$$H_A\colon \nu < 32.0$$

where ν is the median compilation time for all programs run at a certain computer installation.

a. Using the material which led to Equation (7.2.16) as a guide, devise a test for the above hypothesis. Suppose $n = 6$, and the collected data are 25.1, 49.9, 15.2, 28.2, 31.8, 9.5. Find the OSL value.

b. Suppose the true population density is exponential, with mean 27.7. Find the power of the test, for $n = 10$, 20, and 200 (use an approximation in the last case).

7.12 (M) Suppose objects fall into 10 categories, with p_i being the proportion that fall into Category i. Using the material on contingency tables, devise a test for the hypothesis that the equalities

$$p_1 = p_2 = p_3 = p_4 = p_5$$

and

$$p_6 = p_7 = p_8 = p_9 = p_{10}$$

both hold; that is, give the specific numerical value for d.

7.13 (M) Suppose we roll a die 50 times, to test whether it is fair (i.e., perfectly balanced). We find that we get twelve 1's, four 2's, ten 3's, thirteen 4's, five 5's, and six 6's. Test the hypothesis with $\alpha = 0.05$, and find the OSL value.

CHAPTER 8

Discrete-Event Stochastic Processes

The term **stochastic process** in the title above technically means a collection of subscripted random variables. The typical meaning of the subscript is time. For example, consider a timeshare computer system. Jobs arrive at random times, and they have random time requirements for using the system. These two sources of randomness imply that the state of the system at any given time point is a random variable, where the state of the system is defined to be the number of jobs it is running at the given time. We let X_t denote the state of the system at time t $(0 \leq t < \infty)$.

The term *discrete* in the above chapter title refers to the fact that changes in our system occur abruptly, rather than occurring continuously. For example, consider the timeshare computer system. The state changes only when a new job arrives or an existing job finishes; in between the occurrences of such events, there are no changes in the state of the system, as long as the state is defined as above, as the number of jobs currently in the system. A graph of X_t versus t would be that of a step function; that is, the graph would consist of a series of horizontal line segments, with jumps between consecutive segments.

On the other hand, consider the analysis of weather systems, with X_t denoting (say) temperature at time t. Here the state changes continuously, and the graph of X_t is continuous, with no breaks, and with varying slope. Stochastic processes of this kind are out of the realm of this chapter.

As with other random ("stochastic") phenomena we have studied so far, our analysis is based on both mathematical and simulation approaches.

The relation between these two tools is the same as that described at the end of Section 2.4: Both tools are needed, and they act in a complementary manner. However, there is a difference in the discrete-event setting, because it turns out that both types of analysis—including the simulation programming—are much more challenging in this setting than in those we have studied previously.

8.1 MATHEMATICAL METHODS

We first discuss **Markov chains**. To motivate this discussion, we will concentrate on a simple example:

Example 8.1.1

Consider a **random walk** on the set $\{1,2,3,4,5\}$, in which we move at random in that set, say one move per second. Suppose we move according to the result of rolling a die, as follows. Let i denote our current position. If i is not 1 or 5, then we move to $i - 1$ or $i + 1$, or stay at i, depending on whether the die comes up in the subsets $\{1,2\}$, $\{3,4\}$, or $\{5,6\}$, respectively. If i is 1 or 5, then the next move is to 2 or 4, respectively.

The set $\sigma = \{1, 2, 3, 4, 5\}$ is called the **state space** for this process, and the elements of this set are the possible **states**. Let X_t represent the position of the particle at time t, $t = 0, 1, 2, \ldots$ The random walk is a Markov process. The term *Markov* here has meaning similar to that of the term "memoryless" used for the exponential distribution in Chapter 3, in that we can "forget the past:"

$$P(X_{t+1} = s_{t+1} | X_t = s_t, X_{t-1} = s_{t-1}, \ldots, X_0 = s_0) \\ = P(X_{t+1} = s_{t+1} | X_t = s_t) \qquad \textbf{(8.1.1)}$$

Although Equation (8.1.1) has a very complex look, it has a very simple meaning: *The distribution of our next position, given our current position and all our past positions, is dependent only on the current position.* It is clear that the random walk process above does have this property.

Continuing this example, let p_{ij} denote the probability of going from Position i to Position j in one step. For example, $p_{21} = p_{23} = 1/3$, while $p_{24} = 0$ (we can reach Position 4 from Position 2 in two steps, but not in one step). The numbers p_{ij} are called the **one-step transition probabilities** of the process, and are elements of the 5-by-5 **one-step transition matrix** P for this process:

$$P = \begin{bmatrix} 0 & 1 & 0 & 0 & 0 \\ 1/3 & 1/3 & 1/3 & 0 & 0 \\ 0 & 1/3 & 1/3 & 1/3 & 0 \\ 0 & 0 & 1/3 & 1/3 & 1/3 \\ 0 & 0 & 0 & 1 & 0 \end{bmatrix} \tag{8.1.2}$$

Now let q_{ij} denote the probability of going from Position i to Position j in *two* steps, that is, $P(X_{n+2} = j|X_n = i)$. We can calculate q_{ij} by breaking things down according to what happens at the *first* step, at which we may go to some intermediate state k:

$$q_{ij} = \sum_k p_{ik} p_{kj} \tag{8.1.3}$$

Equation (8.1.3) should look familiar to readers with extensive experience with matrices: Let A and B denote n-by-n matrices, with $C = AB$. Let a_{rs} denote the element in the r-th row and s-th column of A, and define b_{rs} and c_{rs} similarly for B and C. Then

$$c_{ij} = \sum_k a_{ik} b_{kj} \tag{8.1.4}$$

which looks just like Equation (8.1.3).

Thus, we have found that if Q denotes the two-step transition matrix of our process (i.e., Q is the matrix whose elements are the two-step transition probabilities q_{ij}), then $Q = P^2$! In fact, the same reasoning shows that if we let T_n denote the n-step transition matrix, then $T_n = P^n$; that is, the n-step transition matrix turns out to be the n-th power of the one-step transition matrix.

For this example,

$$Q = P^2 = \begin{bmatrix} 1/3 & 1/3 & 1/3 & 0 & 0 \\ 1/9 & 5/9 & 2/9 & 1/9 & 0 \\ 1/9 & 2/9 & 1/3 & 2/9 & 1/9 \\ 0 & 1/9 & 2/9 & 5/9 & 1/9 \\ 0 & 0 & 1/3 & 1/3 & 1/3 \end{bmatrix} \tag{8.1.5}$$

The (4,3) element of this matrix is 2/9, which is indeed the probability of going from Position 4 to Position 3 in two steps, since this can be done in the sequence 4-4-3 (probability 1/9) or 4-3-3 (probability 1/9).

In most applications, we are interested in the *long-run* distribution of the process, for example, the long-run proportion of the time that we are at Position 4. For each state i, define

$$\pi_i = \lim_{t \to \infty} \frac{N_{it}}{t} \tag{8.1.6}$$

where N_{it} is the number of visits the process makes to State i among times $1, 2, \ldots, t$.

In most practical cases, this proportion will exist and be independent of our initial position $m = X_0$. (There are mathematical conditions under which this is guaranteed to occur, but they will not be stated here.) Let π denote the vector of the elements π_i, such as $\pi = (\pi_1, \ldots, \pi_5)$ above. Advanced methods can be used to show that we can calculate π, if it exists, by solving the system

$$\pi = \pi P$$

$$\sum_i \pi_i = 1 \qquad \textbf{(8.1.7)}$$

Thus we can compute π_4 and the other long-run proportions π_i by solving Equation (8.1.7). The solution is $\left[\frac{1}{11}, \frac{3}{11}, \frac{3}{11}, \frac{3}{11}, \frac{1}{11}\right]$, indicating, for example, that in the long run we will spend 3/11 of our time at Position 4.

Some problems may occur in the above approach. First, the long-run proportions may not exist, in various senses. Second, some chains have infinite state spaces, in which case Equation (8.1.7) may be difficult or impossible to solve. By the way, the long-run proportions π_i may exist, and satisfy Equation (8.1.7), without the limit

$$\lim_{t \to \infty} P(X_t = i)$$

existing.

For instance, suppose we alter the above example so that

$$p_{i,i-1} = p_{i,i+1} = 1/2$$

for $i = 2, 3, 4$. In this case, the solution to Equation (8.1.7) is $\left[\frac{1}{8}, \frac{1}{4}, \frac{1}{4}, \frac{1}{4}, \frac{1}{8}\right]$, and in fact this solution is still valid—for example, one can show that we will spend 1/4 of our time at Position 4 in the long run. On the other hand, the limit of $P(X_t = 4)$ will certainly not be 1/4, and in fact will not even exist. To see this, note that for (say) $X_0 = 3$, $P(X_t = 4) = 0$ for any even-numbered value of t.

These problems are rarely encountered in practice, but we mention them so that the reader will be aware that the mathematical theory of Markov chains extends far beyond the short introduction given here. The interested reader can consult the references for further information.

In the above example, the labels for the states consisted of single integers i. In some other examples, convenient labels may be r-tuples, for example 2-tuples (i,j).

Example 8.1.2

Consider a digital communication line, over which we transmit 0-1 **bits** (recall Example 2.3.8). Let $B_1, B_2, B_3, \ldots$ denote the sequence of bits. It is reasonable to assume the B_i to be independent, with $P(B_i = 0)$ and $P(B_i = 1)$ both being equal to 0.5.

The receiver will eventually fail. Suppose that the type of failure is **stuck at 0**, meaning that after failure it will report all future received bits to be 0, regardless of their true value. Once failed, the receiver stays failed, and should be replaced. Eventually the new receiver will also fail, and we will replace it; we continue this process indefinitely.

However, the problem is that we will not know whether a receiver has failed (unless we test it once in a while, which we are not including in this example). If the receiver reports a long string of 0s, we should *suspect* that the receiver has failed, but of course we cannot be sure that it has; it is still possible that the message being transmitted just happened to contain a long string of 0s.

Suppose we adopt the policy that, if we receive k consecutive 0s, we will replace the receiver with a new unit. Here k is a design parameter; what value should we choose for it? If we use a very small value, then we will incur great expense, due to the fact that we will be replacing receiver units at a very high rate. On the other hand, if we make k too large, then we will often wait too long to replace the receiver, and the resulting error rate in received bits will be sizable.

Resolution of this tradeoff between expense and accuracy depends on the relative importance of the two. There are also other possibilities, involving the addition of redundant bits for error detection, such as **parity bits**. For simplicity, we will not consider such refinements here. However, the analysis of more complex systems would be similar to the one below.

A natural state space in this example would be

$$\{(i, j): i = 0, 1, \ldots, k-1; j = 0, 1; \text{case } i = j = 0 \text{ excluded}\}$$

where i represents the number of consecutive 0s that we have received so far, and j represents the state of the receiver (0 for failed, 1 for nonfailed). Suppose the lifetime of the receiver, that is, the time to failure, is geometrically distributed with "success" probability ρ, so that the probability of failing on receipt of the i-th bit after the receiver is installed is $(1 - \rho)^{i-1}\rho$, $i = 1, 2, 3, \ldots$. Then calculation of the transition matrix P is straightforward. For example, suppose the current state is (2,1), and that we are investigating the expense and accuracy corresponding to a policy having $k = 5$ (again, for the sake of simplicity of exposition). What can happen upon receipt of the next bit? The next bit will have a true value of either 0 or 1, with probability 0.5 each. The receiver will change from working to failed status with probability ρ. Thus our next state could be:

- (3,1), if a 0 arrives, and the receiver does not fail;
- (0,1), if a 1 arrives, and the receiver does not fail; or
- (3,0), if the receiver fails.

The probabilities of transitions out of State (2,1) are:

$$p_{(2,1),(3,1)} = 0.5(1 - \rho)$$
$$p_{(2,1),(0,1)} = 0.5(1 - \rho)$$
$$p_{(2,1),(3,0)} = \rho$$

Other entries of the matrix P can be computed similarly.

Formally specifying the matrix P using the 2-tuple notation would be very cumbersome. In this case, it would be much easier to map to a one-dimensional labeling. For example, if $k=5$,the nine states (1,0), . . . , (4,0), (0,1), (1,1), . . . , (4,1) could be renamed States 1, 2, . . . , 9. Then we could form P under this labeling, and the transition probabilities above would appear as

$$p_{78} = 0.5(1 - \rho)$$
$$p_{75} = 0.5(1 - \rho)$$
$$p_{73} = \rho$$

After Equation (8.1.7) is solved, we could find the error rate ϵ, and the mean time (i.e., the mean number of bit receptions) between receiver replacements, μ. We can find both ϵ and μ in terms of the π_i obtained from solving Equation (8.1.7), in the following manner.

For concreteness, continue to assume that $k=5$. The quantity ϵ is the proportion of the time that the true value of the received bit is 1 but the receiver is down, or 0.5 times the proportion of the time spent in states of the form $(i, 0)$:

$$\epsilon = 0.5(\pi_1 + \pi_2 + \pi_3 + \pi_4)$$

Now to get μ in terms of the π_i, note that μ is the mean number of bits between receiver replacements, which is the reciprocal of the proportion of bits that result in replacements. For example, if 5% of the received bits result in replacement of the receiver, then (speaking on an intuitive level) the "average" set of 20 bits will contain one bit which makes us replace the receiver, and there will be an average of 20 bits between replacements. A replacement will occur only from states of the form $(4,i)$, and even then only under the condition that the next reported bit is a 0. In other words, there are three possible ways in which replacement can occur:

a. We are in State (4,0). Here, since the receiver has failed, the next reported bit will definitely be a 0, regardless of that bit's true value. We then will have a total of $k = 5$ consecutive received 0s, and therefore will replace the receiver.

b. We are in the State (4,1), and the next bit to arrive is a true 0. It then will be reported as a 0, and we will replace the receiver, as in (a).

c. We are in the State (4,1), and the next bit to arrive is a true 1, but the receiver fails at that time, resulting in the reported value being a 0. Again we have five consecutive reported 0s, so we replace the receiver.

Therefore,

$$\mu^{-1} = \pi_4 + \pi_9(0.5 + 0.5p)$$

Before leaving this example, it would be instructive to relate it to the material in Chapter 7. The replacement policy used above should remind the reader of a hypothesis test, because that is exactly what it is: Our hypothesis is

H_0: receiver OK

H_A: receiver failed

Our test statistic is M, the number of consecutive reported 0s; and we will "reject" H_0 if $M \geq k$.

We could thus use hypothesis testing customs to choose the value of k, and choose k so that we get the commonly used $\alpha = 0.05$. However, this is clearly inappropriate. Our choice for k should be based on a careful analysis of the tradeoff between expense and accuracy.

In the above examples, the Markov chains are of the **discrete-time** type, so called because the time spent in a state before moving to another state—the **holding time**—is a discrete random variable. For example, consider State 4 in Example 8.1.1. Let T denote the length of time we spend in State 4 after first arriving at it; T can take on the values 1, 2, 3,. . . . The probability that $T = 1$ is 2/3, the probability that $T = 2$ is (1/3)(2/3), and so on. Thus, the holding time T is a discrete random variable, and in fact has a geometric distribution.

However, in most applications of Markov chains, we encounter the **continuous-time** type, in which the holding time at a state before leaving is a continuous random variable. Consider the computer timeshare system mentioned at the beginning of this chapter. A new job can arrive at any time, not just at integer times; the same is true for the times at which a job can finish. Note that the word "discrete" in the phrase "discrete-time" is not related to its usage in the phrase "discrete-event" in the title of this chapter.

In continuous-time chains, the changes still occur as discrete events; that is, they occur abruptly rather than continuously.

In continuous time, we still want our chains to have the Markov property, that is, to be memoryless. A little thought about this then reveals that in order to have this property, the holding time random variables mentioned above must have memoryless distributions. Of course, we have already seen in Chapter 3 that the exponential distribution has this property. In fact, it is shown in Exercise 8.2 that the exponential distribution is the *only* continuous distribution having the memoryless property. (It can be shown similarly that the geometric distribution is the only memoryless *discrete* distribution, so it is no accident that the holding times in discrete-time Markov chains have this distribution.) In continuous-time Markov chains, then, the holding times are exponentially distributed, although the mean of the distribution may vary from state to state.

As before, we use the letter π to denote the vector of long-run proportions of time the system spends in each state (again, these proportions may not exist, but they usually do in practical cases). We can calculate π using the following reasoning. Let b_i denote the value of b in the density be^{-bt} of the holding time of State i. Recall that b_i is the reciprocal of the mean of the distribution; that is, the mean holding time at State i is $1/b_i$. It may also be interpreted as the "rate" at which we leave State i, in the following sense.

Suppose the mean holding time at State i is 0.25 seconds. Then, if we look at all our visits to State i, we stay there an average of 0.25 seconds per visit. Assume, just for the moment, that we immediately return to State i each time we leave it. Then the average of 0.25 seconds per visit would imply $1/0.25 = 4.0$ visits per second, or that we leave State i at the rate of 4.0 times per second.

This interpretation of b_i as a rate has another use: We can use it to derive the continuous-time analog of Equation (8.1.7), the equation used for finding the long-run distribution vector π. First note that our temporary assumption above of immediate return to State i after leaving it is not correct; instead of being at State i 100% of the time, we are there only a proportion π_i of the time. Thus, an observer of this system will record $\pi_i(4.0)$ exits of State i per second, rather than 4.0 exits per second. In general, there will be $\pi_i b_i$ transitions out of State i per unit time.

Now consider the rate at which we *enter* State i from other states j. This rate is equal to

$$\sum_{j \neq i} \pi_j \rho_{ji}$$

where ρ_{ji} is the number of transitions from State j to State i per unit time, during the time that the system is in State j. (Note that $b_r = \Sigma_{s \neq r} \rho_{rs}$.)

Now in the long run, there will be equally many transitions into State i as there are out of it, so we have

$$\pi_i b_i = \sum_{j \neq i} \pi_j \rho_{ji} \quad \textbf{(8.1.8)}$$

for all states i. We can then solve Equation (8.1.8) to obtain the π_i.

Example 8.1.3

Suppose the operations in a factory require the use of a certain kind of machine. The manager has installed two of these machines. This is known as a **gracefully degrading system**: When both machines are working, the fact that there are two of them, instead of one, leads to a shorter wait time for access to a machine. When one machine has failed, the wait is longer, but at least the factory operations may continue. Of course, if both machines fail, the factory must shut down until at least one machine is repaired.

Suppose that the time until failure of a single machine, carrying the full load of the factory, is exponentially distributed with mean 20.0. If both machines are working, the strain on each is lessened, and the time to failure of each machine is exponential with mean 25.0. Suppose the repair time is exponential, too, with mean 8.0. It is assumed that all failure and repair times are independent random variables.

Let the state of the system be the number of working machines, so that the state space is $\{0,1,2\}$. Let us evaluate the parameters b_i and ρ_{ji} in Equation (8.1.8). First consider b_2. Let z denote the time the system enters State 2, and let $z+Y_1$ and $z+Y_2$ be the times at which Machines 1 and 2 next fail, respectively. Due to the memoryless property, the random variables Y_i are exponentially distributed, with mean 25.0. For example, suppose that $z=230.8$, and that we enter State 2 at that time because Machine 1 has just been repaired. Suppose Machine 2 had been repaired at time 212.3. Then at time z, Machine 2 has already been running for $230.8 - 212.3 = 18.5$ time units; however, due to the memoryless property, Machine 2 will act "like new" at time z, that is, its *remaining* lifetime Y_2 will still have an exponential distribution with mean 25.0.

The holding time at State 2 is $H = \min(Y_1, Y_2)$. The distribution of H is determined from the following fact, proved in Exercise 8.3: *Suppose $R_1, \ldots, R_k$ are independent and exponentially distributed, with $ER_i = \nu_i$, and $R = \min(R_1, \ldots, R_k)$. Then R is also exponentially distributed, with mean*

$$ER = \frac{1}{\sum_{i=1}^{k} \frac{1}{\nu_i}}$$

Hence, H has an exponential distribution with mean $(1/25.0 + 1/25.0)^{-1} = 12.5$. The *rate* b_2 is then $1/12.5 = 0.08$.

The computation for b_1 is similar. In State 1, we have one working machine and one failed machine. We will leave this state if the working machine fails, or if the failed machine is repaired. Again, the holding time is $H = \min(Y_1, Y_2)$, where in this case Y_1 is the time until failure of the working machine, and Y_2 is the time until repair of the failed machine. Thus the mean holding time at State 1 is $(1/20.0 + 1/8.0)^{-1} = 5.714$, and $b_1 = 0.175$. Similarly, we find that $b_0 = 1/8.0 + 1/8.0 = 0.25$.

To find the rates ρ_{ji} is easy. For example, consider ρ_{21}. This is the rate of transitions from State 2 to State 1, and since *all* transitions out of State 2 are of that form, we have $\rho_{21} = b_2 = 0.08$. Now consider ρ_{12}. This is the rate of transitions from State 1 to State 2 or the rate at which the failed machine is repaired in a one-working, one-failed setting. This rate is $1/8.0 = 0.125$.

The other parameters are obtained similarly. The results are then used in Equation (8.1.8), yielding the system of equations

$$\begin{aligned} \pi_2(0.08) &= \pi_1(0.125) \\ \pi_1(0.175) &= \pi_2(0.08) + \pi_0(0.25) \\ \pi_0(0.25) &= \pi_1(0.05) \end{aligned} \tag{8.1.9}$$

(Of course, we also have the relation $\pi_2 + \pi_1 + \pi_0 = 1$.) The solution is $\pi_2 = 0.566$, $\pi_1 = 0.362$, and $\pi_0 = 0.072$.

Thus, 7.2% of the time, there will be no machine available at all. It is interesting to see how this compares to the availability figure that would result if only one machine had been purchased, rather than two. This machine would alternate between working states, having mean duration 20.0, and failed states, having mean duration 8.0. In such a case the system would be unavailable for a proportion $8.0/(20.0 + 8.0) = 0.286$ of the time, an intolerably high value.

As with discrete-time Markov chains, such as the one in Example 8.1.2, state labels in continuous-time chains can be r-tuples.

Example 8.1.4

Consider a computer disk system with two disk drives and a shared **channel**. The times between arrivals for a given drive are independent and exponentially distributed with mean 5.0. When a request arrives, the drive must first perform a **seek** operation, which we will assume to have an exponentially distributed time requirement with mean 1.25. The drive is then ready to perform the actual data transfer, which requires the use of the channel. If the channel is currently busy, the drive must wait. Once it acquires the channel, the transfer is assumed to take exponential time with mean 0.25. Only one job may be active (i.e., either in seek phase, or waiting for or using the channel) at a given drive at one time.

Call the two drives D1 and D2. Then we can set up a state space for this system, with states of the form (i, j, k, l), as follows. The component i is the number of D1 jobs in the system, regardless of phase (waiting for seek, performing seek, waiting for transfer, performing transfer). The component j is analogous for D2. The component k indicates the status of the D1 job, if any, with respect to the channel:

- k = 0 means that D1 currently has no use for the channel.
- k = 1 means that D1 is currently waiting for the channel.
- k = 2 means that D1 is currently using the channel.

Component l is then similar for D2. By the way, the reader should keep in mind that the state space developed here is not unique. There are many other equivalent formulations.

Let us compute some of the parameters $\rho_{(i,j,k,l),(a,b,c,d)}$ for this system. For example, consider the state (4,2,0,0). Here both drives have seeks in progress, and the channel is idle. To see this, look at D1. The 4 in (4,2,0,0) indicates that there are four jobs at D1, but the first 0 indicates that none of these four have any use for the channel. That must mean that one of the four jobs is currently in its seek phase, and the other three are queued.

What are the possibilities for the next state? One possibility is that a new job could arrive, say at D2. Since this happens at rate 1/5.0, we have

$$\rho_{(4,2,0,0),(4,3,0,0)} = 0.2$$

Another event that could occur is that one of the two seeks may finish, say the one at D1. This happens at rate 1/1.25, so that

$$\rho_{(4,2,0,0),(4,2,2,0)} = 0.8$$

where the second 2 in (4,2,2,0) indicates that a job at D1 (the one that just finished its seek) is using the channel.

The other rates $\rho_{(i,j,k,l),(a,b,c,d)}$ can be determined similarly, so we present only one more. Suppose the current state is (3,6,1,2). Here a D2 job is using the channel, and a D1 job is waiting for it. No seeks are in progress, but as soon as the D2 job finishes, one of the five queued jobs at D2 will begin its seek (and the D1 job which was waiting for the channel will begin its use of the channel). The rate for this transition is 1/0.25, so

$$\rho_{(3,6,1,2),(3,5,2,0)} = 4.0$$

We may set up the system (8.1.8) in this manner. However, unlike in Example 8.1.3, we may have real difficulty in solving that system, since there are infinitely many equations in this case (the state space σ will have infinitely many elements). In most cases, it is not clear how to obtain a solution for an infinite system of equations. We will discuss two methods here.

(a) Truncation. We can try to find an approximate solution by *truncating* the problem. For example, we could place a limit of 10 jobs, queued, or executing, for each drive in Example 8.1.4, with any job arriving to a full queue being lost. Then Equation (8.1.8) would be a system of approximately 900 equations. Although we need to be extremely careful when solving large systems of equations (e.g., with regard to roundoff error arising during the solution), this is at least a *finite* system, which we know how to solve. Intuitively, we could then test the accuracy of this approximation by looking at the sizes of the probabilities $\pi_{(10,j,k,l)}$ and $\pi_{(i,10,k,l)}$, which indicate the proportion of lost jobs in the truncated system. If this proportion were small, it would be intuitively clear that the truncated system is a good approximation to the original one. Another way to test the accuracy of this approximation would be to try a limit of 15 jobs, and see whether the results change much. Of course, we can always use simulation as a check, too.

(b) Birth/Death Processes. As noted in Example 8.1.2, the use of r-tuples as state labels is convenient for describing some systems, but becomes cumbersome when solving the equations for finding the steady-state probabilities. In such cases we may wish to map the r-tuple labels to single subscripts, as was done in Example 8.1.2.

In some lucky cases, there exists some single-subscript labeling such that ρ_{ij} is nonzero only for $j = i - 1$ and $j = i + 1$. Such systems are known as being of the **birth/death** type. The term comes from population genetics applications, in which State k represents a population size of k.

Through some algebraic manipulations, one can show that the solution to Equation (8.1.8) for birth/death processes is

$$\pi_n = \pi_0 r_n \ (n > 0) \tag{8.1.10}$$

and

$$\pi_0 = \frac{1}{1 + \sum_{i=1}^{\infty} r_i}$$

where

$$r_i = \prod_{k=1}^{i} \frac{\lambda_{k-1}}{\mu_k}$$

$$\lambda_k = \rho_{k,k+1}$$

$$\mu_k = \rho_{k,k-1}$$

Example 8.1.5

Suppose we have a system consisting of a single, unbreakable machine. The state space is then the set of nonnegative integers, with State i meaning that there are i jobs either waiting for the machine or using it. This system is known as an **M/M/1 queue**. This notation means that the interarrival times are "Markovian," or exponentially distributed; the service times are also Markovian, and there is one server, that is, one machine.

Suppose the mean time between arrivals is 4.0, and the mean service time is 2.5. Then $\lambda_k = 1/4.0$ and $\mu_k = 1/2.5$. After evaluating Equation (8.1.10), we find that $\pi_i = (0.375)(0.625)^i$ for $i = 0, 1, 2, \ldots$. Thus, the number of jobs in the system in the long run has a geometric distribution (starting at 0 instead of 1) with success probability 0.375. The mean number of jobs in the system is then $1/(1 - 0.625) - 1 = 1.667$. Note that we can also find the mean queue length from this: The mean number of jobs actually using the machine is the probability of the machine being in use, since the number using the machine is either 0 or 1. This probability is $1 - \pi_0 = 0.625$. Thus the mean number of jobs waiting to use the machine is $1.667 - 0.625 = 1.042$.

It is also of interest to know the mean time a job spends in the system, that is, waiting time plus service time. This could be computed from the π_i, but it is more easily found using **Little's Rule**:

Little's Rule

Think of objects entering a box from the left side at random times, spending some random time in the box, and exiting from the right side. Let α be the mean number of arrivals per unit time, μ the mean time spent in the box, η the mean number of objects in the box, and τ the throughput, that is, the mean number of objects exiting per unit time. Assume that the capacity of the box is not strained, so that $\tau = \alpha$. Then

$$\eta = \alpha\mu = \tau\mu \tag{8.1.11}$$

The following is an informal proof of Little's Rule, addressing the case in which objects exit the box in the order of their arrival. Fix attention on a "typical" object, A, at the time it exits. The objects that were in the box at the time A arrived will have all exited the box by the time A exits. Thus, at the time A exits, the only objects in the box are those which arrived during the time that A was in the box. The expected value of the time A was in the box is μ, and during a period of length μ, an average of $\alpha\mu$ objects arrive; that is, the expected value of the number of objects when A leaves is $\alpha\mu$. But since A is "typical," the mean number of objects in the box at the time it exits should be η. Therefore $\eta = \alpha\mu$.

We can use Little's Rule in Example 8.1.5 to find the mean time a job spends in the system. The "box" here consists of the machine together with the waiting area. We then have $\alpha = 1/4.0$ and $\eta = 1.667$, so μ is 6.667, by Equation (8.1.11). Also, since μ in this context is equal to mean queueing time plus a mean service time of 2.5, we see that on the average, a job waits for $6.667 - 2.5 = 4.167$ units of time before the machine is free.

Example 8.1.3 is also a birth/death process (with most of the λ_k and μ_k values equal to 0; e.g., $\lambda_k = 0$ for all $k \geq 2$). Unfortunately, Example 8.1.4 is not a birth/death process.

It should be emphasized again that in a continuous-time Markov chain, all activities are assumed to have exponential distributions. This point deserves some discussion. Let us consider queueing models and reliability analyses, the two most common types of Markov systems encountered in engineering work. In queueing models, such as Example 8.1.4, we are mainly concerned with the distributions of two classes of duration variables, interarrival times and service times. In reliability analyses, such as Example 8.1.3, the quantities of concern are typically lifetimes and repair times.

Empirical studies have found that interarrival times are quite often exponentially distributed. This is very fortunate, because although some queueing theory has been developed that relaxes the assumption of exponential service times, the theory still usually assumes exponential interarrivals. One point which should be mentioned, however, is that often the mean of the exponential interarrival time will vary cyclically through time. For example, consider arrivals to a bank. The peak arrival rate might occur around noon, with the slowest rate being just before closing time. In such cases, the resulting analysis, say of an M/M/1 queue, is valid only for time periods in which the arrival rate is approximately constant.

Lifetimes, or times until failure, have also been found to be exponentially distributed; in many settings. For example, electronic parts often have exponential lifetimes, at least if they are tested for a **burn-in period** before being actually used (during the burn-in time, the parts with major defects are quickly discovered and screened out). Another instance of exponential lifetimes was seen in Example 6.4.2, regarding times until failure for air conditioners.

However, service times and repair times are usually *not* exponentially distributed. One important example of this is in computer networks. Here the "jobs" to be served are messages to be transmitted, and the "servers" are the computers in the network (which act as receiver/transmitters). In many networks, the messages are of constant length, so that the service time is constant too. In this case, the exponential assumption is not even close to correct.

What then can be done? One simple solution is to assume exponential distributions even when they are not justified. This has been found to be relatively safe in a number of situations, as verified through the use of simulation. Of course, we can always use simulation itself as the primary tool of analysis, and we shall discuss this important tool in the following sections. However, we still would prefer to have some mathematical results if possible. There are two ways in which this can be done.

(a) The Method of Stages. It can be shown that the distribution of any positive random variable can be approximated by combinations of exponential random variables. In this way, in theory any system of the type discussed in this chapter can be analyzed approximately as a Markov chain. This theoretical result is sometimes feasible in practice, sometimes not. The full details of the method will not be given here, but it is important to at least introduce it.

Recall the Erlang distribution from Section 3.3. It was mentioned there that such a distribution can be represented as the sum of n independent exponential random variables having the same parameter value b. Also from that section, we know that the mean of an exponential variable with parameter b is $1/b$, and that the variance is $1/b^2$. Thus from the Central Limit Theorem, the Erlang distribution is approximately normal with mean n/b and variance n/b^2. The ratio of the standard deviation to the mean would then be $1/\sqrt{n}$. Of course, this ratio is small if n is large, which means that an Erlang random variable acts like a constant for large n.

This observation has profound implications for the applicability of Markov chains. For example, consider an M/G/1 queue, which is defined identically to an M/M/1 queue, except that service times are general (G) rather than restricted to being exponentially distributed. Suppose service time in the M/G/1 queue is the constant c. Actually, there are exact mathematical results for this system (see below), but for the sake of presenting this method, let us suppose that these are not available. As discussed above, suppose we approximate c by an Erlang variable. Recall again that the Erlang variable can be represented as a sum $W_1 + \cdots + W_n$ of exponential variables. We can thus think of the service time being broken down into n successive "stages" of service, with durations $W_1, \ldots, W_n$, although these stages might not have any physical meaning. However, due to the exponential durations of the stages, we now have a Markov process. The state space is now $\{(i, j) : i = 0,1,2,3,\ldots; j = 0,1,2,\ldots, n\}$, where i is the number of jobs in the system and j is the "stage" of service ($j = 0$ if $i = 0$). We can then do a Markov analysis as usual.

This method works for nonconstant service times, too. For example, again consider the M/G/1 system. Let S be the service time. The random variable S can be approximated by a discrete random variable S' which places probability $1/m$ on each of m points: Partition $(0,\infty)$ into m subintervals of possibly different lengths, in such a way that the area under the density f_S is equal to $1/m$ in each subinterval. Let α_i denote the left endpoint of the i-th subinterval. Then let S' take on the values $\alpha_1, \ldots, \alpha_m$ with probability $1/m$ each.

S' will have approximately the same cumulative distribution function as S. Now note that as above, each α_i, being a constant, can then in turn be approximated by an Erlang random variable, which is a sum of exponentially distributed random variables. In this way, we again have a Markov process, with state space $\{(i,j,k) : i=0,1,2,3,\ldots; j=0,1,2,\ldots, n; k=1,2,\ldots,m\}$. Here, i and j are as above, while k indicates that $S'=\alpha_k$.

Thus, any distribution can be approximated by mixtures of sums of exponentials, and thus, in theory, any system can be analyzed as a Markov process. This may not be as simple as it sounds; for one thing, the addition of all these artificial stages may make the state space enormous. For example, in the M/G/1 queue analysis outline in the last paragraph, suppose we look at a truncated version $\{(i,j,k) : i = 0,1,2,3,\ldots, l; j = 0,1,2,\ldots, n; k = 1,2,\ldots,m\}$. This state space has size $(l+1)(n+1)m$, which can very easily amount to thousands of states. Then the system (8.1.8) may consist of thousands of equations, and thus might place severe strains on time and memory resources in the computer used for solution.

(b) Transform Methods. Recall the concept of moment generating functions, and related functions such as Laplace transforms, introduced in Section 4.3. In some queueing systems, these tools can be used to derive long-run distributional quantities. For example, in the M/G/1 queue, the mean number of jobs in the system can be shown to be

$$\frac{U^2(1+C^2)}{2(1-U)}+U \qquad \textbf{(8.1.12)}$$

where the **utilization** U is the proportion of time the server is busy, and C is the **coefficient of variation**

$$\frac{\sqrt{\text{Var}(S)}}{E(S)}$$

of the service time S (the derivation of this is beyond the scope of this text). Little's Rule implies that that

$$U=\frac{E(S)}{E(A)} \qquad \textbf{(8.1.13)}$$

where A is the interarrival time; neither A nor S need be exponentially distributed for Equation (8.1.13) to hold (Exercise 8.6).

It is important to note that Equation (8.1.12) involves Var(S), not just

E(S). If two machines have the same mean service times, but one has larger variance, then that machine will produce a longer mean waiting time.

The key point in the derivation of Equation (8.1.12) is to notice that, even though the M/G/1 queue itself is not Markovian, the number of jobs in the system at the end of each service completion *does* form a discrete-time Markov chain. The Markovian nature of this chain stems from the exponential interarrival times in the original process. The derivation is not really difficult, but it is beyond the scope of this text. The interested reader may consult the references.

8.2 DISCRETE-EVENT SIMULATION

While mathematical methods play a vital role in the analysis of discrete-event systems, it is clear from the material in Section 8.1 that the mathematical approach definitely has its limitations. First consider Markov chains. One problem is that of solving for the long-run state probabilities π_i. It is often unclear how to solve for these in an infinite system, and even in the finite case, the system may be so large that serious computational problems arise. Of course, the other major problem with Markov chains is that all distributions must be assumed to be exponential. Some mathematical results for nonexponential systems have been derived, but such derivations are quite difficult, and they cover only a small proportion of the various systems arising in practice.

Simulation is a very valuable alternative tool for analysis, just as it has been throughout our study in earlier chapters. We introduce methods for discrete-event simulation programming in this section, and methods for accuracy analysis (analogous to that of Section 6.5) in Section 8.3. As usual, we begin with a simple example.

Example 8.2.1

Suppose we have a computer timeshare system, consisting of five terminals and one CPU (Central Processing Unit, the execution part of the computer). Assume a busy system, so that there is always someone at each terminal; when one user leaves, another arrives immediately. After a user's job finishes, he or she will enter a "thinking period" of duration T before sending in the next job. The mean thinking period is 15 seconds. Let the CPU time requirement of a job be denoted by S, and suppose $E(S) = 2.4$ seconds. Jobs execute according to a **Round Robin** policy, which means that they take turns executing, with each turn lasting a specified amount of time, typically a fraction of a second. A job takes a turn, is idle until its next turn, takes another turn, is idle again, and so on, until it finishes its CPU time requirement. (This is done so that short jobs will not have to wait for a very long job to complete before they can start.)

There is a **context switch** time involved each time a turn passes from one job to another. Suppose on this particular system, the size of a turn is small relative to 2.4, the mean processing time requirement, and suppose the context-switch time is small relative to the turn size. Under these circumstances, job execution alternates so quickly that it would *appear* to an outside observer as if all jobs were simultaneously running, even though they are actually taking turns. This form of round-robin scheduling is called **Processor Sharing**. In this setting, it appears that each job is running continuously at rate $1/k$ normal speed, where k is the number of jobs currently at the CPU ($1 \leq k \leq 5$).

If thinking time and processing time are both exponentially distributed, this problem can be solved as a Markov chain, in fact as a birth/death process: Simply let the state of the system be the number of jobs at the CPU, so that the state space is $\{0,1,2,3,4,5\}$. Consider λ_k, for $0 \leq k < 5$. This is the rate of a transition from the state in which there are k jobs at the CPU, to the state in which there are $k + 1$ jobs. For such an event to occur, one of the $5 - k$ "thinking" terminal users must submit a job to the CPU. This happens for a single such user at the rate of 1/15, so the aggregate rate for the $5 - k$ users is $(5 - k)(1/15)$; that is, $\lambda_k = (5 - k)/15$. The quantity μ_k is also easy to determine. This is the rate at which we move from State k to State $k - 1$, an event which occurs when a job completes its service at the CPU. For the moment, suppose there were only one job at the CPU, rather than k. This job would complete at the rate of 1/2.4 in this case; but with k jobs present, it completes at a rate of only $1/k$ of full speed, or $1/(2.4k)$. However, there are k jobs doing this, so the combined rate is $k[1/(2.4k)] = 1/2.4$, and this is the value of μ_k.

Thus, a Markov analysis of this example is easy. From Equation (8.1.10), we find that the solution vector is $\pi = (0.38,0.30,0.19,0.09,0.03,0.01)$. Note that the CPU is idle 38% of the time. This suggests that we could easily add some more terminals to the system without seriously inconveniencing the present users.

Although the Markov analysis is easy to perform, it may not be sufficient, and a simulation analysis may be needed, for a variety of reasons:

- The assumption that the thinking times and CPU times are exponential may be incorrect, even as approximations.
- We may be specifically interested in investigating the effect of varying the size of a job's turn at the CPU. There is a tradeoff

here: If the turn size is too small, the **overhead** due to context switching may consume a large part of the CPU's resources, that is, most of the CPU's time will be devoted to nonproductive context switching. On the other hand, if the turn size is too large, we lose the round robin policy's advantage of giving shorter jobs higher priority (note that if the turn size is infinite, then the round robin reduces to a first-come, first-served policy). At any rate, the above Markov analysis becomes invalid: The assumption that the context-switch time is much smaller than the turn size, which itself is much smaller than the mean CPU time required for a job, will be inappropriate as we vary the turn size to very small or very large values.

- The above analysis does not make allowance for input/output (I/O) activities. When a program's execution reaches an I/O statement (e.g., a "readln" in Pascal), the program's turn ends. The reason for this stems from the fact that I/O is much slower than the CPU. Therefore, rather than having the CPU wait idle until the I/O action is completed, it makes much more sense to end this program's turn, and start a turn for another program. Our analysis should take this fact into account. Moreover, the I/O actions have another effect on our analysis, in that they may slow down the effective speed of the CPU, if CPU access to memory is occasionally blocked due to memory access by a high-speed I/O device (this is called **memory cycle stealing**). We could try to incorporate these considerations into the Markov model above, but again we might have severe problems with assumptions; for example, the I/O time might not be exponentially distributed.
- Another problem is that even if the Markov assumptions are valid, some quantities of interest are not directly available from the long-run probabilities π_i. For example, in the timeshare system here, we may be primarily interested in the mean response time, that is, the mean time between submission and completion of a job. This quantity is not an immediately obvious function of the π_i. There are indirect ways of determining the mean response time in this case (see Exercise 8.7). However, one often encounters cases in which it is not at all clear how to find certain quantities of interest from the steady-state probabilities.

Some of these problems can be solved by expert queueing analysts, using very deep and intricate mathematics. However, even the best experts can only solve *some* of the problems mathematically. Thus while mathematical analysis is quite useful in certain contexts, it is certainly not powerful enough to cover general applications. For this reason, simulation analysis is an extremely valuable tool.

Let us look into a simulation-based analysis of Example 8.2.1. This will be done in Program 8.2.1 below. The program has been kept simple, since its main purpose for us is to introduce discrete-event simulation programming. However, keep in mind the ways in which a modified version of the program would address the shortcomings cited above for the mathematical Markov analysis:

a. The program could handle nonexponential distributions; for example, we could replace

```
Expon(MeanCPU)
```

by

```
2.0*MeanCPU*Rnd
```

in Lines 88 - 89 below, to model a uniform distribution with the specified mean.

b. We could add code to the program to account for the effect of context-switch time.

c. We could add code to account for I/O effects.

d. We could add code to determine other quantities than the π_i. In fact, the program below already does this for one such quantity, the mean response time (Line 112).

Note also that numerical parameters of the system are entered when the user runs the program, rather than being set by assignment or Pascal "const" statements within the program. For example, the number of terminals NTerm is entered by the user, in Lines 58–59, rather than with a statement such as

```
const NTerm = 5;
```

This way, if we wish to explore the effects of adding more terminals to the system, we do not need to recompile the program each time we investigate a different number of terminals.

Here is the program. Following it there is a discussion of its overall structure, the major variables it uses, how the flow of control works, and so on. Thus, at this point you should merely skim through it, and then return later for a more detailed reading.

Program 8.2.1

```
1       program PS(input,output);
2
3          const RunMode = 1;
4                ThinkMode = 2;
```

```
 5
 6        type ActRec = record
 7                        ActType: integer;  {type of
 8                          activity}
 9                        TimeLeft: real;
10                          {time left before current
11                          activity ends}
12                        SbmtTime: real;
13                          {time job was submitted to
14                          CPU}
15                      end;
16
17       var SimTime,  {current simulated time}
18           Delta,  {time increment size}
19           TimeLim  {simulated time for ending run}
20             : real;
21           Seed: integer;  {for random number
22                             generation}
23           NCurrRun: integer;  {number of jobs
24                      currently running at CPU}
25           SumResp: real;  {total of response time for
26                      all jobs completed so far}
27           NComplt: integer;  {total number of jobs
28                      completed so far}
29           MeanThnk,MeanCPU: real;
30             {mean thinking and cpu times}
31           NTerm: integer;  {number of terminals}
32           TrmActRec: array[1..50] of ActRec;
33             {activity records, one for each terminal}
34           Pi: array[0..50] of real;
35             {counts for estimating the long-run
36               probabilities}
37           Trm: integer;
38
39       function Rnd: real;
40       begin
41          Seed := (25173*Seed + 13849) mod 32768;
42          Rnd := Seed/32768.0
43       end;
44
45       function Expon(Mu: real): real;
46          var U: real;
47       begin
48          U := Rnd;
49          if U < 0.00001 then U := 0.00001;
50          Expon := -Mu*ln(U)
51       end;
52
53       procedure Init;
```

```
54             var I: integer;
55          begin
56             writeln('enter Delta,TimeLim,Seed');
57             readln(Delta,TimeLim,Seed);
58             writeln('enter NTerm');
59             readln(NTerm);
60             SimTime := 0.0;
61             writeln('enter MeanThnk,MeanCPU');
62             readln(MeanThnk,MeanCPU);
63             NComplt := 0;
64             NCurrRun := 0;  {all terminals are in 'think'
65                              mode}
66             SumResp := 0.0;
67             for I := 1 to NTerm do
68                begin
69                Pi[I] := 0.0;
70                TrmActRec[I].ActType := ThinkMode;
71                TrmActRec[I].TimeLeft := Expon(MeanThnk)
72                end
73          end;
74
75
76       begin
77          Init;
78          repeat
79             SimTime := SimTime + Delta;
80             Pi[NCurrRun] := Pi[NCurrRun] + 1.0;
81             for Trm := 1 to NTerm do
82                with TrmActRec[Trm] do case ActType of
83                   ThinkMode: begin
84                                 TimeLeft:=TimeLeft - Delta;
85                                 if TimeLeft <= 0.0 then
86                                    begin
87                                    ActType := RunMode;
88                                    TimeLeft := Expon
89                                                (MeanCPU);
90                                    SbmtTime := SimTime;
91                                    NCurrRun := NCurrRun + 1
92                                    end
93                                 end;
94                   RunMode: begin
95                               TimeLeft := TimeLeft - Delta/
96                                           NCurrRun;
97                               if TimeLeft <= 0.0 then
98                                  begin
99                                  ActType :=ThinkMode;
100                                 TimeLeft:=Expon(MeanThnk);
101                                 SumResp :=SumResp +
102                                        SimTime - SbmtTime;
```

```
103                                  NComplt := NComplt + 1;
104                                  NCurrRun := NCurrRun - 1
105                                  end
106                              end
107                 end
108        until SimTime >= TimeLim;
109        writeln('long-run probabilities:');
110        for Trm := 0 to NTerm do
111           writeln(Trm, '    ', Pi[Trm]*Delta/SimTime:1:3);
112        writeln('mean response time = ',
113           SumResp/NComplt:1:3)
114     end.
```

The program works in the following way. First, consider the major variables and data structures, declared at the beginning of the program. There is a record structure ActRec (Lines 6–15), to describe the current activity of a terminal. This record contains all the current information for this terminal:

- Mode: Is the job from this terminal currently running on the CPU, or is the user thinking?
- Time left before the next change in mode for this terminal; for example, if the mode currently is ThinkMode, the length of time before the user submits his/her next job.
- If there is a job running from this terminal, the submission time of the job. We need to know this so that we can record the response time for this job when it finishes.

Note that there is an entire array TrmActRec (Lines 32–33) of these activity records, one record for each of the NTerm terminals.

There is a variable SimTime which represents the current simulated time. We simulate the operation of the system until time TimeLim. Thus the loop in the main program (Lines 78–108), which has the form

```
repeat
   SimTime := SimTime + Delta;
   ...
until SimTime >= TimeLim
```

is analogous to the loop

```
for RepNum := 1 to TotReps do
    begin
    ...
    end
```

in Program 2.4.3, and the larger we set TimeLim, the more accurate will be our estimate of the long-run performance of the system, just as TotReps plays a similar role in the accuracy of the output of Program 2.4.3. Also, note that simulated time advances in small increments of size Delta; the smaller the value of Delta, the more accurate our results.

The variable NCurrRun records the number of jobs currently running at the CPU. This is used in two ways. One of these ways occurs in Line 95:

```
TrmActRec[Trm].TimeLeft:=
     TrmActRec[Trm].TimeLeft - Delta/NCurrRun;
```

Recall that if k jobs are currently at the CPU, then each job runs at only $1/k$ normal speed. This is the reason for dividing by NCurrRun above.

NCurrRun is also used in Line 80:

```
     Pi[NCurrRun] := Pi[NCurrRun] + 1.0;
```

The array Pi records counts of the number of time periods in which the system was in State i, where as before State i means that i jobs are at the CPU. Since NCurrRun does in fact record the current state of the system, the above incrementation statement does update the array Pi in the desired manner.

As mentioned above, simulated time starts at 0.0 and runs through time TimeLim with the "clock" moving in "ticks" of size Delta. After each tick of the clock, each of the activities has its TimeLeft field updated (Lines 84 and 95). If the result is zero, then an activity (either thinking or running) has completed, in which case the new activity must be started. For example, if a user at a terminal has just completed the thinking process and submits a job to the CPU, we must then simulate starting the job. This is done in Lines 87–91.

Program 8.2.1 demonstrates the major ideas behind discrete-event simulation. However, it is of course not a general program, and does not include some features which are very common in many programs of this type. For example, the system being analyzed in Program 8.2.1 does not involve any queueing, but many other systems do, such as that in Example 8.1.4. Also, in Program 8.2.1, there are a fixed number (NTerm) of activity records, but in many other programs, the number may vary; again Example 8.1.4 is an instance of this, because the number of jobs in the system varies.

It is of obvious benefit to have a *general* program that allows for queueing, varying numbers of activity records, and so on. Furthermore, such a general program should have debugging facilities, since simulation programs tend to be difficult to debug.

Program 8.2.2 below addresses these issues. It is actually not a complete program, but rather a program *template*. It includes generally useful data structures, procedures and a main program. However, it is filled with numerous gaps, labeled "** start ASC **" and "** end ASC **," which indicate where code specific to the application at hand ("application-specific code") must be added. In other words, one uses Program 8.2.2 as the *foundation* for one's program, adding code in various places, as required by the system to be simulated. This format can be used quite conveniently by a programmer at a terminal, since he or she can simply use text editor commands to search for all instances of the string "** start ASC **."

Program 8.2.2 is much larger than those previously presented in this text, and it thus may have something of an intimidating effect on the reader. However, in spite of its formidable appearance, it is basically just a generalization of Program 8.2.1, and it is actually easy to use after the programmer applies it to one or two specific systems. As with all large learning projects, gains in understanding are incremental. In the next few pages, the reader will find that he or she will not immediately understand the entire program, but will gradually understand more and more aspects. Then, after the reader has actually used this template to write a program simulating a specific system, he or she should feel quite comfortable in applying it to other systems.

The following advice should be very helpful: Most of the program consists of procedures which manipulate linked lists. The author highly recommends that, at least at the beginning, the reader not concern himself or herself with the details of such procedures; just learn what their *purposes* are. In particular, for the procedures Delete, AddToTail, AddToQ, and DeleteHeadOfQ, simply read the comments which explain the purposes of these procedures; do not look at the details.

With these points in mind, let us begin to develop a rough idea of how the template is organized. It will be helpful to have a specific example:

Example 8.2.2

Suppose we are simulating a large automobile manufacturing plant. which is set up on an "assembly line" basis. A car will be routed to several different stations in the plant, such as welding, painting, and so on. For simplicity, assume that job orders to build cars arrive at Point O in the plant, and that finished cars exit the plant at Point F.

With this specific example in mind, let us proceed to discuss the program template appearing below, Program 8.2.2. Keep in mind that this is a general template, to be used for all settings; the system in Example 8.2.2 is only one possible application.

First, we will look at the data structures used in the program. The basic record for an activity has type ActRec (Lines 37–59), just as in Program 8.2.1. However, in this case, there are many more fields in the record. The field Facil can be used to indicate where the job is in the system. In Example 8.2.2, this field would allow us to keep track of the current location of the car in the plant; the facilities will be various stations on the factory floor (e.g., a welding station, a painting station, etc.). In Example 8.1.4, the facilities would be the two disk drives and the channel.

The field StartWt is used whenever a job is waiting in a queue. When a job joins a queue, the procedure AddToQ (Line 219) will be called, and StartWt will be set to the current time (Line 224) to indicate that the queue wait started at that time. Then when the job leaves the queue and starts service (Lines 273–276), the total waiting time for all jobs having service for that facility can be incremented by the wait time for this job (Lines 274–275). In this way, at the end of the simulation run, the program can report the mean queue wait at each of the facilities in the system (Lines 524–525).

The field StrtTime (Line 45) indicates when the job first entered the system as a whole. In Example 8.2.2, this would be the time that the job order arrived at Point O. When the car finally gets to Point F, we can use StrtTime to find the total time the car has spent in the system (often called **residence time**), as in Lines 343–344. This is often the main quantity of interest in a given application: for example, the mean time that it takes a customer to obtain an airline reservation, measured from the time she dials the telephone through the time that the agent finishes the final details of the making of the reservation. This is after all the nature of queueing analysis—people do not like to wait!

The field ActNum (Line 50) is convenient for debugging purposes, and in some applications is convenient if there is communication between different activities.

The fields FwdPtr (forward pointer) and BackPtr (backward pointer) in Line 51 are used in connecting activity records together into a linked list. As mentioned above, in many applications the number of activity records in the system will vary with time. Thus instead of having an array of records, as in Program 8.2.1, we will have a chain of records, of varying length. Variables ActListHead and ActListTail (Line 71) point to the head and tail of this chain.

Lines 54–58 then provide space for the user to include his or her own fields in the record, according to the needs of the specific system being simulated (application-specific code). In Example 8.2.2, we would probably have a number of such fields, indicating the type of car, color to be painted, and so on. These fields would be used by the program to decide to which station a car should be routed next after completing service at a particular station on the factory floor. For example, one field would be used to determine whether the car should be routed next to the station for automatic transmissions, or the one for manual transmissions.

The program maintains information on the status of each facility (Lines 79–87). For example, there is a queue for each facility, again implemented as a chain, with the head and tail of the chain for Facility *i* being pointed to by QHead[i] and QTail[i]. The length of that queue is given by QLength[i], and InUse[i] indicates whether Facility *i* is currently in use or free.

Here is an overview of the roles of the procedures in the program: Group I procedures (Lines 155–209) handle the basic operations of deleting an item from a chain, or adding an item to one. They are concerned strictly with the disconnection or connection operations involved; that is, they are concerned only with Pascal pointer manipulation, not with the simulation environment itself.

On the other hand, Group II procedures (Lines 213–287) are directly concerned with simulating the start and completion of jobs in the simulated system. For example, look at procedure AddToQ (Line 219) in Group II. Consider a situation in which a job is ready to use a certain facility, but can not do so because the facility is busy (Lines 254–257). In this case, the job must be added to the queue for that facility, which is done by calling AddToQ. AddToQ calls on AddToTail (Line 225) to do the actual work of connecting the job's activity record to the facility's queue, and it calls on Delete to do the corresponding disconnection from the activity chain (since the job will be queued, for the moment it does not correspond to an activity). In other words, the Group II procedure AddToQ, which is directly involved in simulating the given system, calls on the Group I procedure AddToTail to do the nonsimulation work of pointer manipulation. However, AddToQ also does additional "bookkeeping" associated with queueing the job: It records the start of the queue wait time for the job (Line 224), and updates the variable which records the length of the queue (Line 226).

The other procedures in Group II perform similar operations. For example, consider the situation in which a job has just completed service at a facility. The facility will now be released for possible use by a job which had been waiting at the facility's queue. This is done by procedure Release (Lines 260–278).

The Group III procedures (Lines 291–358) check for changes in the state of a system, and take appropriate action when changes are found. For instance, for Example 8.2.2, procedure CheckForChange might discover that the current "tick" of the simulation clock SimTime has resulted in the completion of the welding operation. Procedure TakeAction would then route the car to the next appropriate assembly line station.

At this point, the reader should browse through the template, paying special attention to the comments in each procedure. Then we will look at complete implementations for two examples.

Program 8.2.2

```
1   {Throughout the code, there are comments labeled
2    "ASC"(application-specific code). Usually an ASC
3    section is bracketed by comments "** start
4    ASC **" and "** end ASC **." These indicate that
5    the user should insert code which is specific to
6    the particular application.}
7
8
```

```
 9   program ActSim(input,output);
10
11
12       {** start ASC **}
13       {Put here comments regarding various activity
14        types, including the conditions under which
15        they are created and destroyed:}
16
17       {** end ASC **}
18
19
20       const {** start ASC **}
21             MaxAry = ;  {maximum value array size}
22             {** end ASC **}
23             {** start ASC **}
24             {declare constants here for
25              activity type names, e.g, Arriv = 1}
26
27             {** end ASC **}
28
29       {** start ASC **}
30       {add comments here explaining the system
31        of facility numbers, e.g, Facil = 3 means
32        the CPU}
33
34       {** end ASC **}
35
36       type ActRecPtr = ActRec;
37            ActRec = record  {activity record structure}
38                       ActType: integer; {type of activity}
39                       Facil: integer;  {facility}
40                       StartWt: real;
41                          {time current queue wait began}
42                       TimeLeft: real;
43                          {time left before current
44                           activity ends}
45                       StrtTime: real;
46                          {time this activity or job began;
47                           used in applications in which
48                           one is interested in the mean
49                           time a job spends in the system}
50                       ActNum: integer;  {activity number}
51                       FwdPtr,BackPtr: ActRecPtr;
52                          {links to next and
53                           previous records}
54                       {** start ASC **}
55                       {application-specific
56                        fields, if any, go here}
57
```

```
 58                       {** end ASC **}
 59                    end;
 60
 61
 62        var
 63            {time variables}
 64            SimTime,  {current simulated time}
 65            Delta, {time increment size}
 66            TimeLim  {time limit at which
 67                       simulation run ends}
 68                    : real;
 69
 70            {pointer variables for activities}
 71            ActListHead, ActListTail,
 72               {point to head and tail of activities list}
 73            CurrPtr,
 74               {points to activity currently being handled
 75                in the main program}
 76            NextPtr
 77                    : ActRecPtr;
 78
 79           {variables showing current status of facilities}
 80            QHead,QTail
 81               {arrays pointing to heads and tails
 82                of facility queues}
 83                    : array[1..MaxAry] of ActRecPtr;
 84            InUse  {indicates whether facilities are busy}
 85                    : array[1..MaxAry] of boolean;
 86            QLength: array[1..MaxAry] of integer;
 87               {array of queue lengths}
 88
 89            {variables keeping track of
 90             statistics for facility queues}
 91            NJobs  {array of numbers of jobs handled so far}
 92                    : array[1..MaxAry] of integer;
 93            SumLng,
 94               {cumulative sums of queue lengths, for
 95                finding mean queue lengths}
 96            SumBusy,
 97           {cumulative busy time, for finding utilizations}
 98            SumWait  {cumulative queue wait time}
 99                    : array[1..MaxAry] of real;
100
101            {miscellaneous}
102            Seed,  {for random number generation}
103            NActs,  {number of activities created so far}
104            NFcls,  {number of facilities}
105            NChange,  {number of changes with this ''tick''
106                        (see Change below)}
```

```
107          Fcl,SaveAct: integer;
108          SumResTime: real;  {total system residence
109                          time, for all jobs (may need to
110                          create several of these, if break
111                          down time according to job
112                          categories)}
113          Change, {indicates that in this ''tick'' of the
114                          clock (of size Delta), a change
115                          in the system has occurred, e.g.,
116                          a service period has completed}
117          Debug
118             {indicates whether user wants debugging
119                          info}
120             : boolean;
121          CommandChar  {for reading one-character
122                          commands}
123             : char;
124          ActName  {names of the activities}
125             : array[1..MaxAry] of array[1..8] of char;
126
127          {** start ASC **}
128          {application-specific declarations go here}
129
130          {** end ASC **}
131
132
133      {************************************
134       Random Number Generation Subprograms.
135       Other distributions may be generated
136       from combinations of these, or by the
137       user adding his or her own functions}
138
139      function Rnd: real;
140      begin
141         Seed := (25173*Seed + 13849) mod 32768;
142         Rnd := Seed/32768.0
143      end;
144
145      function Expon(Mu: real): real;
146         var U: real;
147      begin
148         U := Rnd;
149         if U < 0.00001 then U:=0.00001;
150         Expon := -Mu*ln(U)
151      end;
152
153
154      {****************************************
155       Group I Simulation Subprograms:
```

```
156          Linked-List Manipulation.
157          These do the low-level work of adding or
158          deleting records to or from the activities
159          list or facility queues. Called mainly by
160          Group II subprograms.}
161
162         procedure Delete(Delptr: ActRecPtr;
163                          var ListHead,ListTail: ActRecPtr);
164            label 999;
165         begin
166            {deletes Delptr from the list whose
167             head and tail are ListHead and ListTail}
168            if ListHead = ListTail then
169               begin
170               ListHead := nil;
171               ListTail := nil;
172               goto 999
173               end;
174            if Delptr = ListHead then
175               begin
176               ListHead := ListHead^.FwdPtr;
177               ListHead^.BackPtr := nil;
178               goto 999
179               end;
180            if Delptr = ListTail then
181               begin
182               ListTail := ListTail^.BackPtr;
183               ListTail^.FwdPtr := nil;
184               goto 999
185               end;
186            Delptr^.BackPtr^.FwdPtr := Delptr^.FwdPtr;
187            Delptr^.FwdPtr^.BackPtr := Delptr^.BackPtr;
188            999:
189         end;
190
191         procedure AddToTail(AddPtr:ActRecPtr;
192                     var Listhead, ListTail: ActRecPtr);
193         begin
194            {adds the record pointed to by AddPtr to
195             the list whose last record is pointed
196             to by ListTail}
197            if ListTail <> nil then
198               begin
199               ListTail^.FwdPtr := AddPtr;
200               AddPtr^.BackPtr := ListTail
201               end
202            else
203               begin
204               Listhead := AddPtr;
```

```
205            AddPtr^.BackPtr := nil
206            end;
207         ListTail := AddPtr;
208         AddPtr^.FwdPtr := nil
209      end;
210
211
212      {**********************************
213       Group II Simulation Subprograms:
214       Job Initiation and Completion.
215       Do the low-level work of starting
216       and completing jobs. Called by
217       Group III subprograms.}
218
219      procedure AddToQ(Pt: ActRecPtr; Fcl: integer);
220      begin
221         {deletes Pt from the activity list, and
222          adds it to queue for facility Fcl}
223         Delete(Pt,ActListHead,ActListTail);
224         Pt^.StartWt := SimTime;
225         AddToTail(Pt,QHead[Fcl],QTail[Fcl]);
226         QLength[Fcl] := QLength[Fcl]+1
227      end;
228
229      procedure DeleteHeadOfQ(Fcl: integer;
230                              var Pt: ActRecPtr);
231      begin
232         {deletes head of the queue for facility Fcl,
233          and adds that record back to the activities
234          list}
235         {also, the deleted record is pointed to by Pt}
236         Pt := QHead[Fcl];
237         Delete(QHead[Fcl],QHead[Fcl],QTail[Fcl]);
238         QLength[Fcl] := QLength[Fcl] - 1;
239         AddToTail(Pt,ActListHead,ActListTail)
240      end;
241
242      procedure Start(Pt: ActRecPtr; Fcl: integer);
243      begin
244         InUse[Fcl] := true;
245         Pt^.Facil := Fcl;
246         NJobs[Fcl] := NJobs[Fcl] + 1
247      end;
248
249      procedure StartOrQ(Pt: ActRecPtr; Fcl: integer);
250      begin
251         {tries to start the job pointed to by Pt at
252          facility Fcl, adding it to queue if Fcl
253          is busy}
```

```
254             if not InUse[Fcl] then
255                Start(Pt,Fcl)
256             else
257                AddToQ(Pt,Fcl)
258          end;
259
260          procedure Release(Fcl: integer);
261             var Pt: ActRecPtr;
262          begin
263             {releases facility Fcl, and starts
264              the next job in the queue, if any}
265             InUse[Fcl] := false;
266             if QLength[Fcl] > 0 then
267                begin
268                {** start ASC **}
269                {if not first-come, first-served policy,
270                 then change the lines below,
271                 e.g. DeleteHeadOfQ}
272                {** end ASC **}
273                DeleteHeadOfQ(Fcl,Pt);
274                SumWait[Fcl] := SumWait[Fcl] + SimTime - Pt^
275                                   .StartWt;
276                Start(Pt,Fcl)
277                end
278          end;
279
280          procedure MakeNewAct(var Pt: ActRecPtr);
281          begin
282             {creates a new activity}
283             NActs := NActs + 1;
284             new(Pt);
285             Pt^.ActNum := NActs;
286             AddToTail(Pt,ActListHead,ActListTail)
287          end;
288
289
290          {*********************************
291           Group III Simulation Subprograms.
292           High-level simulation mechanisms,
293           called by the main program.}
294
295
296          procedure CheckForChange;
297          begin
298             {** start ASC **}
299             {look at the activity currently being
300              considered, i.e., the one pointed to
301              by CurrPtr, update it according to
```

```
302          whatever effects the current tick of the
303          clock has, and then check to see
304          whether a ''discrete event'' change
305          has occurred in the system, if so
306          setting Change to true}
307        {in most programs, this procedure will be simply
308
309           Change := false;
310           CurrPtr^.TimeLeft:=CurrPtr^.TimeLeft - Delta;
311           if CurrPtr^.TimeLeft <= 0.0 then
312              Change := true
313
314        }
315        Change := false;
316
317        {** end ASC **}
318     end;
319
320     procedure TakeAction;
321        var Pt: ActRecPtr;
322     begin
323        {** start ASC **}
324        {takes whatever action is needed when
325         a change occurs, i.e., when CurrPtr^.TimeLeft
326         reaches 0.0}
327        {below are listed some possibilities:
328            (a)  an arrival from outside the system
329                 has occurred, so should have a
330                 sequence like
331                    MakeNewAct(Pt);
332                    Pt^.ActType:=***;
333                    Pt^.TimeLeft:=***;
334                    Pt^.StrtTime:=SimTime;
335                    set application-specific fields
336                    StartOrQ(Pt,***);
337                    CurrPtr^.TimeLeft:=***;  for setting
338                       up next external arrival
339            (b)  a job has finished at a facility; if it
340                 now exits the system, should have a
341                 sequence like
342                    Release(CurrPtr^.Facil);
343                    SumResTime := SumResTime + SimTime
344                                    -CurrPtr^.StrtTime;
345                    Delete(CurrPtr,ActListHead,
346                       ActListTail);
347                    dispose(CurrPtr)
348                 if not, it should be forwarded
349                 to its next facility, so should have a
```

```
350                    sequence like
351                       Release(CurrPtr^.Facil);
352                       CurrPtr^.ActType:=***;
353                       CurrPtr^.TimeLeft:=***;
354                       StartOrQ(CurrPtr,***);}
355          case CurrPtr^.ActType of
356
357          {** end ASC **}
358       end;
359
360
361       {************************
362        Initialization Procedure.}
363
364       procedure Init;
365          var Fcl: integer;
366              Pt: ActRecPtr;
367       begin
368          writeln('enter Delta,TimeLim,Seed');
369          readln(Delta,TimeLim,Seed);
370          {** start ASC **}
371          {insert code here for determining NFcls,
372           by assignment or readln statements}
373
374          {** end ASC **}
375          for Fcl := 1 to NFcls do
376             begin
377             QHead[Fcl] := nil;
378             QTail[Fcl] := nil;
379             InUse[Fcl] := false;
380             QLength[Fcl] := 0;
381             SumLng[Fcl] := 0.0;
382             SumBusy[Fcl] := 0.0;
383             SumWait[Fcl] := 0.0;
384             NJobs[Fcl] := 0
385             end;
386          ActListHead := nil;
387          ActListTail := nil;
388          SimTime := 0.0;
389          SumResTime := 0.0;
390          NActs := 0;
391          writeln('debug?');
392          readln(CommandChar);
393          Debug := (CommandChar = 'y');
394          {** start ASC **}
395          {initialize the array 'ActName',
396           e.g., ActName[Arriv] := 'Arriv'}
397
```

```
398        {** end ASC **}
399        {** start ASC **}
400        {other application-specific initialization
401         goes here, e.g. reading in parameter values,
402         and initializing values of variables using
403         assignment statements, e.g. initializing
404         sums to 0}
405
406        {** end ASC **}
407        {** start ASC **}
408        {initialize activities list here, calling
409         MakeNewAct; don't forget to call Start if
410         appropriate; a typical example is the
411         following, in which an arrival
412         activity is initiated:
413               MakeNewAct(Pt);
414               Pt^.ActType := Arriv;
415               Pt^.TimeLeft := Expon(MeanArrivTime);}
416
417        {** end ASC **}
418     end;
419
420
421     {*******************
422      Debugging Procedures.}
423
424     procedure WriteFclName(Fcl: integer);
425     begin
426        {** start ASC **}
427        {writes out facility name, during debugging}
428
429        {** end ASC **}
430     end;
431
432     procedure PrintActList;
433        var Pt: ActRecPtr;
434            Fcl: integer;
435     begin
436        {prints out activity list, during debugging}
437        writeln('activity list:');
438        Pt := ActListHead;
439        repeat
440           write(Pt^.ActNum:1);
441           write('  ');
442           write(ActName[Pt^.ActType]);
443           write('  ');
444           Fcl := Pt^.Facil;
445           if Fcl > 0 then WriteFclName(Fcl);
```

```
446              write('  ');
447              write(Pt^.TimeLeft:1:2);
448              write('  ');
449              {** start ASC **}
450              {write out application-specific info}
451
452              {** end ASC **}
453              writeln;
454              Pt := Pt^.FwdPtr
455           until Pt = nil
456        end;
457
458        procedure PrintFclQs;
459           var Pt: ActRecPtr;
460               Fcl: integer;
461        begin
462           {prints out facility queues,
463            during debugging}
464           for Fcl := 1 to NFcls do
465              begin
466              Pt := QHead[Fcl];
467              if Pt <> nil then
468                 begin
469                 writeln;
470                 write('queue for ');
471                 WriteFclName(Fcl);
472                 writeln;
473                 while Pt <> nil do
474                    begin
475                    write(Pt^.ActNum:1);
476                    write('  ');
477                    {** start ASC **}
478                    {write out application-specific info}
479
480                    {** end ASC **}
481                    writeln;
482                    Pt := Pt^.FwdPtr
483                    end
484                 end
485              end
486        end;
487
488        procedure DbgPro;
489        begin
490           {for debugging}
491           writeln; writeln;
492           writeln('**********************');
493           writeln;
494           writeln('SimTime = ',SimTime:1:2,'     ');
```

```
495         writeln('activity causing change : ',ActName
496                           [SaveAct]);
497         writeln;
498         PrintActList;
499         PrintFclQs
500      end;
501
502
503      {*******************************
504       Procedures for Reporting Results}
505
506      procedure StdStat;
507         var Fcl: integer;
508             MeanLngth,MeanWait: real;
509      begin
510         {writes out standard statistics}
511         writeln; writeln;
512         write('fcl       util       ');
513         writeln('mean q.lng.  mean wait');
514         for Fcl := 1 to NFcls do
515            begin
516            WriteFclName(Fcl);
517            write('  ');
518            write(SumBusy[Fcl]*Delta/TimeLim:1:2);
519               {utilization}
520            write('      ');
521            MeanLngth := SumLng[Fcl]*Delta/TimeLim;
522            write(MeanLngth:1:2);  {mean queue length}
523            write('          ');
524            MeanWait := SumWait[Fcl]/NJobs[Fcl];
525            writeln(MeanWait:1:2)
526               {mean wait for jobs at this facility}
527            end;
528         writeln; writeln
529      end;
530
531      procedure ASCStat;
532      begin
533         {** start ASC **}
534         {application-specific statistics}
535
536         {** end ASC **}
537      end;
538
539
540   begin
541      Init;
542      if Debug then
543         begin
```

```
544          writeln('************************');
545          writeln;
546          writeln('SimTime = 0.00');
547          PrintActList
548          end;
549       repeat
550          SimTime := SimTime + Delta; {advance the clock}
551          {update the statistics}
552          for Fcl := 1 to NFcls do
553             begin
554             SumLng[Fcl] := SumLng[Fcl] + QLength[Fcl];
555             if InUse[Fcl] then SumBusy[Fcl] := SumBusy
556                         [Fcl] + 1;
557             {** start ASC **}
558             {increment application-specific counts
559              for this facility}
560
561             {** end ASC **}
562             end;
563          {** start ASC **}
564          {increment application-specific counts not
565           associated with particular facilities}
566
567          {** end ASC **}
568          {now examine each activity in the activities
569                         list}
570          CurrPtr := ActListTail;
571          NChange := 0;
572          repeat
573             NextPtr := CurrPtr^.BackPtr;  {save pointer}
574             CheckForChange;  {see if this activity has
575                         generated a change}
576             if Change then
577                begin
578                NChange := NChange + 1;
579                SaveAct := CurrPtr^.ActType;
580                TakeAction
581                end;
582             CurrPtr := NextPtr
583          until CurrPtr = nil;
584          if (NChange > 0) and Debug then DbgPro
585       until SimTime > TimeLim;
586       writeln('want standard statistics report?');
587       readln(CommandChar);
588       if CommandChar = 'y' then StdStat;
589       ASCStat
590    end.
```

Now let us look at an example of how the "ASC" sections in the above program template would be filled in for a specific application.

Example 8.2.3

Consider Example 8.1.3, but suppose there is only one machine (the program would be only slightly more complicated if there were two machines). The machine can only serve one job at a time, so any arrivals during a busy period must join the queue. Mean service and interarrival times are 2.5 and 4.0, respectively, and the mean lifetime and repair times are 20.0 and 8.0. Assume for simplicity that all distributions are exponential; again, we could easily simulate other distributions by using functions other than Expon. Assume also that if the machine breaks down in the middle of service for a job, the job does not have to be restarted from the beginning when the machine returns to working status; whatever work was done on it before the machine broke down does not need to be redone. We would like to calculate quantities such as the mean time a job spends in the system, what proportion of jobs are interrupted by server breakdowns, and so on.

Here are application-specific code segments which, when inserted into Program 8.2.2, will result in a complete program for simulating the system in Example 8.2.3.

Lines 12–17:

```
{** start ASC **}
{Put here a list of various activity types,
 including the conditions under which they
 are created and destroyed:}
{The Arriv activity is created initially,
 and continues to run, producing the supply
 of arriving jobs.
 The Serve activity indicates that the
 machine is currently serving a job.
 There is another activity, created initially,
 which alternates between Working and
 Repair, indicating the current status
 of the machine.}
{** end ASC **}
```

Lines 20–22:

```
const {** start ASC **}
      MaxAry = 4;  {maximum value array size}
      {** end ASC **}
```

Lines 23–27:

```
{** start ASC **}
      {declare constants here for
       activity type names, e.g., Arriv = 1}
      Arriv = 1;
      Serve = 2;
      Working = 3;
      Repair = 4;
      {** end ASC **}
```

Lines 29-34:

```
   {** start ASC **}
   {add comments here explaining the system
    of facility numbers, e.g., Facil = 3 means the CPU}
   {Facil = 1 means the machine}
   {** end ASC **}
```

Lines 54–58:

```
                  {** start ASC **}
                  {application-specific
                   fields, if any, go here}
                  {none}
                  {** end ASC **}
```

Lines 127–130:

```
      {** start ASC **}
      {application-specific declarations go here}
      Down: boolean;  {'true' means machine is down}
```

```
NIntrpt: integer;  {number of interrupted jobs}
{the next 4 variables are parameters to be entered
 by the user at execution time (see insert for
 Lines 394-398 below); as in Program 8.2.1,
 reading in these variables allows us to explore
 different sets of parameter values, without
 recompiling the program}
MeanArriv,  {mean interarrival time, 4.0 in example}
MeanSrv,  {mean service time, 2.5 in example}
MeanUp,  {mean up time, 20.0 in example}
MeanDown  {mean down time, 8.0 in example}
    : real;
{** end ASC **}
```

Lines 268–272:

```
{** start ASC **}
{if not first-come, first-served policy, then
 change the lines below, e.g. DeleteHeadOfQ}
{no change}
{ ** end ASC **}
```

Lines 298–317:

```
{** start ASC **}
{original comments deleted here, to save space}
Change := false;
if not (Down and (CurrPtr^.ActType = Serve)) then
   CurrPtr^.TimeLeft := CurrPtr^.TimeLeft-Delta;
if CurrPtr^.TimeLeft <= 0.0 then
   Change := true
{** end ASC **}
```

Lines 323–357:

```
case CurrPtr^.ActType of
   Arriv: begin
             MakeNewAct(Pt);
             Pt^.ActType := Serve;
             Pt^.TimeLeft := Expon(MeanSrv);
             Pt^.StrtTime := SimTime;
             StartOrQ(Pt,1);
```

```
                CurrPtr^.TimeLeft := Expon(MeanArriv)
                end;
        Serve: begin
                release(CurrPtr^.facil);
                SumResTime := SumResTime + SimTime
                             -CurrPtr^.StrtTime;
                Delete(CurrPtr,ActListHead,ActListTail);
                dispose(CurrPtr)
                end;
        Working: begin
                  Down := true;
                  CurrPtr^.ActType := Repair;
                  CurrPtr^.TimeLeft := Expon(MeanDown);
                  if InUse[1] then
                     NIntrpt := NIntrpt + 1
                  end;
        Repair: begin
                 Down := false;
                 CurrPtr^.ActType := Working;
                 CurrPtr^.TimeLeft := Expon(MeanUp)
                 end
        end
    {** end ASC **}
```

Lines 370–374:

```
    {** start ASC **}
    {insert code here for determining NFcls,
     by assignment or readln statements}
    NFcls := 1;
    {** end ASC **}
```

Lines 394–398:

```
    {** start ASC **}
    {initialize the array 'ActName',
     e.g., ActName[Arriv] := 'arriv'}
    ActName[Arriv] := 'arriv';
    ActName[Serve] := 'serve';
    ActName[Working] := 'working';
    ActName[Repair] := 'repair';
    {** end ASC **}
```

Lines 399-406:

```
{** start ASC **}
{other application-specific initialization
 goes here, e.g. reading in parameter values,
 and initializing values of variables using
 assignment statements, e.g. initializing
 sums to 0}
 Down := false;  {machine initially up}
 NIntrpt := 0;
 writeln('enter MeanArriv,MeanSrv,MeanUp,MeanDown');
 readln(MeanArriv,MeanSrv,MeanUp,MeanDown);
{** end ASC **}
```

Lines 407–417:

```
{** start ASC **}
{comments deleted to save space}
MakeNewAct(Pt);
Pt^.ActType := Arriv;
Pt^.TimeLeft := Expon(MeanArriv);
{start the machine aging/repair process}
MakeNewAct(Pt);
Pt^.ActType := Working;
Pt^.TimeLeft := Expon(MeanUp)
{** end ASC **}
```

Lines 426–429:

```
{** start ASC **}
{writes out facility name, during debugging}
write('machine')
{** end ASC **}
```

Lines 449-452:

```
{** start ASC **}
{write out application-specific info}
```

```
{none}
{** end ASC **}
```

Lines 477–480:

```
{** start ASC **}
{write out application-specific info}
{none}
{** end ASC **}
```

Lines 533-536:

```
{** start ASC **}
{application-specific statistics}
write('mean system residence time per job = ');
writeln(SumResTime/NJobs[1]:1:2);
write('proportion of interrupted jobs = ');
writeln(NIntrpt/NJobs[1]:1:2)
{** end ASC **}
```

Lines 557–561:

```
{** start ASC **}
{increment application-specific counts
 for this facility}
{none}
{** end ASC **}
```

Lines 563–567:

```
{** start ASC **}
{increment application-specific counts not
 associated with particular facilities}
{none}
{** end ASC **}
```

Look at the insert for Lines 298–317, which form the Application-Specific Code to go into procedure CheckForChange. In most applications, we simply update CurrPtr^.TimeLeft and then check the new value for nonpositiveness, as in Lines 85 and 97 of Program 8.2.1, and as suggested in the comment in Lines 307-314 of Program 8.2.2. However, in this particular application, we do *not* update CurrPtr^.TimeLeft if this activity corresponds to a job that is halted due to the machine having broken; in this case, CurrPtr^.TimeLeft represents the amount of work left to do for this job, and at present no progress is being made in decreasing this work.

Now look at the insert for Lines 323–357, which will go into procedure TakeAction. Recall that this procedure is called when a change is discovered in procedure CheckForChange. The action then taken depends on what type of activity produced the change, hence the "case" statement in Line 355 of the template. Let us consider the various activity types.

In case Arriv, a new job has arrived to the system. We now must both process this job and set up the next arrival for the future. Here it is important to keep in mind that the ActRec record currently pointed to by CurrPtr^ is the record representing the continuing arrival activity; it is permanently associated with that activity. Thus, we need a *separate* activity record for the particular job that just arrived. This record is created by the statement MakeNewAct(Pt). This record is then initialized; its service time is generated and placed into the TimeLeft field, and then StartOrQ will try to start the service. If the machine is currently free, this will be done; otherwise, this job will be added to the machine's queue (Lines 249–258 of Program 8.2.2).

Case Serve has mostly been discussed above, in connection with procedure Release. However, the calls to procedures Delete and "dispose" (the latter being a built-in Pascal library procedure) deserve comment. Since the job has finished and will exit the system, the record representing the job should be removed from the simulation system. This would not occur in most situations in Example 8.2.2, since a car leaving one assembly line station usually has another to go to next. However, after the car left the last station, we would remove the corresponding activity record from the system. On the other hand, for Example 8.2.1, we would never dispose a record; when a job finishes at the CPU, it simply returns to the terminal to enter a "think" period, rather than leaving the system.

The cases Working and Repair should be fairly clear. In the first of these, the event that precipitated the call to TakeAction was that the TimeLeft field for an activity of type Working reached zero; that is, the machine failed. Thus Down should be set to true, and so on.

You are very strongly urged to run the program for Example 8.2.3, using the "debug" mode (note the user prompt on Line 391 of Program 8.2.2). Although, as the name implies, this mode is normally used for debugging purposes, it is also extremely useful as a learning tool. It reports the step-by-step progress of the simulated system, facilitating understanding of the program itself. For best results, try running this program in the debug mode (say with TimeLim equal to 100.0) before proceeding to the next example.

Example 8.2.4

Consider a computer network arranged in a ring structure. There are *n* stations on the ring. Times between successive messages generated by a station are exponentially distributed (again, the program could easily be changed for other distributions). The transmission time for each message is a constant, XmitTime. Each station sends messages to all the other stations with equal frequency. Note that the "facilities" here are not the stations, but rather the links between stations. In other words, the links are the job servers, and jobs queue for them, rather than for the stations. There are *n* links on the ring, connecting each station to the next. Also, suppose there is an additional station at the center of the ring, called the "hub." There is a link from each outer station to the hub, and from the hub to each outer station. The hub acts as a relay node only; it does not generate any messages of its own.

When a message is first generated, say at Station i, that station makes a routing decision. It can either route the message to Station $i + 1$, or it can send the message to the hub. Suppose the policy is to send a fixed proportion of messages to Station $i + 1$, and the rest to the hub. It is clear that this proportion, which we will call HubProb, should be neither too small nor too large. If it is too small, very few messages will be able to take advantage of the "shortcut" that the hub provides. On the other hand, too large a value will result in congestion along the hub links, negating the shortcut advantage. By running the simulation a number of times, once for each of many different values of HubProb, we can find the optimal value, based, say, on the criterion of mean total transit time from source to destination.

Here is the Application-Specific Code to be inserted into the template Program 8.2.2 for this problem:

Lines 12–17:

```
{** start ASC **}
{Put here a list of various activity types,
 including the conditions under which they
 are created and destroyed:}

{There are two activity types, Arriv
 and Send, which are self-explanatory.}
{** end ASC **}
```

Lines 20–22:

```
const {** start ASC **}

      MaxAry = 60;  {maximum value array size}
      {** end ASC **}
```

Lines 23–27:

```
      {** start ASC **}
      {declare constants here for
       activity type names, e.g., Arriv = 1}
      Arriv = 1;
      Send = 2;
      {** end ASC **}
```

Lines 29–34:

```
 {** start ASC **}
 {add comments here explaining the system
  of facility numbers, e.g., Facil = 3 means the CPU}
 {There are n stations on the ring, and an extra
  relay station at the hub. Facil = i means Link i.
  For i=1, . . . , n, Link i connects Station i to
  Station i + 1 (except for the case i = n, where
  Station n is connected to Station 1). For i =
  n + 1, . . . , 2n, Link i connects Station i to
  Station n + 1, the hub. For i = 2n + 1, . . . , 3n,
  Link i connects Station n + 1 to Station i.}
 {** end ASC **}
```

Lines 54–58:

```
{** start ASC **}
       {application-specific
         fields, if any, go here}
         Source: integer; {number of station at which
                           message originates}
         Dest: integer;  {destination station}
         EndLink: integer {link leading directly to
```

```
                              destination of message}
   {** end ASC **}
```

Lines 127–130:

```
  {** start ASC **}
  {application-specific declarations go here}
  MeanArriv,  {mean interarrival time}
  XmitTime,  {send time}
  HubProb  {probability of being routed through
              the hub}
     : real;
  N: integer;  {number of stations on the ring}
  {** end ASC **}
```

Lines 268–272:

```
    {** start ASC **}
    {if not First-Come, First Served policy, then
     change the lines below, e.g. DeleteHeadOfQ}
    {no change}
    {** end ASC **}
```

Lines 298–317:

```
{** start ASC **}
{comments deleted to save space}
Change := false;
CurrPtr^.TimeLeft := CurrPtr^.TimeLeft-Delta;
if CurrPtr^.TimeLeft <= 0.0 then
   Change := true

{** end ASC **}
```

Lines 323–357:

```
{** start ASC **}
{comments deleted to save space}
case CurrPtr^.ActType of
```

```
Arriv: begin
         MakeNewAct(Pt);
         Pt^.ActType  := Send;
         Pt^.TimeLeft  := XmitTime;
         Pt^.StrtTime  := SimTime;
         Pt^.Source  := CurrPtr^.Source;
         repeat
            Dest  := trunc(N*Rnd)+1
         until Dest <> Pt^.Source;
         if Rnd < HubProb then  {route through hub}
            begin
            Fcl  := N+Pt^.Source;
            Pt^.EndLink  := 2*N+Dest
            end
         else  {route around ring}
            begin
            Fcl  := Pt^.Source;
            if Dest = 1 then
               Pt^.EndLink  := N
            else
               Pt^.EndLink  := Dest-1{can't use 'mod'}
            end;
         StartOrQ(Pt,Fcl);
         CurrPtr^.TimeLeft  := Expon(MeanArriv)
         end;
Send:  begin
         Release(CurrPtr^.Facil);
         CurrPtr^.TimeLeft  := XmitTime;
         if CurrPtr^.Facil = CurrPtr^.EndLink then
            {journey has ended}
            begin
            SumResTime  := SumResTime+SimTime
                           -CurrPtr^.StrtTime;
            Delete(CurrPtr,ActListHead,ActListTail);
            dispose(CurrPtr)
            end
         else
            {must forward to next link}
            if CurrPtr^.Facil <= N then
               {travel along link}
               if CurrPtr^.Facil = N then
                  StartOrQ(CurrPtr,1)
               else
                  StartOrQ(CurrPtr,CurrPtr^.Facil+1)
            else
               {through the hub;
                must already be at hub}
               StartOrQ(CurrPtr,CurrPtr^.EndLink)
         end
```

```
   end
{** end ASC **}
```

Lines 370–374:

```
{** start ASC **}
{insert code here for determining NFcls,
 by assignment or readln statements}
writeln('enter N');
readln(N);
NFcls := 3*N;
{** end ASC **}
```

Lines 394–398:

```
{** start ASC **}
{initialize the array ActName,
 e.g., ActName[Arriv] := 'arriv'}
ActName[Arriv] := 'arriv';
ActName[Send] := 'send';
{** end ASC **}
```

Lines 399–406:

```
{** start ASC **}
{other application-specific initialization
 goes here, e.g. reading in parameter values,
 and initializing values of variables using
 assignment statements, e.g. initializing
 sums to 0}
writeln('enter MeanArriv,XmitTime,HubProb');

readln(MeanArriv,XmitTime,HubProb);
{** end ASC **}
```

Lines 407–417:

```
{** start ASC **}
{comments deleted to save space}
```

```
for Fcl := 1 to N do  {have arrival activity at each
                        of the n stations}
   begin
   MakeNewAct(Pt);
   Pt^.ActType := Arriv;
   Pt^.TimeLeft := Expon(MeanArriv);
   Pt^.Source := Fcl
   end
{** end ASC **}
```

Lines 426–429:

```
{** start ASC **}
{writes out facility name, during debugging}
write('link ',Fcl:1)
{** end ASC **}
```

Lines 449–452:

```
   {** start ASC **}
   {write out application-specific info}
   write('src = ',Pt^.Source:1);
   if Pt^.ActType = Send then
      write('  ','endlink = ',Pt^.EndLink:1);
   {** end ASC **}
```

Lines 477–480:

```
         {** start ASC **}
         {write out application-specific info}
         write('src = ',Pt^.Source:1);
         if Pt^.ActType = Send then
            write('  ','endlink = ',Pt^.EndLink:1);
         {** end ASC **}
```

Lines 533–536:

```
{** start ASC **}
{application-specific statistics}
```

```
Sn := 0;
for I := 1 to N do
   Sn := Sn+NJobs[I];
write('mean message transit time = ');
writeln(SumResTime/Sn:1:2)
{** end ASC **}
```

Lines 557–561:

```
{** start ASC **}
{increment application-specific counts
 for this facility}
{none}
{** end ASC **}
```

Lines 563–567:

```
{** start ASC **}
{increment application-specific counts not
 associated with particular facilities}
{none}
{** end ASC **}
```

Again, the reader is encouraged to try running the above program using the debug mode before going on to the next example.

Example 8.2.5

Consider a roadway intersection, to be referred to below as I. The intersection I has the shape of the letter "T," with each road being one lane only and one direction only. The stem of the T heads north into the intersection, and the bar portion heads east.

The traffic signal at I is green for 75-second periods for traffic that arrives along the bar, while the corresponding value for stem traffic is only 30 seconds. However, a right turn on a red light at I is allowed for stem traffic under safe conditions, which in this case means that no car will arrive at I along the bar within the next 5 seconds. Traffic along the bar reaches a point 300 feet from I with exponential interarrival times having mean 5 seconds, with 10 seconds being the analogous value for stem traffic.

Cars generally travel at a speed of 35 miles per hour. If a moving car sees a red light, it begins a constant deceleration, timed so that it will come to rest just behind the last waiting car (cars are 15 feet long). If a car that is decelerating sees the light turn green, it accelerates back toward 35 miles per hour, at a rate of 7 miles/hour/second. When the light turns green, cars that had been waiting at the red light can move; each car waits 1.5 seconds after the car in front of it starts, before moving itself; it then starts a constant acceleration of 5 miles/hour/second, again up to 35 miles per hour. The intersection is 100 feet wide.

There are a number of questions of interest here; for example: What is the mean wait for cars at the intersection? What is the average number of cars waiting? What is the average speed of a car between the intersection and the point 300 feet preceding it? Also, we may be interested in seeing the effects of changing some of the parameters, such as the durations of the green-light periods.

We will not include a listing of the ASC sections for this example. The program is very complicated, and it would consume too much space. This is true in spite of the fact that the above specification of the system has been deliberately oversimplified. Writing this program would be a larger task than those typically given for homework in a classroom setting; instructors may wish to assign it as a group project.

Thus, we will only discuss the program in broad outline form, omitting the details. First, we could set up two arrival activities, one for the bar and one for the stem, and another activity representing the current status of the traffic signal. Each time a car arrives in the simulated region, that is, arrives at a point 300 feet from the intersection, an activity record for that car would be created; the ASC fields of this record would contain information on the car's current position, speed, and acceleration.

When a car stops for a red light, it would join a queue; in this case, the "server" could be the traffic signal itself. However, some thought about this situation will reveal that some other aspects of queueing in Program 8.2.2 may need to be altered slightly. For example, you might wish to allow a car's activity record to remain in the activity chain when the car joins the queue, which would require alteration of Line 223.

Code would also need to be written to account for the fact that stem traffic can turn right at a red signal. There would be a number of ways to simulate this aspect. Probably the simplest method would be to create another permanent activity, whose sole purpose is to continuously check whether there is a car eligible for the right turn.

Before leaving this section, we will discuss two further software aspects. The first is software validation. How can we check that our simulation code is correct? There is no automatic method for this, but there are steps we can take to help ensure the validity of our code. First, in some cases we can construct a Markov model of the system. For instance, even though the distributions involved in our system may not be exponential, we could temporarily assume that they are, substituting exponential distributions everywhere in our program. We could then solve the corresponding Markov chain, using truncation if necessary, and see if the output from our simulation program matches the results of the Markov analysis. If so, this would be at least a partial validation of our program, and we could then run our program with the nonexponential distributions.

In some other examples, there is an easy Markov analysis available for a simplified version of the system model we are using in our simulation. In this situation, an *approximate* match between the results of the Markov and simulation analyses would serve as partial confirmation of the correctness of the simulation.

Another validation method involves simply running the simulation program under the debug mode, checking to see that all actions occur as intended. This method is generally effective, but in using it we must make sure that we run the program long enough to observe a large number of different situations. In other words, some programming errors may occur only in certain special combinations of situations, and thus may only occur if the program is run for a very long time.

The other software issue that should be discussed is that of alternatives to the approach taken in Program 8.2.2. Actually, there is an enormous variety of simulation software available, some of it quite different from what has been presented here. The interested reader is referred to the "Further Reading" section at the end of this chapter for details, but we can at least provide an introduction here.

Program 8.2.2 runs more slowly than it should. In most "ticks" of the simulation clock, no change to the system has occurred. Yet we must check the system at every tick. It is clear that this wastes a great deal of time.

One way to improve this would be to make the simulated time **asynchronous**, instead of taking the **synchronous** approach used here. What this means is that instead of always advancing the simulation clock with increments of constant size Delta (Line 550 of Program 8.2.2), we advance the clock by an amount equal to the smallest TimeLeft value among the records in the activity chain. Of course, code such as that in Line 310 would have to be modified accordingly.

With these modifications, we may wish to alter the code in Lines 570–584, too. This code checks all activities to see if the advancing of the simulation clock has produced a system change (a "discrete event") associated with that activity. Now, with the modifications described in the last paragraph, we know that there will definitely be at least one such activity; in fact, there may be more than one such change. However, in most applications there will only be a single activity for which a change occurs. If so, the "repeat" loop in Lines 572–583 also represents wasted effort, since it checks *all* of the activities. Thus, we could replace this loop by code that calls TakeAction *only* for the particular activity that had the smallest TimeLeft field, that is, the one discussed above for advancing the simulation clock.

The result of all these modifications would be a simulation program using what is called the **event-oriented** method, as opposed to the **activity-scan** method used in this chapter. Instead of continuously scanning the list of all activities, the event-oriented approach essentially keeps the activities in a chain ordered by their TimeLeft fields. Because of this ordering, we always know which activity has the smallest TimeLeft value, so the clock advance described above is easy.

In practice, there are some further refinements to the event-oriented approach. For example, it is helpful to replace the TimeLeft field by a field EventTime, which is the scheduled absolute time of occurrence of the event, rather than the occurrence time relative to SimTime. Also, a single chain of events ordered by EventTime is still somewhat wasteful, since at any given time we are only interested in which one of the events will occur next. Instead, a data structure known as a **heap** can be used to save time, since it does not need to constantly maintain a complete ordering.

A third general category of discrete-event simulation methodology is known as the **process-oriented** approach. Here, each type of activity is represented in the program by its own procedure. Each call to a procedure creates a **process**, roughly equivalent to an activity. The difference, is that these procedure calls are not like those in Pascal or FORTRAN, in which the calling program suspends action until the called procedure finishes execution. In languages which support processes, such as Simula and Modula-2, the calling program continues its own execution without waiting for the called process to finish. At this point, you can see that the process-oriented approach can get very complicated. Details are beyond the scope of this book. However, you are encouraged to look further into this approach, because many simulation analysts believe that it is actually the clearest of the three methods.

It should also be mentioned there are many **simulation packages** available commercially. These usually have an internal structure based on one of the above three methods, but have an external user interface that makes their use very convenient in many applications. The user specification for a system may be quite simple with such a package. For example, the user might write something like

```
X; INF; ARRIVE EXP 5; SERVE EXP 2; ROUTE Y.
Y; 10; SERVE EXP 3; EXIT.
```

Except for X and Y, all the items above would be reserved words (i.e., command names) built into the package. The two lines above might mean: X is a server with an infinite waiting area; interarrival times are exponentially distributed with mean 5; service times are exponential with mean 2; and after completion of service, the job is routed to Y, as long as there is still room in Y's waiting area, which can only accommodate 10 jobs. Service times at Y are also exponential, with mean 3, and after service at Y, the job exits the system.

The question arises as to which type of simulation software is best. The answer involves considerations of convenience, availability, speed, and generality. For example, the packages can be extremely convenient in many applications, but they may be difficult or even impossible to use in simulating some particular systems. On the other hand, we have already mentioned that the activity-scanning method, perhaps the most general of the methods, is the slowest, especially in its synchronous-time version. However, in some particular systems, such as those in which several simultaneous events interact, it may be the easiest to use. Moreover, it is the most natural method if the system to be simulated is not purely of the discrete-event type; Example 8.2.5 is an instance of this, because of the fact that the speeds change continuously.

The author has found that the synchronous activity-scanning approach is much easier to teach than the event-oriented approach, especially for the more complex applications. Furthermore, in addition to facilitating the teaching of simulation programming, it also helps in presenting concepts involved with the formation of confidence intervals for discrete-event systems, which will be the subject of Section 8.3.

The process approach is also good, but has the distinct disadvantage that languages such as Simula and Modula-2 are not nearly so commonly available, or so widely known, as Pascal. This limits their usefulness as a teaching tool. The same comment applies to the packages, in addition to the above-mentioned problem of occasional lack of flexibility.

In any case, the experience gained by the reader in this chapter should enable him or her to become proficient fairly easily in the use of any other kind of simulation software.

8.3 CONSTRUCTION OF CONFIDENCE INTERVALS

As we saw in Section 6.5, simulation can be viewed as a sampling process, exactly in the sense of Chapter 5. The finiteness of our simulation run time, in the form of a finite value for TotReps in Program 2.4.3 and for TimeLim in Program 8.2.2, produces sampling error. The larger our simulation run time, the more likely this sampling error is to be small.

Once we realize that this sampling error exists, we must find a method for quantifying it. This was fairly straightforward in Section 6.5, because the sampling process of Program 2.4.3 was exactly that covered in the earlier sections of Chapter 6.

The situation is not so simple for Chapter 8. To illustrate this, consider Example 8.1.5. Suppose we run a simulation of this system with a TimeLim value of 200.0, with a goal of finding the long-run average queue length μ. [We have a completely mathematical analysis available for this system in the case of exponential interarrival and service times, but if neither distribution is exponential, we may need to do a simulation analysis.]

Suppose we use Program 8.2.2 as the basis of our simulation, with Delta $=0.5$. Then the average queue length reported by the program will be

$$\overline{X}=\frac{X_1 + \cdots + X_{400}}{400} \tag{8.3.1}$$

where the sample size 400 arises from the fact that we have sampled the system $200.0/0.5=400$ times (in the program, the sampling would be done in Line 554). This expression is similar to that of Equation (5.3.1), but the conditions here are completely different. In Equation (5.3.1), the observations X_i were independent identically distributed random variables. However, in Equation (8.3.1), the random variables X_i are neither independent nor identically distributed.

In fact, far from being independent, the X_i are quite highly correlated. If we sample the system at a certain time and observe that there are eight jobs in the queue, then it is fairly likely that if we sample again a little while later, the length of the queue will not be greatly different from eight. In fact, if the value of Delta were very small, say 0.01, the probability would be quite high that two consecutive observations of the queue length yield the same report, that is, $X_{i+1} = X_i$. Thus, the observations are indeed highly correlated.

Nor are the X_i identically distributed. For example, suppose in procedure Init in Program 8.2.2 we make the system initially empty, that is, with no jobs either waiting or being served. Then $X_1 = 0$ with high probability, and $E(X_1)$ is near 0. On the other hand, $E(X_{400})$ is near μ, because by that time the system is near steady state. (Note, however, that X_{400} itself is not necessarily near μ. The phrase "steady state" for a stochastic process refers to distributions: If G_i denotes the cumulative distribution function of X_i (recall Chapter 3), then

$$\lim_{i\to\infty} G_i$$

exists, and the mean of this distribution is μ. On the other hand,

$$\lim_{i\to\infty} X_i$$

does *not* exist.)

Thus, the X_i are neither independent nor identically distributed. (You should note that the same problems would occur if the X_i were observations on a real system, rather than data from our simulation.) Let us consider the second problem first—the fact that the X_i in Equation (8.3.1) are not identically distributed. The resulting effect on $\overline{X}$ is clear from Equation (5.3.2)—there will be a bias induced in $\overline{X}$ as an estimator of μ. If, for example, we start with an empty system as above, so that EX_i is near 0 for the smaller values of i, then $\overline{X}$ will be biased downward, that is, there will be a tendency for $\overline{X}$ to underestimate μ.

One solution for this may be to throw out the first few observations, estimating μ by, say, the quantity

$$\frac{X_{51} + \cdots + X_{400}}{350}$$

One problem with this is that it is not clear how many of the early observations to throw out. For example, in the above expression, we are throwing out the first 50 observations; is this enough? Another problem is that we are increasing the variance of our estimator when we throw out observations: Even though the relation

$$\mathrm{Var}(\overline{X}) = \frac{\mathrm{Var}(X_1)}{n}$$

does not hold any more due to lack of independence, $\mathrm{Var}(\overline{X})$ will still be inversely related to n. If we decrease n from 400 to 350, $\mathrm{Var}(\overline{X})$ will probably increase, though the decrease in bias may make it all worthwhile.

Another method would be to try to hasten the convergence to steady state, by coding procedure Init in such a way as to approximate steady state conditions. For example, we may believe, say from some approximate mathematical analysis, that $\mu \approx 3$. We then could write Init to set the initial queue length of the system to 3.

Through one of these methods, we could make sure that the X_i used to estimate μ are at least approximately identically distributed. Assume this has been done, and now let us consider the problem of the lack of independence of the X_i.

One approach would be to parallel the discussion in Equation (5.3.3), using Equation (4.2.6) to adjust for the correlation between observations. We then could estimate quantities $\mathrm{Cov}(X_i, X_{i+k})$, just as we estimated $\mathrm{Var}(X_i)$ in Equation (6.3.5) and in Example 6.5.2. However, there would be a great many such terms to estimate, and the accuracy of estimation might not be very good.

Another approach would be that of **batch means**. This method relies on the fact that even though observations that are close in time, such as X_i and X_{i+2}, are highly correlated, observations spaced very distantly in time are nearly independent. To illustrate this method, consider the above example and partition $X_1, \ldots, X_{400}$ into groups of, say, 25, and find the sample mean of each group:

$$\overline{Y}_1 = \frac{X_1 + \cdots + X_{25}}{25}$$

$$\overline{Y}_2 = \frac{X_{26} + \cdots + X_{50}}{25}$$

and so on through $\overline{Y}_{16}$.

Our estimator of μ would then be

$$\overline{W} = \frac{1}{16} \sum_{i=1}^{16} \overline{Y}_i \tag{8.3.2}$$

which, of course, is identical to $\overline{X}$ in Equation (8.3.1). However, we will give it a different name. We will treat the 16 terms in Equation (8.3.2) as uncorrelated, for the following reason. Most of the observations in Batch 1 have very little correlation with those in Batch 2; for example, $\mathrm{Cov}(X_{12}, X_{40})$ should be fairly small. This in turn implies that $\mathrm{Cov}(\overline{Y}_1, \overline{Y}_2)$ should be small, so that we make an approximation that this covariance is 0.

We can now form an ordinary confidence interval for μ using Equation (6.3.5), where the values for $\overline{X}$, n, and S are defined appropriately: $\overline{X}$ is $\overline{W}$ from Equation (8.3.2) above, n is 16, and S is defined by

$$S^2 = \frac{1}{15} \sum_{i=1}^{16} (\overline{Y}_i - \overline{W})^2$$

Of course, the difficulty with this method is that it is not clear how to choose the batch sizes (the batch size of 25 used above was just for illustration purposes). Although no definitive rules for choosing the batch size have been developed, some heuristic guidelines for this choice have been proposed. The interested reader may consult the references at the end of this chapter.

Another method for dealing with the nonindependence of the X_i is the **regenerative approach**. This method is very useful in some settings, but its full description is beyond the scope of this text. Again, you can learn more from some of the references listed at the end of this chapter, but we can at least introduce the method here.

The central idea is that of a **regeneration point**, which is related to the "memorylessness" concept from Markov chains. To explain, let us first consider a system which is Markovian, or completely memoryless. As our example, we will take the single-server queue discussed earlier in this section, for the case in which both interarrival and service times are exponentially distributed.

Suppose in procedure Init we initialize the system so that there are no jobs in the system, either queued or in service. We can imagine that a job has just now finished service and exited the system. Let us call such a situation State 0, meaning that there are 0 jobs in the system. By the

Markov property, the system "starts over" each time we reach State 0. Future events will be independent of what occurred before we entered State 0. For example, the time until the next change of state (the arrival of a new job) will be completely independent of the length of time it took to get to State 0 from State 1.

Let us keep track of all returns to State 0 during the simulation run in the following way. Suppose the first return to State 0 occurs between the simulation "clock ticks" corresponding to X_{22} and X_{23}, the second return comes between X_{30} and X_{31}, and so on. In this way, batches are being formed, as above. However, the batch sizes are now variable, for example, sizes 22 and 8 for the first two batches in this case.

We form the batch means as before:

$$\overline{Y}_1 = \frac{X_1 + \cdots + X_{22}}{22} \tag{8.3.3}$$

and so on. Due to memorylessness, *these batch means are now independent*. Note that they are completely independent, not just approximately uncorrelated as in the original batch means method. Moreover, the batch means $\overline{Y}_i$ are exactly identically distributed, too, though to fully appreciate this, the reader must note that the "sample sizes" (in our example, 22, 8, . . .) are themselves independent and identically distributed random variables.

We could then take the grand mean of all the $\overline{Y}_i$, as we did in Equation (8.3.2), and again form an interval like that of Equation (6.3.5). However, although Equation (8.3.2) would work, that is, it would provide correct confidence levels, it would not be best. The reason for this lack of efficiency is that it weights all batch means $\overline{Y}_i$ equally, which should not be done in the case of varying batch sizes. For example, if the sizes of the first two batches are 22 and 8, we should give the first batch mean more weight. This, in fact, is precisely what is done, but the details will not be given here.

Even though we cannot include a full treatment of the regenerative method, there remains one important point to be discussed. The last few paragraphs have assumed that the system is Markovian, which requires both the interarrival and service times to be exponentially distributed. Suppose now that only the interarrival times are exponential. In this case, the system is not completely memoryless, but in a sense it is partially so: State 0 is still a regeneration point. Once a job being served in the presence of an empty queue leaves the system, everything does start over, with past history being irrelevant. The time until the next system change occurs, that is, the time until the next arrival, will be independent of the history of the system before the entry into State 0. Thus, the batches will again be independent, and so on. Note, however, that this would not be true if the interarrivals were nonexponential. Suppose that the interarrival distribution were $U(0,1)$. If we knew that the job whose completion of service resulted in entry into

State 0 arrived 0.88 units of time before that completion, then we would know that the next system change would occur within 0.12 units of time, making successive batches nonindependent.

At any rate, it is clear from the above discussion that the regenerative method is applicable to many systems other than Markov chains, as long as *some* key distribution is exponential. In some complicated systems, identification of a regeneration point may not be easy, and there may be other problems as well, such as exceedingly long times between visits to the regeneration point. However, the regenerative method does provide considerable hope for solution to the difficult statistical problems discussed in this section.

FURTHER READING

Jerry Banks and John Carson, *Discrete-Event System Simulation*, Prentice-Hall, 1984.

Roy Billinton and Ronald Allan, *Reliability Evaluation of Engineering Systems: Concepts and Techniques*, Pitman, 1983.

Paul Bratley, Bennett Fox, and Linus Schrage, *A Guide to Simulation*, Springer-Verlag, 1983.

George Fishman, *Principles of Discrete Event Simulation*, Wiley, 1978.

Wayne Graybeal and Udo Pooch, *Simulation: Principles and Methods*, Winthrop, 1980.

Donald Gross and Carl Harris, *Fundamentals of Queueing Theory*, Wiley, 1985.

Charles Haan, *Statistical Methods in Hydrology*, Iowa State University Press, 1977.

Jeremiah Hayes, *Modeling and Analysis of Computer Communications Networks*, Plenum Press, 1984.

Samuel Karlin and Howard Taylor, *A First Course in Stochastic Processes*, Academic Press, 1975.

Leonard Kleinrock, *Queueing Theory (Vol. I: Theory; Vol. II: Computer Applications)*, Wiley, 1976.

Hisashi Kobayashi, *Modeling and Analysis: An Introduction to System Performance Evaluation Methodology*, Addison-Wesley, 1978.

Edward Lazowska, John Zahorjan, C. Scott Graham, and Kenneth Sevcik, *Quantitative System Performance: Computer System Analysis Using Queueing Network Models*, Prentice-Hall, 1984.

Stephen Lavenberg, *Computer Performance Modeling Handbook*, Academic Press, 1983.

Averill Law and David Kelton, *Simulation Modeling and Analysis*, McGraw-Hill, 1982.

M. MacDougall, *Simulating Computer Systems*, MIT Press, 1987.

Edward McNair and Charles Sauer, *Elements of Practical Performance Modeling*, Prentice-Hall, 1985.

I. Mitrani, *Simulation Techniques for Discrete Event Systems*, Cambridge University Press, 1982.

Emanuel Parzen, *Stochastic Processes*, Holden-Day, 1962.

Charles Sauer and K. Mani Chandy, *Computer Systems Performance Modeling*, Prentice-Hall, 1981.

Mischa Schwartz, *Telecommunication Networks*, Addison Wesley, 1986.

Kishor Trivedi, *Probability and Statistics, with Reliability, Queueing and Computer Science Applications*, Prentice-Hall, 1982.

Bernard Zeigler, *Multifaceted Modelling and Discrete Event Simulation*, Academic Press, 1984.

EXERCISES

8.1 (S) Consider Example 8.1.1. It was found that the long-run proportion of the time we will spend at position 4 is 3/11. Verify this using simulation. Also, verify that this long-run proportion is 1/4 for the modified version of that Markov chain.

8.2 (M) Consider a continuous random variable X having the memoryless property:

$$P(X > s + t | X > s) = P(X > t)$$

for all s, $t > 0$. Let $G(u) = P(X > u)$. Write the above equation in terms of G, and evaluate its derivative with respect to s at $s = 0$. This will set up a differential equation for G. Solve this equation, and thus demonstrate that the only continuous distribution that is memoryless is the exponential.

8.3 (M, S) Suppose $X_1, \ldots, X_n$ are independent random variables, with X_i having density

$$f_i(t) = b_i e^{-b_i t}$$

Let $W = \min(X_1, \ldots, X_n)$. Show that W has density

$$f(t) = be^{-bt}$$

where $b = b_1 + \cdots + b_n$, and that

$$EW = \frac{1}{\sum_{i=1}^{n} \frac{1}{EX_i}}$$

Use simulation to verify the expression for EW, with parameters b_i of your choice.

8.4 (M, S) It is sometimes claimed that Markovian analysis is often robust with respect to the exponential assumption. We will investigate this claim here, for the distribution of the service time S in M/G/1 queues. Assume that interarrival times are exponentially distributed with mean 4.0, and mean service time is 2.5. Suppose we are interested in ν, the mean number of jobs in the system.

a. Suppose the service time has a $U(0,5)$ density, but that we assume an exponential distribution. Under this assumption, we would have $\nu = 1.667$, as calculated in Example 8.1.5. How much of a mistake would we be making? [Answer this by using Equation (8.1.12), and verify by simulation.]

b. Repeat Part (a) for the case in which the service time has an Erlang distribution with parameters $r = 2$ and $b = 0.8$; that is, $S = X_1 + X_2$, where the terms in this sum are independent exponential random variables having mean 1.25.

c. Repeat Part (a) for the case in which S is equal to W_1 or W_2, with probability 0.5 each, where these are independent exponential random variables having means 2.2 and 2.8. (Hint: Review the material on conditional expectation and variance at the end of Section 4.1.)

8.5 (M) Investigate the feasibility of the truncation method described in Section 8.1 in the following way. Consider the M/M/1 queue, the truncated version of which is denoted M/M/1/N (Markov interarrival and service times, one server, maximum queue length of N). Of course, there is no need to use the truncated system to analyze the untruncated system, since we already have an explicit solution to Equation (8.1.8) from Equation (8.1.10). However, it is precisely this fact that will enable us to see how well the truncation method works. To do this, find expressions for the mean queue length, for both systems, in terms of the mean interarrival and service times. Evaluate for specific numerical values of these means and N, and comment on how well the truncated system approximates the untruncated one, for various values of N.

8.6 (M) Using Little's Rule on the appropriate "box," derive Equation (8.1.13).

8.7 Suppose we have the following **finite source** queueing system, which occurs in many types of applications: There are σ sources and one server. Each source can submit only one job at a time to the server; that is, after a source submits a job, it can not submit another until the first is done. For instance, this type of system exists in Example 8.2.1.

Let ER denote the mean response time, that is, the mean time between submission and completion of a job. Also, let ES be the mean service time, U the utilization proportion of the server, and EW the mean time a source waits before submitting a new job after the previous one has completed (in Example 8.2.1, this was the mean "thinking time").

a. (M) Use Little's Rule on two or more appropriately chosen "boxes,"getting several equations, to show that

$$ER = \frac{\sigma ES}{U} - EW$$

b. (S) One of the advantages of using Program 8.2.1 over Markov chain analysis was said to be that the program could compute mean response time, which was "not an immediately obvious function of the π_i." However, the formula in Part (a) solves this problem. Find ER using the π_i found in Example 8.2.1, and then verify by running Program 8.2.1.

c. (M) Without rerunning the program, estimate the increase in mean response time that would occur if a sixth terminal were to be added to the system. Will your estimate err on the side of being too low or too high?

8.8 (M, S) If interarrival times and service times are exponential, which is better: having two machines of ordinary speed, or one machine of double the speed? Be sure to indicate which criterion you are using as the basis of your comparison.

8.9 (M) Derive Equation (8.1.10) by first showing that

$$\pi_k = \frac{\lambda_{k-1}}{\mu_k} \pi_{k-1}$$

8.10 (M) Using the "flow in = flow out" argument that led to Equation (8.1.8), derive Equation (8.1.7).

8.11 (S) Consider a **multiprocessor** computer system, in which several CPUs share common memory. Suppose there are two CPUs and two memory modules. The memory modules can be accessed simultaneously, but if both CPUs simultaneously request the same module, one of them must wait in a queue. Suppose that we decide to store program code in Module 0 and data code in Module 1, and that this results in 65% of all memory requests being for Module 0. Assume that the module numbers of the requests coming from a particular CPU are independent; for example, the probability of two consecutive requests being for Module 1 is $(0.65)^2$. (This assumption is fairly reasonable in this context, but not in some others. See Exercise 8.12.)

The system is somewhat similar to that of Example 8.2.1: A

CPU will submit a memory access request, wait for it to be completed, and then enter the analog of the "thinking period," which will be the time that the CPU is performing internal work not needing a memory access. Suppose the thinking time is exponentially distributed with mean 400 nanoseconds, that is, 400 billionths of a second. Memory access time is a constant, 100 nanoseconds.

Use simulation to find the mean residence time of a request at the memory system (waiting time plus actual access time). Also, find the percentage of requests that have to wait. Can the approach in Exercise 8.7 be used here?

8.12 Consider the multiprocessor system in Exercise 8.11, but with **interleaved** memory. This means that memory words with consecutive addresses are stored in consecutive modules; that is, if Word i is in Module j, then Word $i+1$ will be in Module $j+1$ mod 2. The reason for this arrangement is that most memory accesses tend to be sequential. (There are other considerations here, but we will omit them in the interests of keeping things simple.) Suppose that the pattern of memory access by a CPU is as follows: If the last module accessed by that CPU was Module 0, then the next one will be Module 1 with probability 0.7; otherwise, the next access will be to Module 0 again. Similarly, if we just finished accessing Module 1, the probability is 0.9 that we next request Module 0, and 0.1 that the next access is to Module 1.

a. (M, S) Use discrete-time Markov chain analysis to find the long-run proportion of requests for Module 0, among all requests, and validate using simulation.

b. (S) Determine the information asked for in Exercise 8.11 (mean wait time, proportion of requests that must wait) in this context.

8.13 (M, S) Orders arrive at a warehouse in each day's afternoon mail. Suppose either 0, 1, 2, or 3 orders arrive, with probabilities 0.20, 0.35, 0.30, and 0.15. The stock on hand at the warehouse is replenished at the constant rate of one item per day, with the item being delivered in the morning. An exception to this is that if there is already a stock on hand of 4 items, no new item is delivered. If an order arrives on a day on which the item is out of the stock, the customer is notified by telephone. Assume that in such cases, the customer purchases the item from another company, representing lost business. Use discrete-time Markov chain analysis to determine the proportion of orders that will be lost in this way, and validate using simulation.

8.14 (S) Carry out the details of the highway simulation in Example 8.2.5.

8.15 (S) A typical example of **polling** in computer networks is as follows: τ terminals, $T_1, \ldots, T_\tau$, are connected to a computer through one common communications link, such as a single phone line. Since they cannot all use the link at the same time, an **access protocol**, a set of rules for deciding which terminal can use the line at which time, is needed.

The "polling" protocol works in the following way. When a user at a terminal finishes typing a line and hits the carriage return key, a buffer in the terminal saves this line of characters. In fact, the user can "type ahead," typing several lines, and they will all be stored in the terminal's buffer (assumed here to be of infinite size, for simplicity). The terminal then waits until it receives an "invitation" (a poll) from the computer, allowing that terminal to use the link to send the user's typed lines to the computer. The last character sent by the terminal will be a special "End of Message" character (if the terminal has nothing to send, it will send only this character). When the computer receives the message, it will then send an invitation to the next terminal, cycling through all τ terminals repeatedly: $T_1, \ldots, T_\tau, T_1, \ldots, T_\tau, \ldots$. Assume that service is exhaustive, that is, that when a terminal's turn to send comes, it is allowed to send all the messages stored so far, *plus* any messages typed by the user while the stored messages are being sent.

Suppose the link speed is 150 characters per second. This is the rate at which characters can be put onto the link by a terminal or the computer. (Technically, another type of speed, the propagation delay of a signal from its source to its destination, should also be taken into account, but we will assume that the distances here are so short that this delay is negligible.) Assume that each user types lines at random times, with the time between successive submissions being exponentially distributed with mean 35.2 seconds, and that the number of characters per line is uniform on the set $\{1,2,3,\ldots,10\}$. Assume that the basic poll is 4 characters long, and that when a poll is sent by the computer to a terminal, the computer includes any messages it has saved to send to that terminal (most of these messages are responses to the messages that the user submitted); suppose the number of characters in these messages has a geometric distribution with $p = 0.05$.

Find the mean delay here, that is, the mean time elapsed between a user typing a carriage return for a line, and the receipt by the computer of the last character in the line. Do this for several values of τ, such as 2, 5, 10, 20, and so on.

8.16 (M, S) Recall the bin-packing problem in Exercise 3.42, but with a few changes: First, there is a continuing stream of items, not just n, of lengths $L_1, L_2, L_3, \ldots$. Second, the bins are of size 5, and the lengths L_i are either 1, 2, 3, or 4, with probability 1/4 each. Use discrete-time Markov chain analysis to find the long-run proportion of wasted space.

(Hint: Let the current state of the system at time n be the number of empty spaces left in the box currently being filled, after we have packed the nth object. For example, suppose $L_1=3$, $L_2=1$ and $L_3=2$. Then the state of the system at time 1 will be 2, since we just put the first object in the first bin, and there are still $5-3=2$ units of space left. The second object will fit into this space, leaving $2-1=1$ unit of space, so $X_2=1$. The third object will be too big to fit into this, so we will have to put it into the next bin, leaving $5-2=3$ units of space, so $X_3=3$. You may find it helpful to review Example 8.1.2, especially the expression for μ^{-1}, before starting this problem.)

8.17 (S) In Example 8.1.5, suppose the service time for a job is known when the job arrives, and that Shortest Time First scheduling is used; that is, when the server finishes one job and is ready to select another job for service from among those waiting, it chooses the one with shortest service time. However, as a protection against the possibility that some long jobs may wait indefinitely, the server gives priority to any job that has been waiting for longer than c units of time. If there is more than one such job, then again the one with shortest service time—among these long jobs—is selected for service. Find the mean system residence time (waiting time plus service time) for jobs of length more than 3.8 units of time, for various values of c.

8.18 (S) In Example 8.1.5, suppose that the mean interarrival time is 8.6, and that the machine automatically shuts down if it has been idle for d units of time, in order to save power and wear; when a job then arrives, the machine takes 1 unit of time to power up again. Find the mean system residence time for various values of d.

8.19 Consider a discrete-time Markov chain with a finite state space σ. In this exercise we will develop techniques for analysis beyond that presented in the text.

a. (M) Let ν_{ij} denote the mean time that it takes to reach State j, if we start at State $i (i, j \in \sigma)$. Set up a system of linear equations for the ν_{ij}, similar to Equation (8.1.7), as follows: Fix i, and suppose we are at State i at time 0. Let the random variable N denote the time at which we first reach State j. Apply Equation (4.1.15a), with N playing the role of Y, and with X being our state at time 1.

b. (M, S) Use (a) in Example 2.4.3 to find the mean number of rolls needed to get three consecutive 4s. Verify using simulation.

c. (M) Let q_{ijk} be the probability that it takes exactly k steps to get from State i to State j. Develop a system of equations for these quantities. Then use these equations to solve Example 2.4.3. (In order to keep the number of equations small, solve a scaled-down version of the problem; for example, find the probability that six rolls are needed.)

8.20 (M, S) In Example 8.1.5, suppose that the role of the machine is to manufacture an item, so that the service time is the manufacturing time (which varies from one item to another, since there are different kinds of items). There is a 15% chance that the item will turn out defective, in which case the machine must remake the item before moving on to the next waiting order. Find the mean system residence time for an order. (Hint: Review the material on conditional expectation and variance at the end of Section 4.1, and use the M/G/1 queue formula.)

8.21 (M) Recall Example 4.1.10. Consider the case in which the distribution of light bulb lifetime is discrete. To review, let X be a random variable representing the lifetime of a bulb, and let $\mu = EX$. At time 0, a new bulb is inserted into the socket. When the bulb burns out, it is immediately replaced by a new one, and when that one burns out, it too is replaced, and so on. Suppose we observe the system at time t. Let W be the accumulated lifetime of the current bulb, that is, the amount of time that the bulb currently in use has been burning since its installation—if the bulb was installed at time $t - i$, then $W = i$. Also, define Z to be the remaining lifetime of that bulb, and let f_i and F_i denote $P(X = i)$ and $P(X \leq i)$, respectively. The intuitive argument in the discrete half of Example 4.1.10 implied that

$$P(Z=i)=\frac{1-F_{i-1}}{\mu}$$

for large t. It turns out that the same result is true with Z replaced by W. Derive this latter result using Markov chain theory, defining State i to mean that the current bulb has $W = i$. (Hint: For $i > 0$, the probability of going from State i to State $i + 1$ is the probability that a bulb will not die in its ith time period, given that it reaches its ith time period.)

APPENDIX A

Pascal Tutorial

Pascal is currently the most widely taught computer language for introductory courses in computer programming, especially for students in engineering and computer science. However, you may have learned programming through another language, probably FORTRAN. If so, this Appendix is for you. It will serve as a quick introduction to Pascal for students who know FORTRAN (whom we will call "FORTRAN speakers").

Our point of view here will be that your main goal is to be able to *read* Pascal programs. This will enable you to understand the programs in this book, while still using FORTRAN for writing your own simulation programs (e.g., those in the exercises marked S at the end of each chapter). If you also wish to *write* Pascal, this would be easy too, but would require spending time on small details of syntax, such as the placement of semicolons and commas, which we will not cover here.

You will find that it will be surprisingly easy to learn to read Pascal. Just a few minutes spent on this Appendix should be quite sufficient for this purpose. In fact, reading Pascal is so easy for a FORTRAN speaker that you probably could understand most of the programs in this book even *without* reading this Appendix. This is because

- Pascal is fairly similar to FORTRAN, and
- one of the major design goals in the original development of the Pascal language was to achieve an English-like clarity (i.e., it was supposed to appear similar to English).

The fastest method for teaching FORTRAN speakers to read Pascal is to display FORTRAN and Pascal versions of the same example program, and show the correspondences. We will take this approach here. The two program versions will

(a) read in an integer n, and then read in n elements of an array,
(b) call a subroutine to compute the mean, the minimum value, and the maximum value of the array elements, and then
(c) write out the results.

Here is the FORTRAN program.

```
      REAL X(100),MEAN,MIN,MAX
      INTEGER I,N

      READ(5,*)N
      READ(5,*)(X(I),I=1,N)
      CALL CALC(X,N,MEAN,MIN,MAX)
      WRITE(6,*)MEAN,MIN,MAX
      END


      SUBROUTINE CALC(Z,K,ME,MI,MA)
      REAL Z(K),ME,MI,MA,TMP
      INTEGER J

      ME=Z(1)
      MI=Z(1)
      MA=Z(1)
      DO 10 J=2,K
         TMP=Z(J)
         ME=ME+TMP
         IF (TMP.LT.MI) THEN
            MI=TMP
         ELSE IF (TMP.GT.MA) THEN
            MA=TMP
         END IF
10    CONTINUE
      ME=ME/K
      RETURN
      END
```

Here is a Pascal version of the same program (we have added line numbers at the left margin, for convenience, but they would not actually be in the program file).

```
1        program Example(input,output);
2
```

```
 3         type AryType = array[1..100] of real;
 4
 5         var X: AryType;
 6            Mean,Min,Max: real;
 7            I,N: integer;
 8
 9         procedure Calc(Z: AryType; K: integer;
10            var Me,Mi,Ma: real);
11            var Tmp: real;
12              J: integer;
13         begin    {beginning of this procedure}
14            Me:= Z[1];
15            Mi:= Z[1];
16            Ma:= Z[1];
17            for J:= 2 to K do
18               begin  {beginning of this 'for' loop}
19               Tmp:= Z[J];
20               Me:= Me + Tmp;
21               if Tmp < Mi then
22                  Mi := Tmp
23               else
24                  if Tmp > Ma then Ma:= Tmp
25               end; {end of this 'for' loop}
26            Me:= Me/K
27         end  {end of this procedure}
28
29      begin  {beginning of main program}
30         read(N);
31         for I:= 1 to N do
32            read(X[I]);
33         Calc(X,N,Mean,Min,Max);
34         write(Mean,Min,Max)
35      end.  {end of the main program}
```

Here are the major items one needs to know in order to read Pascal programs, as appearing in the program above:

(a) Line 1 is just a "header," giving the program a name ("Example") and declaring that both input and output from the terminal will be done.

(b) In Line 3, a user-defined type is being declared. Pascal has types similar to those of FORTRAN (i.e., integer, real, etc.), but it also gives the programmer the capability to define new types. Here, we are inventing a type which we will name "AryType," which will be an array of 100 real variables.

(c) Keep in mind that in Line 3 we merely defined the *type* AryType.

Now in Line 5 we will use it, declaring X to be of this type, just as the line

```
REAL X(100)
```

did in the FORTRAN version. We also define the other variables, just as we did in the FORTRAN version, though Pascal's syntax is a little different.

(d) Pascal's analog of a SUBROUTINE is called a "procedure." (The Pascal analog of a FUNCTION is also called a "function.") In Pascal, the subprograms (procedures and functions) always appear before the main program, not after it (Line 9).
Parameters (also called "arguments") are declared in a way similar to those of FORTRAN. One difference is that the type is given immediately after the parameter (e.g., "Z: AryType" and "K: integer"). Also, those parameters that will be changed within the procedure (e.g., "Me" but not "K") must be preceded by "var" (short for "variable").

(e) Pascal comments are indicated by braces: {. . .}. For example, in Line 13, we have inserted a comment which says the procedure computational section begins at that point.

(f) Pascal's analog of FORTRAN's DO loop is the "for" loop, which works the same way. Instead of declaring the scope of the loop through specifying a statement number (e.g., DO 10), the scope is defined through "begin" and "end" delimiters. For example, for the loop starting on Line 17, the scope of the loop is Lines 18 through 25. [If the body of the loop is only one line, these are not needed (Lines 31–32).]
Pascal also has other kinds of loops, such as "repeat . . . until w > 12"; these are self-explanatory.

(g) The if/then/else constructs in Pascal are similar to those of FORTRAN (Lines 21–24). Again, begins and ends can be used for multiple-line code: for example,

```
if A < > B then {<> means "not equal to"}
   begin
   A := A + 1;
   C := B/5
   end
else
   begin
   B := B - 1;
   A := B
   end
```

(h) Pascal's "read" and "write" (and alternate versions, "readln" and "writeln") work essentially the same as FORTRAN's.

(i) Note that, unlike FORTRAN, a procedure call does not use the word "call." To call the procedure, we simply write the procedure name and its parameters (Line 33).

(j) In Pascal, variables declared in the main program (Lines 5–7 here) are "global," that is, accessible from within subprograms as well as the main program. Thus, an alternative version of the procedure Calc could be written, without any parameters as follows:

```
procedure Calc;
   var Tmp: real;
      J: integer;
begin {beginning of this procedure}
   Mean := X[1];
   Min := X[1];
   Max := X[1];
   for J:= 2 to N do
      begin {beginning of this 'for' loop}
      Tmp := X[J];
      Mean := Mean + Tmp;
      if Tmp < Min then
         Min := Tmp
      else
         if Tmp > Max then Max := Tmp
      end;  {end of this 'for' loop}
   Mean := Mean/N
end {end of this procedure}
```

Instead of using parameters, this version of the procedure accesses the main program's variables—X, Mean, and so forth—directly. The main program's procedure call (Line 33) would of course also be parameterless:

```
Calc;
```

(k) Pascal has a set of "intrinsic" functions similar to those offered under most versions of FORTRAN (e.g., sin ( ), cos( ), sqrt( ), etc.). Pascal also includes functions such as sqr( ), which computes the square of the argument, and trunc( ), which computes the "floor" function, which is defined as the greatest integer less than or equal to the argument [e.g., trunc(3.8) = 3 and trunc(3.0) = 3].

Review the above points, and then do a line-by-line comparison of the FORTRAN and Pascal program versions above (again, ignoring details such as semicolons and commas). You then should be comfortable reading the Pascal programs in this book.

APPENDIX B

Tables

Table 1 Cumulative Poisson Distribution Function.

(Tabulated values are $P(Y \leq y) = \sum_{k=0}^{y} \frac{e^{-\mu}\mu^k}{k!}$.)

y \ μ	.1	.2	.3	.4	.5	.6	.7	.8	.9	1.0
0	.9048	.8187	.7408	.6703	.6065	.5488	.4966	.4493	.4066	.3679
1	.9953	.9825	.9631	.9384	.9098	.8781	.8442	.8088	.7725	.7358
2	.9998	.9989	.9964	.9921	.9856	.9769	.9659	.9526	.9371	.9197
3	1.0000	.9999	.9997	.9992	.9982	.9966	.9942	.9909	.9865	.9810
4	1.0000	1.0000	1.0000	.9999	.9998	.9996	.9992	.9986	.9977	.9963
5	1.0000	1.0000	1.0000	1.0000	1.0000	1.0000	.9999	.9998	.9997	.9994
6	1.0000	1.0000	1.0000	1.0000	1.0000	1.0000	1.0000	1.0000	1.0000	.9999

y \ μ	1.1	1.2	1.3	1.4	1.5	1.6	1.7	1.8	1.9	2.0
0	.3329	.3012	.2725	.2466	.2231	.2019	.1827	.1653	.1496	.1353
1	.6990	.6626	.6268	.5918	.5578	.5249	.4932	.4628	.4337	.4060
2	.9004	.8795	.8571	.8335	.8088	.7834	.7572	.7306	.7037	.6767
3	.9743	.9662	.9569	.9463	.9344	.9212	.9068	.8913	.8747	.8571
4	.9946	.9923	.9893	.9857	.9814	.9763	.9704	.9636	.9559	.9473
5	.9990	.9985	.9978	.9968	.9955	.9940	.9920	.9896	.9868	.9834
6	.9999	.9997	.9996	.9994	.9991	.9987	.9981	.9974	.9966	.9955
7	1.0000	1.0000	.9999	.9999	.9998	.9997	.9996	.9994	.9992	.9989
8	1.0000	1.0000	1.0000	1.0000	1.0000	1.0000	.9999	.9999	.9998	.9998

y \ μ	2.1	2.2	2.3	2.4	2.5	2.6	2.7	2.8	2.9	3.0
0	.1225	.1108	.1003	.0907	.0821	.0743	.0672	.0608	.0550	.0498
1	.3796	.3546	.3309	.3084	.2873	.2674	.2487	.2311	.2146	.1991
2	.6496	.6227	.5960	.5697	.5438	.5184	.4936	.4695	.4460	.4232
3	.8386	.8194	.7993	.7787	.7576	.7360	.7141	.6919	.6696	.6472
4	.9379	.9275	.9162	.9041	.8912	.8774	.8629	.8477	.8318	.8153
5	.9796	.9751	.9700	.9643	.9580	.9510	.9433	.9349	.9258	.9161
6	.9941	.9925	.9906	.9884	.9858	.9828	.9794	.9756	.9713	.9665
7	.9985	.9980	.9974	.9967	.9958	.9947	.9934	.9919	.9901	.9881
8	.9997	.9995	.9994	.9991	.9989	.9985	.9981	.9976	.9969	.9962
9	.9999	.9999	.9999	.9998	.9997	.9996	.9995	.9993	.9991	.9989
10	1.0000	1.0000	1.0000	1.0000	.9999	.9999	.9999	.9998	.9998	.9997
11	1.0000	1.0000	1.0000	1.0000	1.0000	1.0000	1.0000	1.0000	.9999	.9999

y \ μ	3.1	3.2	3.3	3.4	3.5	3.6	3.7	3.8	3.9	4.0
0	.0450	.0408	.0369	.0334	.0302	.0273	.0247	.0224	.0202	.0183
1	.1847	.1712	.1586	.1468	.1359	.1257	.1162	.1074	.0992	.0916
2	.4012	.3799	.3594	.3397	.3208	.3027	.2854	.2689	.2531	.2381
3	.6248	.6025	.5803	.5584	.5366	.5152	.4942	.4735	.4532	.4335
4	.7982	.7806	.7626	.7442	.7254	.7064	.6872	.6678	.6484	.6288
5	.9057	.8946	.8829	.8705	.8576	.8441	.8301	.8156	.8006	.7851
6	.9612	.9554	.9490	.9421	.9347	.9267	.9182	.9091	.8995	.8893
7	.9858	.9832	.9802	.9769	.9733	.9692	.9648	.9599	.9546	.9489
8	.9953	.9943	.9931	.9917	.9901	.9883	.9863	.9840	.9815	.9786
9	.9986	.9982	.9978	.9973	.9967	.9960	.9952	.9942	.9931	.9919
10	.9996	.9995	.9994	.9992	.9990	.9987	.9984	.9981	.9977	.9972
11	.9999	.9999	.9998	.9998	.9997	.9996	.9995	.9994	.9993	.9991
12	1.0000	1.0000	1.0000	.9999	.9999	.9999	.9999	.9998	.9998	.9997
13	1.0000	1.0000	1.0000	1.0000	1.0000	1.0000	1.0000	1.0000	.9999	.9999

y \ μ	4.1	4.2	4.3	4.4	4.5	4.6	4.7	4.8	4.9	5.0
0	.0166	.0150	.0136	.0123	.0111	.0101	.0091	.0082	.0074	.0067
1	.0845	.0780	.0719	.0663	.0611	.0563	.0518	.0477	.0439	.0404
2	.2238	.2102	.1974	.1851	.1736	.1626	.1523	.1425	.1333	.1247
3	.4142	.3954	.3772	.3594	.3423	.3257	.3097	.2942	.2793	.2650
4	.6093	.5898	.5704	.5512	.5321	.5132	.4946	.4763	.4582	.4405
5	.7693	.7531	.7367	.7199	.7029	.6858	.6684	.6510	.6335	.6160
6	.8786	.8675	.8558	.8436	.8311	.8180	.8046	.7908	.7767	.7622
7	.9427	.9361	.9290	.9214	.9134	.9049	.8960	.8867	.8769	.8666
8	.9755	.9721	.9683	.9642	.9597	.9549	.9497	.9442	.9382	.9319
9	.9905	.9889	.9871	.9851	.9829	.9805	.9778	.9749	.9717	.9682
10	.9966	.9959	.9952	.9943	.9933	.9922	.9910	.9896	.9880	.9863
11	.9989	.9986	.9983	.9980	.9976	.9971	.9966	.9960	.9953	.9945
12	.9997	.9996	.9995	.9993	.9992	.9990	.9988	.9986	.9983	.9980
13	.9999	.9999	.9998	.9998	.9997	.9997	.9996	.9995	.9994	.9993
14	1.0000	1.0000	1.0000	.9999	.9999	.9999	.9999	.9999	.9998	.9998
15	1.0000	1.0000	1.0000	1.0000	1.0000	1.0000	1.0000	1.0000	.9999	.9999

TABLE 1 Continued

y	μ = 5.5	6.0	6.5	7.0	7.5	8.0	8.5	9.0	9.5	10.0
0	.0041	.0025	.0015	.0009	.0006	.0003	.0002	.0001	.0001	.0000
1	.0266	.0174	.0113	.0073	.0047	.0030	.0019	.0012	.0008	.0005
2	.0884	.0620	.0430	.0296	.0203	.0138	.0093	.0062	.0042	.0028
3	.2017	.1512	.1118	.0818	.0591	.0424	.0301	.0212	.0149	.0103
4	.3575	.2851	.2237	.1730	.1321	.0996	.0744	.0550	.0403	.0293
5	.5289	.4457	.3690	.3007	.2414	.1912	.1496	.1157	.0885	.0671
6	.6860	.6063	.5265	.4497	.3782	.3134	.2562	.2068	.1649	.1301
7	.8095	.7440	.6728	.5987	.5246	.4530	.3856	.3239	.2687	.2202
8	.8944	.8472	.7916	.7291	.6620	.5925	.5231	.4557	.3918	.3328
9	.9462	.9161	.8774	.8305	.7764	.7166	.6530	.5874	.5218	.4579
10	.9747	.9574	.9332	.9015	.8622	.8159	.7634	.7060	.6453	.5830
11	.9890	.9799	.9661	.9467	.9208	.8881	.8487	.8030	.7520	.6968
12	.9955	.9912	.9840	.9730	.9573	.9362	.9091	.8758	.8364	.7916
13	.9983	.9964	.9929	.9872	.9784	.9658	.9486	.9261	.8981	.8645
14	.9994	.9986	.9970	.9943	.9897	.9827	.9726	.9585	.9400	.9165
15	.9998	.9995	.9988	.9976	.9954	.9918	.9862	.9780	.9665	.9513
16	.9999	.9998	.9996	.9990	.9980	.9963	.9934	.9889	.9823	.9730
17	1.0000	.9999	.9998	.9996	.9992	.9984	.9970	.9947	.9911	.9857
18	1.0000	1.0000	.9999	.9999	.9997	.9993	.9987	.9976	.9957	.9928
19	1.0000	1.0000	1.0000	1.0000	.9999	.9997	.9995	.9989	.9980	.9965
20	1.0000	1.0000	1.0000	1.0000	1.0000	.9999	.9998	.9996	.9991	.9984
21	1.0000	1.0000	1.0000	1.0000	1.0000	1.0000	.9999	.9998	.9996	.9993
22	1.0000	1.0000	1.0000	1.0000	1.0000	1.0000	1.0000	.9999	.9999	.9997
23	1.0000	1.0000	1.0000	1.0000	1.0000	1.0000	1.0000	1.0000	.9999	.9999

y	μ = 11.0	12.0	13.0	14.0	15.0	16.0	17.0	18.0	19.0	20.0
0	.0000	.0000	.0000	.0000	.0000	.0000	.0000	.0000	.0000	.0000
1	.0002	.0001	.0000	.0000	.0000	.0000	.0000	.0000	.0000	.0000
2	.0012	.0005	.0002	.0001	.0000	.0000	.0000	.0000	.0000	.0000
3	.0049	.0023	.0011	.0005	.0002	.0001	.0000	.0000	.0000	.0000
4	.0151	.0076	.0037	.0018	.0009	.0004	.0002	.0001	.0000	.0000
5	.0375	.0203	.0107	.0055	.0028	.0014	.0007	.0003	.0002	.0001
6	.0786	.0458	.0259	.0142	.0076	.0040	.0021	.0010	.0005	.0003
7	.1432	.0895	.0540	.0316	.0180	.0100	.0054	.0029	.0015	.0008
8	.2320	.1550	.0998	.0621	.0374	.0220	.0126	.0071	.0039	.0021
9	.3405	.2424	.1658	.1094	.0699	.0433	.0261	.0154	.0089	.0050
10	.4599	.3472	.2517	.1757	.1185	.0774	.0491	.0304	.0183	.0108
11	.5793	.4616	.3532	.2600	.1848	.1270	.0847	.0549	.0347	.0214
12	.6887	.5760	.4631	.3585	.2676	.1931	.1350	.0917	.0606	.0390
13	.7813	.6815	.5730	.4644	.3632	.2745	.2009	.1426	.0984	.0661
14	.8540	.7720	.6751	.5704	.4657	.3675	.2808	.2081	.1497	.1049
15	.9074	.8444	.7636	.6694	.5681	.4667	.3715	.2867	.2148	.1565
16	.9441	.8987	.8355	.7559	.6641	.5660	.4677	.3751	.2920	.2211
17	.9678	.9370	.8905	.8272	.7489	.6593	.5640	.4686	.3784	.2970
18	.9823	.9626	.9302	.8826	.8195	.7423	.6550	.5622	.4695	.3814
19	.9907	.9787	.9573	.9235	.8752	.8122	.7363	.6509	.5606	.4703
20	.9953	.9884	.9750	.9521	.9170	.8682	.8055	.7307	.6472	.5591
21	.9977	.9939	.9859	.9712	.9469	.9108	.8615	.7991	.7255	.6437
22	.9990	.9970	.9924	.9833	.9673	.9418	.9047	.8551	.7931	.7206
23	.9995	.9985	.9960	.9907	.9805	.9633	.9367	.8989	.8490	.7875
24	.9998	.9993	.9980	.9950	.9888	.9777	.9594	.9317	.8933	.8432
25	.9999	.9997	.9990	.9974	.9938	.9869	.9748	.9554	.9269	.8878
26	1.0000	.9999	.9995	.9987	.9967	.9925	.9848	.9718	.9514	.9221
27	1.0000	.9999	.9998	.9994	.9983	.9959	.9912	.9827	.9687	.9475
28	1.0000	1.0000	.9999	.9997	.9991	.9978	.9950	.9897	.9805	.9657
29	1.0000	1.0000	1.0000	.9999	.9996	.9989	.9973	.9941	.9882	.9782
30	1.0000	1.0000	1.0000	.9999	.9998	.9994	.9986	.9967	.9930	.9865
31	1.0000	1.0000	1.0000	1.0000	.9999	.9997	.9993	.9982	.9960	.9919
32	1.0000	1.0000	1.0000	1.0000	1.0000	.9999	.9996	.9990	.9978	.9953
33	1.0000	1.0000	1.0000	1.0000	1.0000	.9999	.9998	.9995	.9988	.9973

Source: Computed by D. K. Hildebrand from *Statistical Thinking for Behavioral Scientists* by David K. Hildebrand. Copyright © 1986 by PWS Publishers.

Table 2 Cumulative Normal Distribution Function.

(Tabulated values are $P(Z \leq z) = \int_0^z \frac{1}{\sqrt{2\pi}} e^{-t^2/2} dt.$)

z	.00	.01	.02	.03	.04	.05	.06	.07	.08	.09
0.00	.0000	.0040	.0080	.0120	.0160	.0199	.0239	.0279	.0319	.0359
0.10	.0398	.0438	.0478	.0517	.0557	.0596	.0636	.0675	.0714	.0753
0.20	.0793	.0832	.0871	.0910	.0948	.0987	.1026	.1064	.1103	.1141
0.30	.1179	.1217	.1255	.1293	.1331	.1368	.1406	.1443	.1480	.1517
0.40	.1554	.1591	.1628	.1664	.1700	.1736	.1772	.1808	.1844	.1879
0.50	.1915	.1950	.1985	.2019	.2054	.2088	.2123	.2157	.2190	.2224
0.60	.2257	.2291	.2324	.2357	.2389	.2422	.2454	.2486	.2517	.2549
0.70	.2580	.2611	.2642	.2673	.2704	.2734	.2764	.2794	.2823	.2852
0.80	.2881	.2910	.2939	.2967	.2995	.3023	.3051	.3078	.3106	.3133
0.90	.3159	.3186	.3212	.3238	.3264	.3289	.3315	.3340	.3365	.3389
1.00	.3413	.3438	.3461	.3485	.3508	.3531	.3554	.3577	.3599	.3621
1.10	.3643	.3665	.3686	.3708	.3729	.3749	.3770	.3790	.3810	.3830
1.20	.3849	.3869	.3888	.3907	.3925	.3944	.3962	.3980	.3997	.4015
1.30	.4032	.4049	.4066	.4082	.4099	.4115	.4131	.4147	.4162	.4177
1.40	.4192	.4207	.4222	.4236	.4251	.4265	.4279	.4292	.4306	.4319
1.50	.4332	.4345	.4357	.4370	.4382	.4394	.4406	.4418	.4429	.4441
1.60	.4452	.4463	.4474	.4484	.4495	.4505	.4515	.4525	.4535	.4545
1.70	.4554	.4564	.4573	.4582	.4591	.4599	.4608	.4616	.4625	.4633
1.80	.4641	.4649	.4656	.4664	.4671	.4678	.4686	.4693	.4699	.4706
1.90	.4713	.4719	.4726	.4732	.4738	.4744	.4750	.4756	.4761	.4767
2.00	.4772	.4778	.4783	.4788	.4793	.4798	.4803	.4808	.4812	.4817
2.10	.4821	.4826	.4830	.4834	.4838	.4842	.4846	.4850	.4854	.4857
2.20	.4861	.4864	.4868	.4871	.4875	.4878	.4881	.4884	.4887	.4890
2.30	.4893	.4896	.4898	.4901	.4904	.4906	.4909	.4911	.4913	.4916
2.40	.4918	.4920	.4922	.4925	.4927	.4929	.4931	.4932	.4934	.4936
2.50	.4938	.4940	.4941	.4943	.4945	.4946	.4948	.4949	.4951	.4952
2.60	.4953	.4955	.4956	.4957	.4959	.4960	.4961	.4962	.4963	.4964
2.70	.4965	.4966	.4967	.4968	.4969	.4970	.4971	.4972	.4973	.4974
2.80	.4974	.4975	.4976	.4977	.4977	.4978	.4979	.4979	.4980	.4981
2.90	.4981	.4982	.4982	.4983	.4984	.4984	.4985	.4985	.4986	.4986
3.00	.4987	.4987	.4987	.4988	.4988	.4989	.4989	.4989	.4990	.4990

z	area
3.50	.49976737
4.00	.49996833
4.50	.49999660
5.00	.49999971

Source: Computed by P. J. Hildebrand from *Statistical Thinking for Behavioral Scientists* by David K. Hildebrand. Copyright © 1986 by PWS Publishers.

Table 3 Student Distribution.
(Tabulated values are $t(a; df)$.)

df	a=.1	a=.05	a=.025	a=.01	a=.005	a=.001
1	3.078	6.314	12.706	31.821	63.657	318.309
2	1.886	2.920	4.303	6.965	9.925	22.327
3	1.638	2.353	3.182	4.541	5.841	10.215
4	1.533	2.132	2.776	3.747	4.604	7.173
5	1.476	2.015	2.571	3.365	4.032	5.893
6	1.440	1.943	2.447	3.143	3.707	5.208
7	1.415	1.895	2.365	2.998	3.499	4.785
8	1.397	1.860	2.306	2.896	3.355	4.501
9	1.383	1.833	2.262	2.821	3.250	4.297
10	1.372	1.812	2.228	2.764	3.169	4.144
11	1.363	1.796	2.201	2.718	3.106	4.025
12	1.356	1.782	2.179	2.681	3.055	3.930
13	1.350	1.771	2.160	2.650	3.012	3.852
14	1.345	1.761	2.145	2.624	2.977	3.787
15	1.341	1.753	2.131	2.602	2.947	3.733
16	1.337	1.746	2.120	2.583	2.921	3.686
17	1.333	1.740	2.110	2.567	2.898	3.646
18	1.330	1.734	2.101	2.552	2.878	3.610
19	1.328	1.729	2.093	2.539	2.861	3.579
20	1.325	1.725	2.086	2.528	2.845	3.552
21	1.323	1.721	2.080	2.518	2.831	3.527
22	1.321	1.717	2.074	2.508	2.819	3.505
23	1.319	1.714	2.069	2.500	2.807	3.485
24	1.318	1.711	2.064	2.492	2.797	3.467
25	1.316	1.708	2.060	2.485	2.787	3.450
26	1.315	1.706	2.056	2.479	2.779	3.435
27	1.314	1.703	2.052	2.473	2.771	3.421
28	1.313	1.701	2.048	2.467	2.763	3.408
29	1.311	1.699	2.045	2.462	2.756	3.396
30	1.310	1.697	2.042	2.457	2.750	3.385
40	1.303	1.684	2.021	2.423	2.704	3.307
60	1.296	1.671	2.000	2.390	2.660	3.232
120	1.289	1.658	1.980	2.358	2.617	3.160
240	1.285	1.651	1.970	2.342	2.596	3.125
∞	1.282	1.645	1.960	2.326	2.576	3.090

Source: Computed by P. J. Hildebrand from *Statistical Thinking for Behavioral Scientists* by David K. Hildebrand.

Table 4 Chi-Square Distribution.
(Tabulated values are $x^2(a; df)$.)

df	a=.999	a=.995	a=.99	a=.975	a=.95	a=.9
1	.000002	.000039	.000157	.000982	.003932	.01579
2	.002001	.01003	.02010	.05064	.1026	.2107
3	.02430	.07172	.1148	.2158	.3518	.5844
4	.09080	.2070	.2971	.4844	.7107	1.064
5	.2102	.4117	.5543	.8312	1.145	1.610
6	.3811	.6757	.8721	1.237	1.635	2.204
7	.5985	.9893	1.239	1.690	2.167	2.833
8	.8571	1.344	1.646	2.180	2.733	3.490
9	1.152	1.735	2.088	2.700	3.325	4.168
10	1.479	2.156	2.558	3.247	3.940	4.865
11	1.834	2.603	3.053	3.816	4.575	5.578
12	2.214	3.074	3.571	4.404	5.226	6.304
13	2.617	3.565	4.107	5.009	5.892	7.042
14	3.041	4.075	4.660	5.629	6.571	7.790
15	3.483	4.601	5.229	6.262	7.261	8.547
16	3.942	5.142	5.812	6.908	7.962	9.312
17	4.416	5.697	6.408	7.564	8.672	10.09
18	4.905	6.265	7.015	8.231	9.390	10.86
19	5.407	6.844	7.633	8.907	10.12	11.65
20	5.921	7.434	8.260	9.591	10.85	12.44
21	6.447	8.034	8.897	10.28	11.59	13.24
22	6.983	8.643	9.542	10.98	12.34	14.04
23	7.529	9.260	10.20	11.69	13.09	14.85
24	8.085	9.886	10.86	12.40	13.85	15.66
25	8.649	10.52	11.52	13.12	14.61	16.47
26	9.222	11.16	12.20	13.84	15.38	17.29
27	9.803	11.81	12.88	14.57	16.15	18.11
28	10.39	12.46	13.56	15.31	16.93	18.94
29	10.99	13.12	14.26	16.05	17.71	19.77
30	11.59	13.79	14.95	16.79	18.49	20.60
40	17.92	20.71	22.16	24.43	26.51	29.05
50	24.67	27.99	29.71	32.36	34.76	37.69
60	31.74	35.53	37.48	40.48	43.19	46.46
70	39.04	43.28	45.44	48.76	51.74	55.33
80	46.52	51.17	53.54	57.15	60.39	64.28
90	54.16	59.20	61.75	65.65	69.13	73.29
100	61.92	67.33	70.06	74.22	77.93	82.36
120	77.76	83.85	86.92	91.57	95.70	100.62
240	177.95	187.32	191.99	198.98	205.14	212.39

Table 4 Continued

a=.1	a=.05	a=.025	a=.01	a=.005	a=.001	df
2.706	3.841	5.024	6.635	7.879	10.83	1
4.605	5.991	7.378	9.210	10.60	13.82	2
6.251	7.815	9.348	11.34	12.84	16.27	3
7.779	9.488	11.14	13.28	14.86	18.47	4
9.236	11.07	12.83	15.09	16.75	20.52	5
10.64	12.59	14.45	16.81	18.55	22.46	6
12.02	14.07	16.01	18.48	20.28	24.32	7
13.36	15.51	17.53	20.09	21.95	26.12	8
14.68	16.92	19.02	21.67	23.59	27.88	9
15.99	18.31	20.48	23.21	25.19	29.59	10
17.28	19.68	21.92	24.72	26.76	31.27	11
18.55	21.03	23.34	26.22	28.30	32.91	12
19.81	22.36	24.74	27.69	29.82	34.53	13
21.06	23.68	26.12	29.14	31.32	36.12	14
22.31	25.00	27.49	30.58	32.80	37.70	15
23.54	26.30	28.85	32.00	34.27	39.25	16
24.77	27.59	30.19	33.41	35.72	40.79	17
25.99	28.87	31.53	34.81	37.16	42.31	18
27.20	30.14	32.85	36.19	38.58	43.82	19
28.41	31.41	34.17	37.57	40.00	45.31	20
29.62	32.67	35.48	38.93	41.40	46.80	21
30.81	33.92	36.78	40.29	42.80	48.27	22
32.01	35.17	38.08	41.64	44.18	49.73	23
33.20	36.42	39.36	42.98	45.56	51.18	24
34.38	37.65	40.65	44.31	46.93	52.62	25
35.56	38.89	41.92	45.64	48.29	54.05	26
36.74	40.11	43.19	46.96	49.65	55.48	27
37.92	41.34	44.46	48.28	50.99	56.89	28
39.09	42.56	45.72	49.59	52.34	58.30	29
40.26	43.77	46.98	50.89	53.67	59.70	30
51.81	55.76	59.34	63.69	66.77	73.40	40
63.17	67.50	71.42	76.15	79.49	86.66	50
74.40	79.08	83.30	88.38	91.95	99.61	60
85.53	90.53	95.02	100.43	104.21	112.32	70
96.58	101.88	106.63	112.33	116.32	124.84	80
107.57	113.15	118.14	124.12	128.30	137.21	90
118.50	124.34	129.56	135.81	140.17	149.45	100
140.23	146.57	152.21	158.95	163.65	173.62	120
268.47	277.14	284.80	293.89	300.18	313.44	240

Source: Computed by P. J. Hildebrand from *Statistical Thinking for Behavioral Scientists* by David K. Hildebrand.

Table 5 *F*-Distribution. (Tabulated values use the upper-*a* points of an *F*-distribution having (dF_1, dF_2) degrees of freedom.)

df_2	a	df_1 = 1	2	3	4	5	6	7	8	9	∞
1	.25	5.83	7.50	8.20	8.58	8.82	8.98	9.10	9.19	9.26	9.85
	.10	39.86	49.50	53.59	55.83	57.24	58.20	58.91	59.44	59.86	63.33
	.05	161.4	199.5	215.7	224.6	230.2	234.0	236.8	238.9	240.5	254.3
	.025	647.8	799.5	864.2	899.6	921.8	937.1	948.2	956.7	963.3	1018
	.01	4052	5000	5403	5625	5764	5859	5928	5981	6022	6366
2	.25	2.57	3.00	3.15	3.23	3.28	3.31	3.34	3.35	3.37	3.48
	.10	8.53	9.00	9.16	9.24	9.29	9.33	9.35	9.37	9.38	9.49
	.05	18.51	19.00	19.16	19.25	19.30	19.33	19.35	19.37	19.38	19.50
	.025	38.51	39.00	39.17	39.25	39.30	39.33	39.36	39.37	39.39	39.50
	.01	98.50	99.00	99.17	99.25	99.30	99.33	99.36	99.37	99.39	99.50
	.005	198.5	199.0	199.2	199.2	199.3	199.3	199.4	199.4	199.4	199.5
	.001	998.5	999.0	999.2	999.2	999.3	999.3	999.4	999.4	999.4	999.5
3	.25	2.02	2.28	2.36	2.39	2.41	2.42	2.43	2.44	2.44	2.47
	.10	5.54	5.46	5.39	5.34	5.31	5.28	5.27	5.25	5.24	5.13
	.05	10.13	9.55	9.28	9.12	9.01	8.94	8.89	8.85	8.81	8.53
	.025	17.44	16.04	15.44	15.10	14.88	14.73	14.62	14.54	14.47	13.90
	.01	34.12	30.82	29.46	28.71	28.24	27.91	27.67	27.49	27.35	26.13
	.005	55.55	49.80	47.47	46.19	45.39	44.84	44.43	44.13	43.88	41.83
	.001	167.0	148.5	141.1	137.1	134.6	132.8	131.6	130.6	129.9	123.5
4	.25	1.81	2.00	2.05	2.06	2.07	2.08	2.08	2.08	2.08	2.08
	.10	4.54	4.32	4.19	4.11	4.05	4.01	3.98	3.95	3.94	3.76
	.05	7.71	6.94	6.59	6.39	6.26	6.16	6.09	6.04	6.00	5.63
	.025	12.22	10.65	9.98	9.60	9.36	9.20	9.07	8.98	8.90	8.26
	.01	21.20	18.00	16.69	15.98	15.52	15.21	14.98	14.80	14.66	13.46
	.005	31.33	26.28	24.26	23.15	22.46	21.97	21.62	21.35	21.14	19.32
	.001	74.14	61.25	56.18	53.44	51.71	50.53	49.66	49.00	48.47	44.05
5	.25	1.69	1.85	1.88	1.89	1.89	1.89	1.89	1.89	1.89	1.87
	.10	4.06	3.78	3.62	3.52	3.45	3.40	3.37	3.34	3.32	3.10
	.05	6.61	5.79	5.41	5.19	5.05	4.95	4.88	4.82	4.77	4.36
	.025	10.01	8.43	7.76	7.39	7.15	6.98	6.85	6.76	6.68	6.02
	.01	16.26	13.27	12.06	11.39	10.97	10.67	10.46	10.29	10.16	9.02
	.005	22.78	18.31	16.53	15.56	14.94	14.51	14.20	13.96	13.77	12.14
	.001	47.18	37.12	33.20	31.09	29.75	28.83	28.16	27.65	27.24	23.79
6	.25	1.62	1.76	1.78	1.79	1.79	1.78	1.78	1.78	1.77	1.74
	.10	3.78	3.46	3.29	3.18	3.11	3.05	3.01	2.98	2.96	2.72
	.05	5.99	5.14	4.76	4.53	4.39	4.28	4.21	4.15	4.10	3.67
	.025	8.81	7.26	6.60	6.23	5.99	5.82	5.70	5.60	5.52	4.85
	.01	13.75	10.92	9.78	9.15	8.75	8.47	8.26	8.10	7.98	6.88
	.005	18.63	14.54	12.92	12.03	11.46	11.07	10.79	10.57	10.39	8.88
	.001	35.51	27.00	23.70	21.92	20.80	20.03	19.46	19.03	18.69	15.75

Table 5 Continued

		df$_1$									
df$_2$	a	1	2	3	4	5	6	7	8	9	∞
7	.25	1.57	1.70	1.72	1.72	1.71	1.71	1.70	1.70	1.69	1.65
	.10	3.59	3.26	3.07	2.96	2.88	2.83	2.78	2.75	2.72	2.47
	.05	5.59	4.74	4.35	4.12	3.97	3.87	3.79	3.73	3.68	3.23
	.025	8.07	6.54	5.89	5.52	5.29	5.12	4.99	4.90	4.82	4.14
	.01	12.25	9.55	8.45	7.85	7.46	7.19	6.99	6.84	6.72	5.65
	.005	16.24	12.40	10.88	10.05	9.52	9.16	8.89	8.68	8.51	7.08
	.001	29.25	21.69	18.77	17.20	16.21	15.52	15.02	14.63	14.33	11.70
8	.25	1.54	1.66	1.67	1.66	1.66	1.65	1.64	1.64	1.63	1.58
	.10	3.46	3.11	2.92	2.81	2.73	2.67	2.62	2.59	2.56	2.29
	.05	5.32	4.46	4.07	3.84	3.69	3.58	3.50	3.44	3.39	2.93
	.025	7.57	6.06	5.42	5.05	4.82	4.65	4.53	4.43	4.36	3.67
	.01	11.26	8.65	7.59	7.01	6.63	6.37	6.18	6.03	5.91	4.86
	.005	14.69	11.04	9.60	8.81	8.30	7.95	7.69	7.50	7.34	5.95
	.001	25.41	18.49	15.83	14.39	13.48	12.86	12.40	12.05	11.77	9.33
9	.25	1.51	1.62	1.63	1.63	1.62	1.61	1.60	1.60	1.59	1.53
	.10	3.36	3.01	2.81	2.69	2.61	2.55	2.51	2.47	2.44	2.16
	.05	5.12	4.26	3.86	3.63	3.48	3.37	3.29	3.23	3.18	2.71
	.025	7.21	5.71	5.08	4.72	4.48	4.32	4.20	4.10	4.03	3.33
	.01	10.56	8.02	6.99	6.42	6.06	5.80	5.61	5.47	5.35	4.31
	.005	13.61	10.11	8.72	7.96	7.47	7.13	6.88	6.69	6.54	5.19
	.001	22.86	16.39	13.90	12.56	11.71	11.13	10.70	10.37	10.11	7.81
∞	.25	1.32	1.39	1.37	1.35	1.33	1.31	1.29	1.28	1.27	1.00
	.10	2.71	2.30	2.08	1.94	1.85	1.77	1.72	1.67	1.63	1.00
	.05	3.84	3.00	2.60	2.37	2.21	2.10	2.01	1.94	1.88	1.00
	.025	5.02	3.69	3.12	2.79	2.57	2.41	2.29	2.19	2.11	1.00
	.01	6.63	4.61	3.78	3.32	3.02	2.80	2.64	2.51	2.41	1.00
	.005	7.88	5.30	4.28	3.72	3.35	3.09	2.90	2.74	2.62	1.00
	.001	10.83	6.91	5.42	4.62	4.10	3.74	3.47	3.27	3.10	1.00

Source: Computed by P. J. Hildebrand from *Statistical Thinking for Behavioral Scientists* by David K. Hildebrand.

Answers to Selected Exercises

Chapter 2

2.1 0.122 **2.3** $\frac{3}{4}$ **2.7** $\frac{1}{5}$ **2.9** $\frac{1}{3}$ **2.10** $\frac{125}{1944}$ **2.15** 0.058
2.19 0.01

Chapter 3

3.1 For example, $p_Z(6) = \frac{1}{9}$ **3.4** 0.44 **3.6** $\frac{1}{9}se^{-s^2/18}$

3.8(a) 0.0228 **3.22** $\frac{a+b}{2}, \frac{(b-a)^2}{12}$ **3.29(b)** $10e^{-12}(1-e^{-4})^2$
3.33 1.39, 2.198

Chapter 4

4.6 23 **4.7** $0.04e^{-0.04t}(e^{0.02t}-1)$, 75, 3125 **4.12** 0.9453
4.13 0.71 **4.26** 2.83 **4.27** $N(k, 2k)$ **4.28(c)** 0.242
4.32 0.16

Chapter 5

5.1(a) $\frac{1}{2}$ **5.2** 0.305 **5.3** 0.464

Chapter 6

6.2 $\dfrac{\overline{X}}{1-\overline{X}}$ **6.4** 430 **6.5** 0.006 **6.9(b)** 4800 **6.10** 152
6.11 $a + b = 1$ **6.13** $2X - 2$ **6.14(a)** 0.33 **(b)** E.g. for $c = 0.5$, the probability is 0.0643 **6.23(a)** $p_N(i) = (1 - \pi)^{i-1}\pi$, $i = 1, 2, \ldots$
(b) $1/N$

Chapter 7

7.5(a) $\chi^2 = 3.01$, for 1 df. OSL is between 0.05 and 0.10. **7.8** 625
7.13 $\chi^2 = 9.6$, for 5 df. OSL is between 0.05 and 0.10.

Chapter 8

8.4(a) $\nu = 1.32$, so the value 1.667 is not terribly wrong. **8.12(a)** $\frac{9}{16}$
8.13 Define the system state to be the number of items in stock after a morning delivery. Then we find that $\pi_1 = 0.62$, $\pi_2 = 0.24$, $\pi_3 = 0.11$, $\pi_4 = 0.03$. The proportion of orders lost is 0.408/1.4, or about 29%.

INDEX